Genetics, Demography and Viability of Fragmented Populations

Habitat fragmentation is one of the most ubiquitous and serious environmental threats confronting the long-term survival of plant and animal species worldwide. As species become restricted to remnant habitats, effective management for long-term conservation requires a quantitative understanding of the genetic and demographic effects of habitat fragmentation, and the implications for population viability. This book provides a detailed introduction to the genetic and demographic issues relevant to the conservation of fragmented populations such as demographic stochasticity, genetic erosion, inbreeding, metapopulation biology and population viability analysis. Also presented are two sets of case studies, one on animals, the other on plants, which illustrate a variety of approaches, including the application of molecular genetic markers, the investigation of reproductive biology and the combination of demographic monitoring and modelling, to examine long-term population viability. This book highlights the value of conducting integrated and inclusive studies for effective conservation management and will be of value to all those working in this crucial area of research.

ANDREW YOUNG is a Senior Research Scientist at the Centre for Plant Biodiversity Research, CSIRO Plant Industry where his research focuses on plant population genetics and ecology. He has published extensively on the genetic consequences of habitat fragmentation and the implications for plant conservation. He is co-editor of *Forest Conservation Genetics* (2000).

GEOFF CLARKE is a Senior Research Scientist at CSIRO Entomology and founder Research Leader of Conservation, Molecular Ecology and Systematics, Australia's first insect conservation biology project. He has worked extensively on insect genetics, particularly on the genetic consequences of habitat fragmentation for insect species. He has been responsible for the preparation of a number of Recovery and Action Plans for threatened invertebrate species in Australia and has also contributed to conservation and biomonitoring programmes throughout the world.

Conservation Biology

Conservation biology is a flourishing field, but there is still enormous potential for making further use of the science that underpins it. This new series aims to present internationally significant contributions from leading researchers in particularly active areas of conservation biology. It will focus on topics where basic theory is strong and where there are pressing problems for practical conservation. The series will include both single-authored and edited volumes and will adopt a direct and accessible style targeted at interested undergraduates, postgraduates, researchers and university teachers. Books and chapters will be rounded, authoritative accounts of particular areas with the emphasis on review rather than original data papers. The series is the result of a collaboration between the Zoological Society of London and Cambridge University Press. The series editor is Professor Morris Gosling, Professor of Animal Behaviour at the University of Newcastle upon Tyne. The series ethos is that there are unexploited areas of basic science that can help define conservation biology and bring a radical new agenda to the solution of pressing conservation problems.

Published Titles

1. *Conservation in a Changing World*, edited by Georgina Mace, Andrew Balmford and Joshua Ginsberg 0 521 63270 6 (hardcover), 0 521 63445 8 (paperback)
2. *Behaviour and Conservation*, edited by Morris Gosling and William Sutherland 0 521 66230 3 (hardcover), 0 521 66539 6 (paperback)
3. *Priorities for the Conservation of Mammalian Diversity*, edited by Abigail Entwistle and Nigel Dunstone 0 521 77279 6 (hardcover), 0 521 77536 1 (paperback)

GCN

Genetics, Demography and Viability of Fragmented Populations

Edited by

ANDREW G. YOUNG
CSIRO Plant Industry

GEOFFREY M. CLARKE
CSIRO Entomology

CAMBRIDGE
UNIVERSITY PRESS

Natural Heritage Trust
Helping Communities Helping Australia

CSIRO

THE ZOOLOGICAL
SOCIETY OF LONDON

PUBLISHED BY THE PRESS SYNDICATE OF THE UNIVERSITY OF CAMBRIDGE
The Pitt Building, Trumpington Street, Cambridge, United Kingdom

CAMBRIDGE UNIVERSITY PRESS
The Edinburgh Building, Cambridge CB2 2RU, UK
40 West 20th Street, New York, NY 10011-4211, USA
10 Stamford Road, Oakleigh, VIC 3166, Australia
Ruiz de Alarcón 13, 28014 Madrid, Spain
Dock House, The Waterfront, Cape Town 8001, South Africa
http://www.cambridge.org

First published 2000

Printed in the United Kingdom at the University Press, Cambridge

Typeset in FF Scala 9.75/13 pt [V N]

A catalogue record for this book is available from the British Library

Library of Congress Cataloguing in Publication data

Genetics demography and viability of fragmented populations/edited by Andrew G.
Young and Geoffrey M. Clarke.
    p. cm. – (Conservation biology)
  ISBN 0 521 78207 4 (hc)
  1. Fragmented landscapes.   2. Population biology.   3. Conservation biology.   I. Young,
Andrew G. (Andrew Graham), 1965–   II. Clarke, Geoffrey M. (Geoffrey Maurice), 1960–
III. Conservation biology series (Cambridge, England)
QH541.15.F73 G46 2000
577.8'8 – dc21   00-029253

ISBN 0 521 78207 4 hardback
ISBN 0 521 79421 8 paperback

Contents

Contributors

DAVID J. AYRE
Australian Flora and Fauna Research
Centre
Department of Biological Sciences
University of Wollongong, NSW 2522
Australia

ANDREW J. BAKER
Philadelphia Zoological Gardens
3400 W. Girard Avenue
Philadelphia, PA 19104
USA

JONATHAN D. BALLOU
Department of Zoological Research
National Zoological Park
Washington DC 20008
USA

FIONA BEYNON
Australian Flora and Fauna Research
Centre
Department of Biological Sciences
University of Wollongong, NSW 2522
Australia

DAVID H. BOSHIER
Oxford Forestry Institute
Department of Plant Sciences
University of Oxford
Oxford OX1 3RB
UK

ANTHONY H. D. BROWN
Centre for Plant Biodiversity Research
CSIRO Plant Industry
GPO Box 1600
Canberra ACT 2601
Australia

PETER F. BRUSSARD
Department of Biology
University of Nevado Reno
Reno, NV 89557-0050
USA

JEREMY J. BURDON
Centre for Plant Biodiversity Research
CSIRO Plant Industry
GPO Box 1600
Canberra ACT 2601
Australia

MARK BURGMAN
School of Botany
University of Melbourne
Parkville, VIC 3052
Australia

GEOFFREY M. CLARKE
CSIRO Entomology
GPO Box 1700
Canberra ACT 2601
Australia

SUSAN J. DANIELS
Department of Biology
Virginia Polytechnic Institute and State
University
Blacksburg, VA 24061-0406
USA

JAMES M. DIETZ
Department of Biology
University of Maryland
College Park, MD 20742
USA

MICHELE R. DUDASH
Department of Biology
University of Maryland
College Park, MD 20782
USA

PHILIP R. ENGLAND
Australian Flora and Fauna Research
Centre
Department of Biological Sciences
University of Wollongong, NSW 2522
Australia

CHARLES B. FENSTER
Department of Botany
Norwegian University of Science and
Technology
Trondheim
Norway

GUSTAVO A. GUTIÉRREZ-ESPELETA
Escuela de Biología
Universidad de Costa Rica
San José
Costa Rica

PHILIP W. HEDRICK
Department of Biology
Arizona State University
Tempe, AZ 85287-1501
USA

KENT E. HOLSINGER
Department of Ecology and
Evolutionary Biology
University of Connecticut
Storrs, CT 06269-3043
USA

STEVEN T. KALINOWSKI
National Marine Fisheries Service
Northwest Fisheries Science Center
Seattle, WA 98112
USA

DAVE KELLY
Plant and Microbial Sciences and
School of Forestry
University of Canterbury
Christchurch 8001
New Zealand

JENNY J. LADLEY
Plant and Microbial Sciences
University of Canterbury
Christchurch 8001
New Zealand

DAVID LINDENMAYER
Centre for Resource and Environmental
Studies and Department of Geography
Australian National University
Canberra ACT 0200
Australia

TANYA LLORENS
Australian Flora and Fauna Research
Centre
Department of Biological Sciences
University of Wollongong, NSW 2522
Australia

CATHY MILLER
Centre for Plant Biodiversity Research
CSIRO Plant Industry
GPO Box 1600
Canberra ACT 2601
Australia

CRAIG MORITZ
Department of Zoology and
Entomology
University of Queensland
St Lucia, QLD 4072
Australia

BRAD R. MURRAY
Centre for Plant Biodiversity Research
CSIRO Plant Industry
GPO Box 1600
Canberra ACT 2601
Australia

BRIAN G. MURRAY
School of Biological Sciences
University of Auckland
Private Bag 92019
Auckland
New Zealand

DAVID A. NORTON
School of Forestry
University of Canterbury
Christchurch 8001
New Zealand

J. GERARD B. OOSTERMEIJER
Institute for Systematics and Ecology
University of Amsterdam
NL-1098 SM Amsterdam
The Netherlands

ROD PEAKALL
Division of Botany and Zoology
Australian National University
Canberra ACT 0200
Australia

HUGH POSSINGHAM
Department of Environmental Science
and Management
University of Adelaide
Roseworthy, SA 5371
Australia

JEFFERY A. PRIDDY
Nicholas School of the Environment
Marine Laboratory
Duke University
Beaufort, NC 28516-9712
USA

CHRISTOPHER M. RICHARDS
Department of Biology
Vanderbilt University
Nashville, TN 37235
USA

ALASTAIR W. ROBERTSON
Department of Ecology
Massey University
Palmerston North
New Zealand

WILLIAM B. SHERWIN
Department of Biological Sciences
University of New South Wales
Sydney, NSW 2052
Australia

SUKAMOL SRIKWAN
Department of Biology
Chulalongkorn University
Bangkok 10330
Thailand

PETER H. THRALL
Centre for Plant Biodiversity Research
CSIRO Plant Industry
GPO Box 1600
Canberra ACT 2601
Australia

JEFFREY R. WALTERS
Department of Biology
Virginia Polytechnic Institute and State
University
Blacksburg, VA 24061-0406
USA

ROBERT J. WHELAN
Australian Flora and Fauna Research
Centre
Department of Biological Sciences
University of Wollongong, NSW 2522
Australia

GEMMA M. WHITE
Cell and Molecular Genetics
Department
Scottish Crop Research Institute
Invergowrie, Dundee DD2 5DA
UK

DAVID S. WOODRUFF
Department of Biology
University of California at San Diego
La Jolla, CA 92093-0116
USA

ANDREW G. YOUNG
Centre for Plant Biodiversity Research
CSIRO Plant Industry
GPO Box 1600
Canberra ACT 2601
Australia

Foreword

When I was a graduate teaching assistant in vertebrate biology 35 years ago, I could take my students to a location a few miles out of town and show them eight of the nine species of lizards that are found in western Nevada. Now gone as natural habitat, the location is now a subdivision bordered by a strip mall. These same lizard species are also very scarce for several miles beyond the subdivisions. There they have been eliminated by the house cats kept by the owners of 20-acre ranchettes. Farther out into the desert it is very hard to find a lizard within a mile or more of a road; commercial collectors have captured most of them and sold them to pet dealers in the eastern United States. Now I show my students specimens of dead lizards, kept in my laboratory in dusty, alcohol-filled jars.

Similar scenarios are played out relentlessly all over the world as the human enterprise expands and natural habitats disappear. None of the lizards of western Nevada is as yet threatened with extinction, but others in adjacent states are. Of the 10–20 million species of plants, animals, and microbes estimated to be extant, perhaps half will become extinct in the next millennium. Most of these will disappear anonymously – unknown and undescribed by science. Others, with proper scientific names, will vanish quietly, their epitaphs simply stating, 'last collected in 1999.' A few charismatic species – primates, colorful birds, butterflies – will be kept from extinction by captive propagation, but most have no hope of ever returning to their long-since fragmented and degraded natural habitats. The survivors of this massive extinction event will be the ecological generalists, the weeds, the invaders.

Why should we care? Some of the species doomed to extinction may hold the cure for an emerging disease in their genetic codes; others might have become nutritious crop plants or useful domesticated animals. Some species may be keystones in certain ecosystems, their importance unrecognized until they disappear and the ecological services that their ecosystems provided then diminish or fail. Our knowledge base will be poorer; we will never know the life histories, ecological roles, or potential utilitarian value of many thousands of extinct species. In aggregate, the global depletion of biodiversity – the genetic information, the species, and the ecosystems in which those species occur – will have negative effects on our economic and physical well-being. Even more important, as each extinction erodes the biological legacy that our descendents will inherit, part of our humanity slips away. Stewardship for the planet and its inhabitants will have been lost, replaced by greed, ignorance, and short-sightedness.

Conservation biology emerged about 20 years ago to provide the underlying science needed to slow the extinction crisis. This relatively new discipline draws on information from the basic sciences of population and evolutionary biology, ecology, biogeography, and genetics and from the more applied domains of wildlife biology, fisheries science, forestry, and rangeland management. The focus of conservation biology is simultaneously broad, seeking to understand the interactions among large landscape elements and the species they contain, and narrow, concentrating on individual species that are immediately imperiled. The metapopulation paradigm bridges these two foci, connecting the spatial structure of landscape elements with the population dynamics of individual species.

This book provides the conceptual framework for examination of the impacts of fragmentation, supported by a number of case studies that integrate both demographic and genetic analyses into the conservation of imperiled species in fragmented and degraded landscapes. The applications of sophisticated molecular tools, population viability analysis, spatial modelling, and classical genetics to conservation problems are well represented in these studies.

Recently, a well-known American resource economist Randal O'Toole stated: 'conservation biology is not a science but a political movement based at least in part on nineteenth-century ideals of what an ecosystem is all about.'[1] If those ideals include the notion that the world's biodiversity is worth saving, O'Toole is at least partly right. But anyone who reads this book will understand clearly that O'Toole is wrong in his contention that conservation biology is not science. The material presented here is testimony to how good science may yet slow, and perhaps eventually arrest, the extinction crisis.

Peter F. Brussard
Past President, Society for Conservation Biology

[1] O'Toole, R. (1999). *Subsidies Anonymous, #36.* The Thoreau Institute, PO Box 1590, Bandon, OR 97411, USA.

Preface

Since the publication of Soulé & Wilcox's *Conservation biology: an evolutionary–ecological perspective* in 1980, there has been a steady string of multi-authored works on conservation biology, most notably the other two Soulé books, *Conservation biology: the science of scarcity and diversity* (1986) and *Viable populations for conservation* (1987) and Schonewald-Cox *et al.*'s *Genetics and conservation: a reference for managing wild animal and plant populations* (1983). In addition, there has been a recent surge in the number of conservation biology textbooks available for both undergraduate and postgraduate teaching.

So we might ask ourselves, is there a need for another work in an already overcrowded field, and what makes this volume different from all the others. Many of the earlier works were written at the time that conservation biology was in the process of defining itself and finding its feet as a rigorous scientific discipline. As a result, and as could be expected, many of the individual contributions to these works were written from the perspective of the individual authors' traditional backgrounds in ecology, genetics and resource management and adapted to suit the growing concern of species decline. In addition, the works were very broad in coverage reflecting the very broad scope of conservation biology, ranging from biodiversity loss on large scales through deforestation to impacts on individual species and populations. These works made an invaluable contribution to promoting both an awareness and acceptance of conservation issues both within academia and land-use management agencies and laid the groundwork for much of modern conservation biology as a science.

Over the last 20 years conservation biology has matured, building on the concepts outlined by these early editions, to become a rigorous scientific discipline with its own theoretical framework based on the traditional fields of ecology, genetics and biogeography and applied to small and declining populations. This volume seeks to reflect this new level of maturity. By focusing on the most ubiquitous and pervasive of all threats to long-term species survival, habitat fragmentation, the book provides the necessary relevance to the bulk of modern conservation efforts. The introductory section provides the theoretical context for explaining and predicting the demographic and genetic consequences of fragmentation. The subsequent two sections present a number of empirical case studies on animal and plant species respectively. These case studies, which integrate ecological, genetic and population biology approaches for assessing impacts of frag-mentation, exemplify the development and coming of age of conservation biology as a science. We believe the book will be equally at home in the classroom as on the desk of the professional conservation biologist.

Like conservation biology itself, this book has matured (and grown) since its original concept. The idea for the volume grew out a symposium of the same title held as part of the 12th Annual Meeting of the Society for Conservation Biology in Sydney in 1998. This symposium attracted enor-mous interest with 11 oral presentations and 20 posters. Many of these contributions form the basis of the empirical studies included in the book. We subsequently decided that these studies needed a theoretical context and thus solicited the contributions included in the introductory section from leading authorities. We have tried to ensure the book is both taxonomically and geographically representative, thus our list of contributors and their study organisms has a global distribution.

The book would not have been possible without the help of our many colleagues. All chapters were peer-reviewed by at least two referees and for this we thank the following: Fred Allendorf, Jon Ballou, Dave Boshier, Dave Coates, Paul Downey, Dick Frankham, Sue Haig, Kringen Henien, Susan Hoebee, Bob Lacy, Gordon Luikart, Georgina Mace, Eric Menges, Neil Mitchell, John Morgan, Phil Nott, David Paetkau, Rod Peakall, Katherine Ralls, Dave Rowell, Kat Shea, Andrea Taylor, Phil Taylor, Pete Thrall, Bob Wayne and Gerry Zegers. Environment Australia provided financial sup-port for the original Society for Conservation Biology symposium. Finally, we thank Alan Crowden and Maria Murphy of Cambridge University Press for their support, encouragement and excellent editorial and production skills.

Geoff Clarke
Andrew Young

Introduction: genetics, demography and the conservation of fragmented populations

GEOFFREY M. CLARKE & ANDREW G. YOUNG

In one of the earliest books on modern conservation biology, Soulé & Wilcox describe the science of conservation biology as being 'as broad as biology itself. It focuses the knowledge and tools of all biological disciplines, from molecular biology to population biology, on one issue – nature conservation' (Soulé & Wilcox, 1980a). Subsequently, in his seminal paper on the nature of conservation biology, Soulé extended this concept to include non-biological sciences such as hazard evaluation and the social sciences (Soulé, 1985). He went on to describe conservation biology as being holistic in nature, in the sense that multidisciplinary approaches will ultimately be the most fruitful. He also stressed that the borders between traditional scientific pursuits and between the 'pure' and 'applied' sciences were artificial in the conservation context.

The truth in these statements with regard the multidisciplinary nature of conservation biology can be revealed by an examination of any issue of the frontline journal *Conservation Biology*. Here you will find papers written by ecologists, resource managers, geneticists, sociologists, political scientists, mathematicians and even politicians. However, for a multidisciplinary approach to be effective, particularly in a crisis discipline such as conservation biology (Soulé, 1985), there must be a high level of integration among the separate fields. It is this integration which we feel has generally been lacking in much of modern conservation research.

There are many possible reasons for this lack of integration. Soulé & Wilcox (1980a) talk of academic snobbery which hampered the acceptance of conservation biology within academic circles, and the subsequent academic elitism (Soulé, 1986a), such that once accepted, it became the 'property' of academia and academic disciplines. While these observations are justified, there is also a degree of discipline rivalry among different areas of scientific pursuit. This rivalry is perpetuated by the structure of university faculties and departments, and the patterns of research funding worldwide. In addition, it reflects the educational and professional background of the first generation of conservation-biology practitioners. Through the period of the 1950s–80s there was a move away from the training of 'field biologists' or 'naturalists', as was common prior to this period, to increasing specialisation. This level of specialisation became extreme in the 1970s and 1980s when graduates were no longer zoologists or geneticists, but rather physiologists, behaviourists, morphologists, or ecologists or population, bacterial, or molecular geneticists. This high degree of specialisation leads to an increased sense of insecurity and fear (and often contempt) of things we don't know, which obviously does not promote integrative collaborative associations. In 1986 Soulé noted that there were no degree programmes in conservation biology, which he viewed as a major impediment to the future of the science. Fortunately this has turned around with many hundreds of broad-based degree and graduate programmes in conservation biology being offered worldwide.

Nowhere has this lack of integration within conservation biology been more evident than between the traditionally somewhat rival fields of ecology and genetics. Over the last 20 years, ecologists have been busily and rigorously investigating the roles of demographic processes such as changes in habitat quality and quantity, population growth rates, breeding structures and migration on species extinction, while population geneticists have been just as industrious looking at loss of genetic diversity, inbreeding and changes in fitness, with little interaction between the two groups. The lack of integration between these two fundamentally related areas of pursuit is somewhat surprising given that one of the earliest and most cited papers in modern conservation biology, by Gilpin & Soulé (1986), clearly details the interaction of demographic and genetic processes in the extinction process.

This disassociation between genetics and demography has been perpetuated in the literature by two very influential papers and has been popularised by articles such as that by Martin Brookes in *New Scientist* (Brookes, 1997). In his commentary Brookes states that 'money spent on conservation genetics would be better spent on either good science or good conservation, rather than a halfway house of nothingness'. He goes on to say that 'while the ship is sinking, conservation geneticists are busy counting the deck chairs. Conservation and genetics, like pop and politics, just don't mix. A swift divorce should leave both science, and what's left of life on Earth, in better shape.'

The first important paper, by Lande (1988) published in *Science*, argued that for wild populations, demographic factors may usually be of more importance than genetic factors in assessing the requirements for long-term species persistence. However, he concludes by saying that there is an immediate practical need in biological conservation for understanding the interaction of demographic *and* genetic factors in the extinction of small populations and that future conservation plans should incorporate both demography and population genetics in assessing the requirements for species survival. These latter statements are often ignored by those wishing to propagate the genetics/demography dichotomy.

The second paper, by Graeme Caughley (1994) and published in the *Journal of Animal Ecology* just before he died, has generated the most controversy and discussion on this issue. In his paper, Caughley made the distinction between what he termed the 'small-population paradigm' and the 'declining-population paradigm'. He argued that the former sought to determine the risk of extinction inherent in low numbers, whereas the latter dealt with the causes of smallness and its cure; essentially representing stochastic and deterministic processes respectively. In what is an intellectually stimulating and elegantly argued hypothesis Caughley states that 'no instance of extinction by genetic malfunction has been reported, whereas the examples of driven extinction are plentiful. Genetic thinking often intrudes where it is not relevant and where it sometimes obscures the real issues. . . .' However, despite these statements, Caughley, like Lande, finishes on a more positive note by saying 'The declining-population paradigm is urgently in need of more theory. The small-population paradigm needs more practice. Each has much to learn from the other. In combination they might enlarge our idea of what is possible.'

The publication of Caughley's article subsequently led to an essay by Hedrick *et al.* (1996) appearing in *Conservation Biology*. In this paper the authors argue that Caughley's paper had created a false dichotomy and also contained a number of misunderstandings about the role of both demography and genetics in extinctions. They argued that 'both the deterministic factors that reduce population size and the stochastic factors that lead to the final extinction of a small population are critical to consider in preventing extinction. Only through an overall and comprehensive effort, which we call inclusive population viability analysis, can extinction processes be understood and mitigated.'

While these papers seem to promote the differences between genetics and demography, essentially they all emphasise the same points, viz. that both genetics and demography and their interactions are important in the extinction process and that only by the integration of these two fields can we hope to achieve effective conservation management and long-term population and species survival. This call for integration has been supported by others (e.g. Nunney & Campbell, 1993; Mills & Smouse, 1994; Schemske et al., 1994; Soulé & Mills, 1998) and brings us to the rationale behind this volume.

Although the debate has distracted many researchers and delayed collaborative and integrated research in some areas, there has been considerable progress in recent years to the extent that we felt there was a need for a synthetic treatment of current activities. Undoubtedly the development and application of highly variable DNA markers such as microsatellites (Bruford et al., 1996), the increase in speed, and reduction in cost, of DNA sequence analysis, and the possibility for non-invasive DNA sampling (Morin & Woodruff, 1996) have led to a rapid expansion of the field of molecular ecology in which any distinction between genetics and ecology is lost. In addition, the advances in mathematical ecology and increases in computing power and speed have led to the development of more realistic models of population dynamics and extinction probabilities, many of which incorporate genetic data. The two fields are thus becoming more inclusive and integrative through their normal advances. This can only benefit the science of conservation biology.

Critical factors in the extinction process, such as population size, breeding structure and dispersal, are now routinely estimated by a combination of genetic and demographic techniques. Genetic data are being used to define units for conservation management and for inferring past and recent changes in population structure and dynamics. In addition, molecular markers can be used to identify and track individuals within populations, which is useful for the development of spatially explicit individual-based models for population persistence.

This volume, which grew out of a symposium of the same name run during the 1998 meeting of the Society for Conservation Biology held in Sydney, Australia, aims to showcase some of the recent and ongoing work which exemplifies attempts to integrate demographic and genetic data in an effort to understand the impacts of habitat fragmentation on population and species survival. Habitat fragmentation is recognised as one of the major environmental factors threatening the survival of populations and species worldwide. Fragmentation has dramatically shaped large areas of temperate and tropical landscapes, forests, heathlands, prairies and grasslands alike into ecosystems that now bear little structural, and probably limited functional, resemblance to the original. For example between 1978 and 1988 the mean rate of deforestation and fragmentation in the Brazilian Amazon basin was estimated to be 53 000 km^2 per year (Skole & Tucker, 1993). For many plants and animals, preservation within relatively intact habitats is no longer an option, and for these, a quantitative understanding of the effects of fragmentation on population processes and viability is now a prerequisite if informed management decisions are to be made for their long-term conservation. However, although the demographic (e.g. Wilcove et al., 1986; Saunders et al., 1991) and genetic (e.g. Young et al., 1996) consequences of fragmentation have been documented, very few attempts have been made to examine these simultaneously and their interaction within single populations.

The first section of this book contains a series of six chapters which each provide an overview of the important genetic and demographic issues relating to conservation biology and provide a framework for the subsequent empirical studies. Sherwin & Moritz (Chapter 2) give an overview of the genetic consequences of fragmentation, with particular emphasis on the loss of genetic diversity. This is followed by Dudash & Fenster (Chapter 3) who focus on perhaps the two most important issues with potential to link genetics and demography, viz. inbreeding and outbreeding depression. Holsinger (Chapter 4) provides a description of the demographic factors relating to extinction in small populations. Given that many fragmented populations exist as a metapopulation, Thrall, Burdon & Murray (Chapter 5) give an overview of metapopulation theory in terms of population structure and dynamics. Burgman & Possingham (Chapter 6) outline the past and future applications of population viability analyses (PVAs) and provide a checklist of what a good PVA should include. Finally Hedrick (Chapter 7) gives some examples of the application of population genetics and molecular markers for providing both genetic and demographic data for threatened species.

These introductory chapters are followed by a series of 12 empirical case studies covering both plants and animals. Each of these studies investigates the genetic and demographic consequences of fragmentation and their interactions in small populations. While some studies are clearly preliminary, others have reached the stage of incorporating genetic and demographic information into quantitative models of population persistence, for example the work of Daniels, Priddy & Walters (Chapter 8) on red-cockaded woodpeckers and Oostermeijer on *Gentiana pneumonanthe* (Chapter 18). Although some chapters are more 'demographic' or more 'genetic' than others, there is nowhere any debate as to the relative importance of one data set over another. Authors have used all the data available to them, regardless of the scientific discipline that provided the tools to generate it, to attempt to understand both the causes and consequences of fragmentation. They provide examples of what can be done in terms of the provision of data important for effective conservation management of threatened species when one ignores the distractions of petty academic snobbery and rivalry. The value of these studies rests on this level of inclusiveness.

Thus we believe this volume represents an overview of the application of modern conservation biology to the issue of habitat fragmentation as Soulé originally imagined it, viz. a holistic, multidisciplinary and integrated science, unified by a common purpose – nature conservation.

PART I
Introductory concepts

Managing and monitoring genetic erosion

WILLIAM B. SHERWIN & CRAIG MORITZ

ABSTRACT

Fragmentation, decline or perturbation of a species can lead to genetic changes. Often these changes can have adverse implications for the conservation of the species, but there is a diversity of responses by different species. Therefore, managers must use a variety of methods to detect, avert or remedy genetic changes which actually affect population viability. The objective should be to maintain optimal fitness in changing conditions, rather than to maintain specific arrays of phenotypes. This effort should be accompanied by monitoring of genetic contributions to fitness, to confirm the effectiveness of the conservation genetic strategy. This approach presumes we have the ability to directly or indirectly manipulate and measure adaptive genetic variants, such as many multilocus (quantitative) traits, or a representative array of single-locus traits associated with fitness. Such analyses are challenging, but are becoming more accessible. It is also important to examine the association between adaptive diversity and surrogates which may be more amenable to monitoring or manipulation, such as neutral DNA variants, size or number of populations, or the range of ecological conditions in which populations of the species are found. In evaluating different types of genetic variation and their surrogates, two important points are the replaceability of the variation (that is, how long it would take for the variation to be replaced) and its utility (likely contribution to adaptation).

INTRODUCTION

Biodiversity conservation targets three interdependent levels: ecosystems, species and genes. This chapter will highlight genetic variation within species, an area which is currently experiencing a wealth of new field, laboratory and statistical methods. A declining or fragmenting species may experience genetic changes including loss of differentiated populations, al-

(a)

A1/A1	A1/A1	A1/A1	A1/A1				
A1/A2	A1/A2	A1/A2	A1/A2	A1/A2	A1/A2	A1/A2	A1/A2
A2/A2	A2/A2	A2/A2	A2/A2				
A1/A3	A1/A3	A1/A3	A1/A3	A1/A3	A1/A3	A1/A3	A1/A3
A2/A3	A2/A3	A2/A3	A2/A3	A2/A3	A2/A3	A2/A3	A2/A3
A3/A3	A3/A3	A3/A3	A3/A3				

(b)

A1/A1	A1/A1	A1/A1	A1/A1	A1/A1	A1/A1	A1/A1	A1/A1	A1/A1
A1/A2	A1/A2	A1/A2	A1/A2	A1/A2	A1/A2	A1/A2	A1/A2	A1/A2
A1/A2	A1/A2	A1/A2	A1/A2	A1/A2	A1/A2	A1/A2	A1/A2	A1/A2
A2/A2	A2/A2	A2/A2	A2/A2	A2/A2	A2/A2	A2/A2	A2/A2	A2/A2

(c)

A1/A1	A1/A1	A1/A1	A1/A1	A1/A1	A1/A1
A1/A1	A1/A1	A1/A1	A1/A1	A1/A1	A1/A1
A2/A2	A2/A2	A2/A2	A2/A2	A2/A2	A2/A2
A2/A2	A2/A2	A2/A2	A2/A2	A2/A2	A2/A2
A3/A3	A3/A3	A3/A3	A3/A3	A3/A3	A3/A3
A3/A3	A3/A3	A3/A3	A3/A3	A3/A3	A3/A3

Fig. 2.1. The difference between inbreeding and lowered genetic variation among individuals. (a) A population in Hardy–Weinberg equilibrium; gene diversity $H = 0.67$, observed heterozygosity $H_0 = 0.67$, allelic diversity $K = 3$, inbreeding coefficient (calculated from the depression of heterozygosity relative to Hardy–Weinberg equilibrium) $F = 0$. (b) A population with lower genetic variation than (a), but no evidence of inbreeding; this population is also in Hardy–Weinberg equilibrium; $H = 0.5$, $H_0 = 0.5$, $K = 2$, $F = 0$. (c) A population with the same genetic variation (H, K) among individuals as (a), but also a history of recent inbreeding which has resulted in deviation from Hardy–Weinberg equilibrium – deficit of heterozygotes. $H = 0.67$, $H_0 = 0$, $K = 3$, $F = 1$.

Gene diversity (H) refers to the chance that random mating would produce a heterozygote at any locus, or the average expected Hardy–Weinberg hetero-zygosity. Allelic diversity (K) refers to the number of alleles at the average locus. We use the word 'inbreeding' to refer to the result of mating between relatives, not as a general term for reduced genetic variation in the population; these often-confused concepts are further clarified in Templeton & Read (1994).

teration of differentiation between populations, loss of variation among members of the same population and changes to the level of inbreeding (Fig. 2.1). These changes are unlikely to be positive for conservation, al-though sometimes they may be of no immediate conservation significance (Lande, 1988). Diminished genetic variation between populations, or loss of distinct populations, reduces the opportunity for adaptive responses to geographically varying local conditions. Lowered genetic variation within populations also reduces the opportunity for adaptation, and may result in reduced reproduction or survival and thereby reduce the viability of the population (Madsen *et al.*, 1996). Inbreeding can also reduce fitness (Ralls *et al.*, 1988). The importance of genetic variation to short-term viability depends upon many aspects of the particular species' biology, including the chromosomal system (James *et al.*, 1991), mating system (Young *et al.*, Chapter 19, this volume) and reproductive potential (Mills & Smouse, 1994). Some species survive with very little variation detectable at a molecular level (Reeve *et al.*, 1990).

Faced with this diversity of responses to genetic change, a conservation manager must use appropriate monitoring to judge whether erosion of genetic variation is actually affecting population viability. When necessary, the manager must also consider how best to avert or remedy erosion of variation. Like most conservation problems, the solutions to genetic problems are easier if action is taken early in the process of decline, when the existence of a number of individuals and populations allows choice of various management strategies. However, managers are often faced with small isolated populations, or even a single small remaining population. Whatever the situation, conservation genetic decisions should be based on the replaceability and utility of genetically determined phenotypic variation. Replaceability depends heavily upon the origin of the variation – the longer it took for the variation to accumulate, the slower its replacement is likely to be. The utility of the variation depends upon its likely importance for current and future adaptation, and therefore its contribution to the viability of populations.

Figure 2.2 shows a classification of overlapping types of genetically determined variation within and between populations. Recently, there has been intense conservation interest in the historical component of genetic diversity (categories 1, 3, 5, 7 in Fig. 2.2), which refers to variants that have accumulated over thousands of generations through the random processes of drift and mutation, and also possibly through selection and adaptation. This historical component is essentially irreplaceable because the circumstances which generated the variation can only be surmised, and the time-scale cannot be replicated in a conservation program (Moritz, 1999*a*). For short-term conservation, a second type of genetic variation is important: variants that affect current adaptation and viability (categories 3, 4, 5, 6 in Fig. 2.2). If lost, phenotypes corresponding to these variants (especially 4 and 6) can potentially be re-created relatively rapidly through selection,

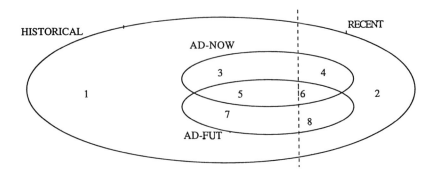

Fig. 2.2. Classification of genetically determined variation, for use in conservation planning. This classification can be applied to variation within or between populations, but the relative sizes of the categories (1 to 8) may differ. HISTORICAL: variation which has accumulated over thousands of generations or longer (categories 1, 3, 5, 7). RECENT: variation of relatively recent origin (categories 2, 4, 6, 8). AD-NOW: variation which is of adaptive significance at present (categories 3, 4, 5, 6). AD-FUT: variation which will be of adaptive significance in the future (categories 5, 6, 7, 8). Any variants, including those not encompassed by AD-NOW or AD-FUT, may have been of adaptive significance at some time in the past.

provided that allelic variation remains at some of the relevant quantitative trait loci, and that the impact of this selection, or other problems, is not so great as to cause immediate extinction of the population(s) (Lynch, 1996; Moritz, 1999b). A third type of genetic variation is also of obvious concern to conservation managers: the variation which will form the basis of continued adaptation to changing conditions in the future (categories 5, 6, 7, 8 in Fig. 2.2). This variation is critical for long-term conservation, but unfortunately, we are unable to specify which genetic variants belong to these categories. Nevertheless, we can make some inferences about the behaviour of each category, and hence the best strategies for genetic conservation.

How, then, can we look after the variation that is important for short- and long-term conservation? We have already suggested that variation in categories 4 and 6 is likely to be restored relatively quickly if lost, as may some of the variation in categories 3 and 5. Variation in categories 7 and 8 is vital for long-term viability, but not of any current adaptive significance, and therefore prone to loss by genetic drift. Once lost, this variation will not be regained as fast as variation in other categories (e.g. category 8 is likely to be slower to recover than 4 or 6). Planning our management strategy would be much easier if we knew which genes determined the variation in each category, but by and large, this is unknown. We can, however, infer that the underlying genetic basis of an adaptive trait will affect its replaceability (see below).

Planning conservation genetic programmes is also aided by considering the likely relative sizes of the categories. The sizes may differ for particular species, and for within- and between-population genetic variation, but some generalisations can be attempted. Because of the time-scales of population genetic processes, we expect that much current genetic variation is in the historical/irreplaceable categories (1, 3, 5, 7). Three of these are important to current or future adaptation (3, 5, 7), and therefore there has been great interest in identifying and maintaining the historical component of variation, both within and between populations (Franklin, 1980; Moritz, 1999b). In order to maintain current population viability, we must also manage variants which are of recent origin, and are currently adaptive (categories 4 and 6). We would also like to manage the variation which is of recent origin, and will be adaptive in the future, but not at present (category 8). Unfortunately, this is the category we know least about, and many managers hope that by concentrating on other aspects, we will serendipitously manage this last category, or (perhaps more likely) that this category will be so small that it will not contribute greatly to long-term conservation. Interestingly, Fitch *et al.* (1997) have shown that we may have hitherto-unsuspected power to make predictions about variation in category 8. For the influenza virus phylogeny, Fitch *et al.* (1997) showed that amongst each year's many newly arisen clades, it is possible to identify which clade is most likely to give rise to the next year's genetically different epidemic.

The next two sections will discuss the effect of genetic variation on population processes, and the relationship between these processes and the types of genetic variation which we can actually measure. We will then consider the erosion of genetic variation within and between populations, before discussing management actions which may forestall or remedy conservation genetic problems. As we have already indicated, in many cases our decisions will have to be guided by incomplete information. We often have to manage or monitor surrogates for the genetic variation that we aim to conserve. For example, we may manage population size instead of gene diversity (Brown *et al.*, 1997), and it is important to examine how well these surrogates reflect the capacity for genetic adaptation.

GENETICS AND CONSERVATION FORECASTING: EFFECT OF GENETIC VARIATION ON POPULATION PROCESSES

To choose appropriate genetic variants for management and monitoring, we must consider the ways in which genetic variation affects conservation outcomes. Does genetic variation affect fitness in managed species, and do changes in fitness alter the chance that a population will persist? For example, one effect of lowered variation is a reduction of heterozygote frequencies, and in *Daphnia*, development of heterozygous clones is better buffered against environmental variation than homozygotes (Deng, 1997). As a result, lowered variation might be expected to affect *Daphnia* population viability in a fluctuating environment. To assess the generality of observations such as this, evidence comes from several lines of investigation: studies of inbreeding depression, studies of single wild populations comprising individuals with differing levels of heterozygosity and studies of multiple conspecific populations with different levels of variation. Pedigree inbreeding (mating between relatives: Templeton & Read, 1994) results in elevated homozygosity at all loci, so that a comparison of more and less inbred individuals can shed light on the likely effects of erosion of genetic variation. In *ex-situ* populations of plants and animals, inbreeding often results in reduced fitness, called inbreeding depression (Gall, 1987; Ralls *et al.*, 1988; Dudash & Fenster, Chapter 3, this volume). Inbreeding depression is also seen in natural populations, though not in all species, and its expression can be environment-dependent, which could lead to underestimation of the impact of inbreeding depression (Stevens & Bougourd, 1988; Jiménez *et al.*, 1994; Keller *et al.*, 1994; Pray *et al.*, 1994). From inbreeding studies, one could infer that members of populations experiencing increased homozygosity due to genetic drift would be likely to show phenomena similar to inbreeding depression. However, we have some reservations about applying the results of inbreeding studies to wild populations with lowered genetic diversity. Firstly, the rate of genetic change (inbreeding) in a population being artificially inbred may often be faster than the rate of genetic change (drift) experienced by a wild population slowly losing genetic variation. The higher rate of change in the inbred population reduces the opportunity for adaptation to homozygosity. Secondly, there are differences of prior variation: some inbreeding studies utilise populations which have a history of inbreeding, which could allow adaptation to homozygosity (Barrett & Charlesworth, 1991; Ouborg & van Treuren, 1994).

As an alternative to inbreeding studies, the search for an association between genetic variation and fitness can utilise comparisons of fitness-related traits among individuals with different levels of gene diversity at marker loci. Unlike the inbreeding studies, lowered (or elevated) gene diversity at marker loci is less likely to indicate the same change in gene diversity at other loci. Despite this, there have been many reports of positive relationships between marker heterozygosity and fitness components. In bighorn sheep *Ovis canadensis*, high heterozygosity is associated with large horn size at sexual maturity, which confers breeding superiority (Fitzsimmons *et al.*, 1995). Multilocus genetic variation appears to be important in adaptation to continually varying parasite infestations in the fish *Poeciliopsis* (Lively *et al.*, 1990). Published studies of captive and wild organisms ranging from shellfish to mammals mostly showed positive relationships between allozyme marker heterozygosity and fitness components such as individual growth rate and body size (Allendorf & Leary, 1986). A caveat to these studies is that usually individual heterozygosity explains only a low proportion of variation in fitness ($r^2 < 0.2$). It must also be noted that there may be a reporting bias against studies which show no association, so their number may be an underestimate (Allendorf & Leary, 1986; Savolainen & Hedrick, 1995). However, there is unlikely to be a bias against studies showing a negative relationship between heterozygosity and fitness components, yet only a few such studies are known, suggesting that negative association is rare (Allendorf & Leary, 1986). In summary, the variety of relationships underlines the importance of studying the association between fitness and heterozygosity in a wide range of managed species. Finally, it is important to note that studies of individuals, whether they are inbred or not, do not fully address the phenomenon of loss of variation in a population. There can be considerable alteration to inbreeding coefficients, and observed levels of heterozygosity, without any change to the amount of gene or allelic diversity available for adaptive evolution (Fig. 2.1).

Studies of multiple populations with different levels of variation provide some support for the contentions that population size affects genetic diversity and that lowered genetic variation affects the viability of populations, through altering characteristics such as growth rate, population size or extinction probability (Frankham, 1995a; Lesica & Allendorf, 1995). There is some direct evidence that the level of genetic variation affects adaptability. Small isolated populations of adders with reduced allozyme variation showed poor recruitment until individuals from other populations were added (Madsen *et al.*, 1996). Bouzat *et al.* (1998a) showed similar results in prairie chickens. In the plant scarlet *Gilia*, individuals from smaller populations have poorer life-history characteristics such as seed size and germination success, and are more susceptible to stress; the partial genetic basis of these effects in small populations can be demonstrated by the improvement of life-history characteristics after transfer of pollen from other populations (Heschel & Paige, 1995). Bottlenecked *Drosophila* populations which have lost variation at microsatellite markers show poorer adaptive response to selection for salt tolerance (Frankham *et al.*, 1999). In most of the cases cited above, erosion of genetic variation and pedigree inbreeding are probably occurring simultaneously, so their effects are confounded.

Thus, a modest number of examples suggests that reduced genetic variation may sometimes affect demography sufficiently to have noticeable effects at the level of the population. There are more studies which indicate that this effect might be mediated by an erosion of gene diversity combined with pedigree inbreeding, and the consequences of both for individual fitness. However, a small proportion of the studies shows discordant results, indicating that there are limits to our ability to predict when adaptive genetic variation is being lost.

TYPES OF GENETIC VARIATION AND THEIR RELEVANCE TO POPULATION PROCESSES

Whereas many discussions allude to 'genetic variation' in general, it is important to recognise that the forms of variation that can be assayed differ with respect to mode of inheritance, accessibility, the nature of information recovered and relevance to population viability and management. Conservation genetic management and monitoring needs to target genes on the basis of their utility in adaptation, their replaceability and their usefulness to other aspects of conservation management. Firstly, there is clearly an interest in genetic diversity within and between populations which is relevant for current or future adaptation (categories 3–8 in Fig. 2.2). Unfortunately, we are presently rarely able to identify any of the appropriate genes, or even when we can do so, their analysis may not be cheap and easy. Thus the use of surrogates is a very important area in conservation genetics. Secondly, managers are also concerned with the replaceability of adaptive variation, and we speculate that replaceability is related to the underlying genetic architecture. After loss due to a bottleneck or other perturbation, variation due to specific alleles at loci of major effect is less likely to recover than variation due to the cumulative effects of alleles at numerous loci of small and equal effects. Therefore, diversity accumulated historically will be of particular importance for variants based on major genes. Finally, genetic analyses can be applied to the assessment of mating systems, gene flow and connectivity (Avise & Hamrick, 1996; Smith & Wayne, 1996), especially as these relate to the efficiency of selection and maintenance of meta-population viability (Endler, 1977; Varvio *et al.*, 1986; Hanski, 1998).

Neutral molecular markers (categories 1 and 2 in Fig. 2.2) have been employed widely in studies of threatened species, to analyse population structure and gene flow, and to assess levels of gene or allelic diversity within populations (Moritz. 1994*a*; Avise & Hamrick, 1996; Smith & Wayne, 1996). Newer laboratory and statistical methods are broadening the array of problems that can be tackled with neutral markers (Luikart & England, 1999). As tracers of gene flow, neutral molecular markers provide valuable information, although care needs to be taken to separate historical connectivity from current gene flow and to test inferences against a solid background of ecological and natural-history information. Maternally inherited loci (mtDNA in most animals; cpDNA in some plants) provide particularly relevant information insofar as the recovery of populations depends on recolonisation by females. More generally, comparisons of uniparentally (principally organellar) vs. biparentally (nuclear) inherited loci can reveal differences in dispersal between sexes, or other aspects of mating behaviour that affect gene flow between locations (e.g. Fitzsimmons *et al.*, 1997; Worthington-Wilmer *et al.*, 1999). Recent developments in the use of relative assignment probabilities based on hypervariable multilocus profiles (Paetkau *et al.*, 1998; Waser & Strobeck, 1998) suggest that, without assuming migration–drift equilibrium, it may be possible to identify individual migrants within populations and to test for sex-bias in dispersal.

It is more complex to analyse portions of the genome that are thought to be directly relevant to survival or reproduction (categories 3–8 in Fig. 2.2). Ideally, we would analyse the genetic components of fitness characters such as survival and fertility, and investigate the response of this genetic variation to different stresses, threatening processes and management regimes. There have been a few successful attempts to do this (Sgro & Hoffmann, 1998), but it is usually necessary to use surrogate measures ranging from analysis of a suite of representative loci to monitoring the distributions of populations. One relatively simple approach is to use neutral molecular markers to assess levels of gene or allele diversity within and between populations, and hope that the behaviour of these genes will reflect that of chromosomal regions which contain genetic variation of adaptive significance (Smith & Wayne, 1996; Lynch *et al.*, 1999).

It may also be possible to analyse molecular markers which are directly relevant to survival or reproduction, such as the major histocompatibility complex (MHC) in vertebrates (Hughes, 1991; Hedrick & Miller, 1994) and the mating incompatibility genes (S loci) in plants (DeMauro, 1993). Patterns of variation within and between species provide strong evidence that MHC loci are under balancing selection (Figueroa et al., 1988; Lawlor et al., 1988; Hughes et al., 1990) and the selective agents proposed include pathogens (Hughes et al., 1990; Hill et al., 1997), disassortative mating (Yamazaki et al., 1976) and maternal–foetal interactions (Hedrick & Thompson, 1988). However, data linking MHC diversity or specific genotypes with individual survival and reproduction are limited. The proposed association between population declines, susceptibility to disease and low MHC diversity in cheetahs (O'Brien et al., 1985) has proved controversial on several counts (Caro & Laurenson, 1994). A further caveat is that some species with evidently flourishing populations exhibit very low diversity at MHC loci (Slade, 1992; Ellegren et al., 1993).

In outcrossing plants, mating compatibility requires mismatch at the incompatibility loci, so polymorphism at these loci is a requirement for successful outcrossing. Within large populations, vast numbers of alleles occur, presumably maintained by balancing selection and, as with MHC, there is a high rate of amino acid evolution as well as retention of allele lineages across related species. Fragmentation and small population size reduce the diversity of S alleles, with demonstrable effects on embryo viability and the mating system (evolution of self-compatibility) (Reinartz & Les, 1994). Numbers of alleles can be analysed through laborious crossing studies, or sometimes by direct molecular assays of S locus variation, which may provide a useful tool for monitoring and determining the cause of reduced fertility.

Other genes that have been characterised at a molecular level, and which may predict response to specific environmental challenges, include insecticide-resistance loci, heat-shock loci and their regulatory genes, as well as loci coding for response to salinity, alcohol tolerance and other specific selective agents (McDonald & Kreitman, 1991; McKechnie et al., 1998; McKenzie & Batterham, 1998).

These studies demonstrate eloquently the process of natural selection at the molecular level, but we currently cannot directly monitor and manage all loci relevant for conservation, for three reasons. Firstly, it is unlikely that the same set of loci would be consistently responsible for adaptive traits in all conditions. Populations should be managed for the full spectrum of environmental influences on survival and reproduction, and any given selective agent will affect a large number of genes in diverse ways. Secondly, although associations of gene diversity and fitness components are often

Table 2.1. Examples of tests for loci under selection

Criterion	Reference
Heterozygote excess/deficit	Hedrick, 1985
Allele frequency distributions	Ewens, 1979; Tajima, 1989; Rand, 1996; Hartl & Clark, 1997
Number of common alleles $p \gg 0.05$	Ewens, 1979; Hartl & Clark, 1997
Recovery of frequencies from perturbation	McKenzie et al., 1994
Family data	Hedrick, 1985
Enzyme activity studies	Watt, 1985
Ratio of neutral to synonymous nucleotide substitution within one population	Muse, 1996
Ratio of neutral to synonymous nucleotide substitution between different taxa	McDonald & Kreitman, 1991; Muse, 1996
Comparison of levels of polymorphism within taxa to levels of variation between taxa or lineages (e.g. duplicated genes)	Fu & Li, 1993; McDonald, 1996

significant and positive, the range of values for the fitness component is often only small, so there is only a low likelihood that any particular locus will have an identifiable effect on fitness. Thirdly, in short-term experiments, it is difficult to distinguish between the effects of heterozygosity at individual loci, and heterozygosity at linked loci in the same chromosomal segment. However, some studies have demonstrated the direct effects of a single locus; by the use of null alleles, Leary et al. (1993) have shown that increased developmental stability of lactate dehydrogenase (LDH) heterozygotes is more likely due to possession of two different alleles at the LDH locus, rather than heterozygosity of linked loci (although the latter may also play a part).

Nevertheless, irrespective of the ability of individual loci to predict total fitness, monitoring of a suite of loci such as MHC or S loci may be useful as indicators of trends in diversity at other genes subject to similar evolutionary processes (e.g. balancing selection). Thus declines in allelic diversity could reflect increasing effects of genetic drift, assuming that the selection pressure has remained reasonably constant. There has recently been an enormous expansion in the range of data and tests available to detect loci under selection (Table 2.1). Some attempts at broadscale surveys for genes under selection have been successful (Endo et al., 1996). For conservation purposes, we wish to target loci and alleles of greatest relevance to population viability: genes that are under strong and consistent selection. Additionally, for these genes with major effects on fitness, diversity accumulated over long periods is unlikely to be readily replaced, so we are especially interested in persistent variation. Such genes can be best identified by using a variety of tests (Table 2.1), each of which has different sensitivity to the window of sequence analysed, time-scale of selection, type of selection, recombination rate and population expansion (Fu & Li, 1993; Simonsen *et al.*, 1995; Endo *et al.*, 1996; McDonald, 1996; Kelly, 1997; Munte *et al.*, 1997). Also, the same locus can apparently be under different selective pressures in different populations of the same species (Hamblin & Aquadro, 1997). Given all these provisos, and the low power of many of the tests, considerable strength and/or long-term consistency of selection is necessary to produce consistently detectable signals in two or more of the tests from Table 2.1. Therefore, this approach favours the detection of loci whose variation is most important for adaptation. As the content of DNA databases increases, these methods could be used more commonly.

A third major class of genes relevant to conservation consists of those that contribute to variation in quantitative traits, such as the great majority of ecological and demographic characteristics relevant to population viability. Falconer & Mackay (1996) have suggested that many, or possibly all, quantitative traits are under direct or indirect balancing selection. If this is true, these traits would be important targets for conservation genetic management and monitoring. Quantitative traits are typically determined by multiple loci with variable levels of effect on the trait, ranging from major loci to those with very small effect. Quantitative traits also display dependence on specific environments (environment–genotype interaction) as well as strong interactions between loci (epistasis) and effects of one locus on multiple traits (pleiotropy). Key issues in understanding the evolutionary dynamics of quantitative traits, and therefore management of genetic diversity, are the number of genes affecting a trait, the distribution of effect sizes, the diversity of alleles maintained at loci of large effect, and the extent of epistasis and pleiotropy. It is fair to say that our understanding of these issues is primitive, except for a few model systems (Falconer & Mackay, 1996; Lynch & Walsh, 1998). In principle, mapped molecular markers can be used to locate quantitative trait loci (QTLs, chromosome segments carrying one or more genes that influence a trait), as well as to compare the magnitude of effects and to predict phenotype value from specific crosses. Studies of model systems (e.g. *Arabidopsis, Drosophila, Mus*) have provided important insights, including the frequent detection of genes with major effect, strong epistasis and variable dependence on specific environments (Mitchell-Olds, 1995). The statistical methods for these analyses have been extended for use in natural populations (Haley *et al.*, 1994), but the expense and effort required preclude the use of QTLs in most circumstances relevant to conservation management.

Traditional analyses of the genetic diversity underlying quantitative traits (Falconer & Mackay, 1996; Lynch & Walsh, 1998) require large numbers of individuals of known relatedness, obtained from experimental crosses or intensively studied field populations (Cheverud *et al.*, 1994; Lynch & Walsh, 1998). There are very few estimates of the genetic component of phenotypic variation in wild populations, and these estimates may be very different between populations of the same species (Blows & Hoffmann, 1993) or between field and laboratory estimates (Prout & Barker, 1989). Despite the difficulty of obtaining precise estimates of additive variance and heritability underlying significant ecological traits, it is an important exercise for meaningful genetic monitoring, and efficient methods for obtaining such information should be pursued. For example, if it is possible to analyse large numbers of individuals spanning a wide range of relatedness, Ritland's (1996) method based on correlation of relatedness (estimated from microsatellite analysis) and phenotypic variation may be applicable to field populations of animals and plants. For many plants, estimation of heritability in natural populations can be achieved by collecting large numbers of half-sibs from individual parent plants, and raising them under controlled conditions, bearing in mind that unless the conditions replicate the wild environment, the heritability estimates are difficult to apply to the field.

EROSION OF GENETIC VARIATION WITHIN POPULATIONS

This section will examine the processes which are likely to result in erosion of variation within populations. The distribution of both single-locus and quantitative genetic diversity, within and among populations, is shaped by the combined action of mutation, genetic drift, migration, selection and mating patterns. Several forms of selection may be important, especially balancing selection and negative correlations between traits (antagonistic pleiotropy) (Barton & Turelli, 1989; Falconer & Mackay, 1996). Spatial and temporal environmental heterogeneity may also be important in maintaining quantitative diversity within populations (Mackay, 1981).

Genetic variation between individuals within populations is likely to be sharply reduced by bottlenecks (periods of small population size), because of random events in the transmission of small numbers of alleles (Wright, 1931; Franklin, 1980; Frankel & Soulé, 1981; Lande & Barrowclough, 1987;

Frankham, 1995b). Changes of genetic variation at the average locus within a population can be characterised in a number of ways; two qualitatively different measures are allelic diversity (the number of different alleles) and gene diversity (the likelihood that random mating would produce a heterozygote). For neutral loci, the rate of loss of gene diversity is dependent upon many demographic factors, including population size and fluctuations, sex ratio and variation in reproductive output between individuals. These factors are summarised as the 'variance effective population size' (N_e), which is inversely related to the rate of loss of two aspects of genetic variation: gene diversity at individual loci, and additive genetic variation for quantitative traits (Falconer & Mackay, 1996). Loss of neutral allelic diversity is a function of the same variables, but usually occurs more rapidly (Lande & Barrowclough, 1987). Most models of loss of genetic variation in small or fragmented populations are based on selectively neutral variation (categories 1 and 2 in Fig. 2.2). Additionally, most experimental investigations in conservation genetics use loci that are likely to be neutral, partly because of their utility for investigation of gene flow, and partly because they are relatively easily analysed. For example, mitochondrial DNA phylogeography probably falls into category 1, while the frequently analysed dinucleotide repeat microsatellite alleles almost certainly belong to category 2. Using methods such as these, a number of studies has shown association between bottleneck size or stable population size and levels of genetic variation at loci which are presumed to be neutral (van Treuren et al., 1991; Frankham, 1995b; Houlden et al., 1996; Madsen et al., 1996; Bouzat et al., 1998a).

The assertion that genetic variants are important in conservation because they are non-neutral (i.e. they affect fitness and adaptation) stands in stark contrast to the extensive modelling and study of selectively neutral genetic variation in conservation biology. For two reasons, neutral variants may be poor representatives of the adaptive genetic variation we wish to manage (categories 3–8 in Fig. 2.2). Firstly, their mutation rate and mode may differ from genes under selection. Evolution of some neutral markers such as microsatellites is best described by the stepwise mutation model, with high mutation rates (Di Rienzo et al., 1994), while genes under selection are more likely to follow the infinite-alleles model, with rates of mutation several orders of magnitude lower (Kimura, 1983). Secondly, neutral genes are not likely to have the same rate of loss of genetic variation as genes under selection. Genetic variants with small selective coefficients, or only occasional periods of selection, may sometimes be reasonably modelled by neutral methods, but this would not be true for variants with higher selective coefficients, which are likely to include many of the

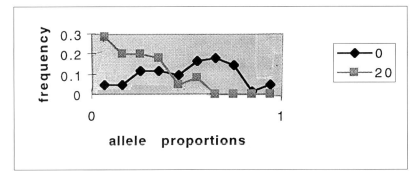

Fig. 2.3. Distribution of allele proportions, for the disadvantageous allele, under directional selection (20) and neutrality (0), modelled using the program POPULUS (Allstad *et al.*, 1993). Sixty diallelic loci were modelled, for 57 generations, with equal allele proportions initially, effective population size $N_e = 100$, selection coefficient against recessive homozygote $s = 0.05$ (additive effect in heterozygote), $4N_es = 20$. Equivalent results are shown for 60 neutral loci, $4N_es = 0$ ($N_e = 100$, $s = 0$).

most important variants to be conserved in a management programme.

Thus it is important that our monitoring should include genes, or genetically determined traits, that are under selection. There should be an adequate sample of these loci or traits to represent the behaviour of all genes under selection (discussed above). This approach should not be confused with suggestions that one or a few loci under selection should be the main focus of conservation genetic efforts (Hughes, 1991). Moreover, it is important to consider the different types of selection that genes may experience, as they have very different consequences in bottlenecked populations. In studies of inbreeding using single- or multilocus traits, inbreeding depression appears to be produced by directional and balancing selection ('recessive lethals' and 'heterosis' models respectively) (Charlesworth & Charlesworth, 1987; Templeton & Read, 1994). The same forms of selection are probably the most relevant for prediction of the effects of genetic erosion. Directional selection occurs when one allelic variant or phenotypic extreme is favoured at the expense of other(s). Directional selection is thought to be important for adaptation to local conditions, and is also sometimes called positive, negative or purifying selection. In a bottleneck, alleles at a local selective advantage would be expected to be retained, resulting in no adverse outcome for conservation. Figure 2.3 shows the results of a simulation in which the selection coefficient (s) and effective population size are both quite high ($N_e = 100$, $s = 0.05$, $4N_es = 20$). In this simulation, the effects of selection can be clearly seen: at many loci the disadvantageous

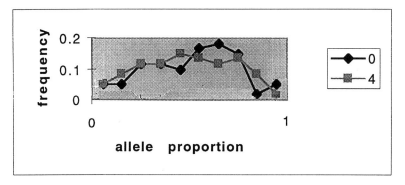

Fig. 2.4. Distribution of allele proportions, for the disadvantageous allele, under directional selection (4) and neutrality (0), modelled using the program POPULUS (Allstad *et al.*, 1993). Sixty diallelic loci were modelled, for 57 generations, with equal allele proportions initially, effective population size $N_e = 100$, selection coefficient against recessive homozygote $s = 0.01$ (additive effect in heterozygote), $4N_e s = 4$. Equivalent results are shown for 60 neutral loci, $4N_e s = 0$ ($N_e = 100$, $s = 0$).

allele is eliminated or decreased in frequency. Also, 14 out of 60 selected loci were fixed (allele proportion < 0.05), compared to four out of 60 neutral loci; these results are significantly different at the 5% level. The 14 selected loci were all fixed for the advantageous allele, which is significantly different from random expectations, at the 5% level. Thus it appears that in a small population of this size, selection is efficient at conserving alleles that are currently advantageous.

However, this relatively optimistic picture breaks down when we look at situations which may be more realistic in conservation management, with lower effective size and weaker selection. Adult population sizes will often be below 1000, leading to effective sizes well below 100. Natural selective coefficients of only 1 or a few percent can be very significant for long-term adaptation, so the allelic variants involved would be worth conserving. Unfortunately, drift can overcome natural selection if there are periods of small effective population size, or low selective coefficients. In this quasi-neutrality, alleles under directional selection will actually be lost as if neutral. Figure 2.4 shows that with $4N_e s = 4$ ($N_e = 100$, $s = 0.01$), the shape of the resultant allele frequency distribution is very similar to neutrality, as is the number of fixations (one selected locus was fixed for the advantageous allele, another was fixed for the disadvantageous allele). In general, selection will overcome the effects of random genetic drift if $4N_e s \gg 1$, whereas drift will prevail if $4N_e s \ll 1$ (Kimura, 1983).

Secondly, in extremely small populations, it is actually possible that the

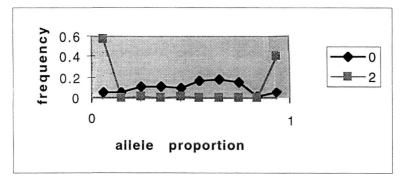

Fig. 2.5. Distribution of allele proportions, for the disadvantageous allele, under directional selection (2) and neutrality (0), modelled using the program POPULUS (Allstad *et al.*, 1993). Sixty diallelic loci were modelled, for 57 generations, with equal allele proportions initially, effective population size $N_e = 10$, selection coefficient against recessive homozygote $s = 0.05$ (additive effect in heterozygote), $4N_e s = 2$. Equivalent results are shown for 60 neutral loci, $4N_e s = 0$ ($N_e = 10$, $s = 0$).

common, advantageous allele could be lost altogether, the cumulative effects leading to 'mutational meltdown'. Figure 2.5 shows the results for $N_e = 10$ (perhaps 100 adults in the population) and $s = 0.05$ (quite a strong selection coefficient). The selection is still evident – significantly more loci are fixed for advantageous alleles than disadvantageous ones. However, in 24 out of 60 loci, the disadvantageous allele has been fixed, and this, accumulated over many loci, could lead to seriously reduced fitness. Lande (1995) and Lynch *et al.* (1995*b*) provide a more thorough treatment of mutational meltdown, in which it becomes apparent that it could be a factor in very small populations, although Gilligan *et al.* (1997) found no evidence for it in *Drosophila*.

Monitoring of the genetic variation required for response to directional selection is problematic. In severely bottlenecked populations, meltdown can be studied through analysis of total fitness through time. Analysis of individual marker loci is not helpful, because in any population there is only a small chance that any particular locus would be involved. In less severely bottlenecked populations, we wish to know whether our conservation measures are maintaining adaptive variation. Alleles with only a small selective advantage could be very important in the long term, and indeed any currently neutral or nearly-neutral allele could be at a selective advantage if the environment changes (categories 7 and 8 in Fig. 2.2). Therefore, in our attempts to manage and monitor variants which respond to directional selection, three possible strategies present themselves: (1) monitor total neutral allelic diversity, (2) monitor a set of genes that are representative of those under directional selection (see above), or (3) perhaps the most appropriate monitoring would be analysis of whether the ability to respond to different environments is being retained (Frankham *et al.*, 1999). This will be discussed further below.

The second major form of selection which may affect the rate of erosion of genetic variation is balancing selection. Balancing selection is a collective name given to a number of forms of selection which actively maintain variation within a population (Hedrick, 1985). Confusingly, balancing selection is sometimes called positive selection. Balancing selection has been identified at the antigen-recognition site of the vertebrate MHC loci (Hughes *et al.*, 1990), in a butterfly glucose-phosphate isomerase locus (*GPI*) (Watt *et al.*, 1983) and other loci (Endler, 1986). In one form of balancing selection, called overdominance, the heterozygote is fitter than both homozygotes, so the fittest genotypes can only be present in a population which has some genetic variation. Hence, maintenance of diversity at loci under balancing selection has considerable relevance to management.

It could be argued that it is so obvious that genes under balancing selection will retain their variation in bottlenecks that they require no monitoring, but this argument is far too simplistic. Firstly, theory shows that whether balancing selection will prevail over drift depends upon the selection intensity and other details. Overdominant balancing selection can dramatically retard the loss of genetic variation by drift for a wide range of population sizes, total selection intensities, and equilibrium allele proportions (Fig. 2.6) (Robertson, 1962).

However, Fig. 2.6 also shows that the loss of variation can actually be accelerated by as much as an order of magnitude, when unequal selection coefficients for the two homozygotes result in equilibrium allele proportions outside the range 0.2 to 0.8. This occurs because the selection holds the allele proportions close to fixation of one allele or the other (Robertson, 1962). Thus alleles which could be particularly useful in long-term adaptation are at a disadvantage in the short term, at effective population sizes which are not unrealistic in conservation management (e.g. $N_e = 80$, $s_1 = 0.02$, $s_2 = 0.18$). Another form of balancing selection, frequency-dependence, is likely to have a wider range of conditions under which selection overcomes the effects of drift, because this form of selection opposes fixation with greatest intensity when an allele is close to fixation. This form of selection may be common under conditions of antagonistic coevolution, such as predator–prey or host–parasite interactions.

Aside from theoretical criticisms of the argument that genes under bal-

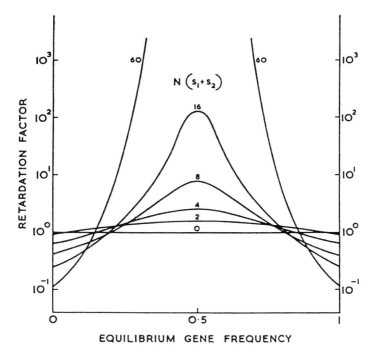

Fig. 2.6. The rate of fixation for a diallelic locus, with heterozygote advantage, in the absence of mutation. The retardation factor is given relative to neutral expectations. N is N_e, the effective population size, and s_1 and s_2 are the coefficients of selection against the two homozygotes. (From Robertson, 1962).

ancing selection will always retain their variation in bottlenecks, this contention is also not supported by a landmark study of some of the best-known genes under balancing selection, the MHC loci. O'Brien *et al.* (1985) showed that cheetahs have very low MHC variation, as well as low variation at other loci. It was suggested that this low variation might have been due to a presumed ancient bottleneck which had affected the non-selected and selected loci equally. Sanjayan *et al.* (1996) showed similar results for pocket gophers. Moreover, it has since been discovered that in cheetahs, some presumably neutral minisatellite loci actually have more variation than MHC loci. This observation conflicts with the prediction that balancing selection should retard loss of variation, and may be a consequence of the higher mutation rate of the minisatellite loci (O'Brien *et al.*, 1985; Hedrick, 1996).

What then can we manage and monitor within a single population, in order to conserve adaptive genetic variation (whether multilocus or single locus)? Because of the difficulties of measuring quantitative genetic variation in the wild (or even under controlled conditions such as in the laboratory), it is usual to use molecular variation as a surrogate for adaptive genetic variation. However, it must always be remembered that molecular variation is only a surrogate. The study of Ouborg & van Treuren (1994) on wild mint showed that levels of inbreeding depression were not related to the allelic diversity at allozyme loci, suggesting that these loci were not representative of the variation at fitness loci in the studied populations. Even if it is possible to find a set of markers whose equilibrium frequencies will reflect potential for adaptation, these markers may have different dynamics to adaptive quantitative traits, making them poor surrogates in the (more usual) non-equilibrium situation. There is theoretical and empirical support for the proposition that quantitative genetic variation, especially that due to many genes of small effect, is either not lost as fast or recovers faster than molecular variation (Cheverud et al., 1994; Lynch, 1996; Lynch et al., 1999). If this is true, then using molecular variation as a surrogate for quantitative variation will probably lead to over-conservative conservation decisions, which is in accordance with the 'precautionary principle' in conservation. Whatever measure of variation we adopt, there will always be the problem of defining the baseline for genetic variation. Studies without data on prior variation run the risk of making quite incorrect assumptions about the baseline against which to compare contemporary variation (Sherwin et al., 1991; Robinson et al., 1993). This problem can sometimes be overcome by amplification of DNA from more numerous populations represented in museum specimens, or congeneric species (Taylor et al., 1994).

Other surrogates for adaptive genetic variation can be considered. A frequently used surrogate is maintenance of large population size. Support for the use of this surrogate might be drawn from a general correlation of population size and genetic diversity (Frankham, 1996), but a number of possible criticisms means that this method must used cautiously. Firstly, there are some exceptions to the correlation of size and diversity (Leberg, 1993). Secondly, it is difficult to calculate effective population size from census size and demographic data (Frankham, 1995c), and there are discrepancies between the results of different methods of estimating N_e from genetic data (Nunney & Elam, 1994; Luikart & England, 1999). Thirdly, even for currently neutral variants (categories 1, 2, 7, 8 in Fig. 2.2), there are uncertainties about the effective population size necessary to maintain a balance between gain of gene diversity through mutation, and loss through drift (Franklin, 1980; Lande, 1995). The previous discussion shows that the necessary size for maintenance of diversity at particular selected loci may be one or more orders of magnitude higher or lower than for neutral loci.

Fourthly, conservation of allelic diversity may be important for long-term conservation, and the appropriate population sizes are many orders of magnitude larger than those required for maintenance of gene diversity (Lande & Barrowclough, 1987). As a result of these uncertainties, an extremely cautious approach should be taken, and if historical population sizes are known, it would have to be questioned whether any reduction of population size is acceptable. Often historical N_e is not known, or N_e may have been depressed by methods which are not immediately apparent from population size, such as alteration of sex ratio through selective harvest (Ginsberg & Millner-Gulland, 1994). Luckily, there are methods of inferring whether a population has been recently bottlenecked, using relatively accessible data (Luikart et al., 1998).

Given the uncertainties of conservation of adaptive variation, it is important to monitor whether a management programme is having the desired effect. Hypervariable markers such as microsatellites will allow us to check whether neutral allelic variation is being retained. Markers under balancing or directional selection are more difficult to devise at present, but the methods are being developed for both single-locus and quantitative variation (see above). There is a need for more studies which check whether programmes based on neutral markers, representative selective markers or other surrogates such as population size, actually conserve variation which aids response to changed selective regimes (Frankham et al., 1999).

EROSION OF GENETIC VARIATION BETWEEN POPULATIONS

A major threat to biological diversity is the reduction in geographic range of species as a consequence of habitat modification, species introductions and so on (Hughes et al., 1997). This phenomenon is particularly stark for Australian mammals, where many species have suffered range declines of more than 90% and often are now restricted to remnant populations on offshore islands. The changed distribution has potential consequences for metapopulation stability. A more insidious threat is the accompanying loss of genetic diversity, which ranges from local extinctions within historically connected populations through to extirpation of entire historically isolated populations, along with their unique and irreplaceable genetic diversity. Another often-suggested threat to geographic diversity is that posed by genetic introgression following restocking.

Evolutionarily significant units (ESUs) may be identified simply as historically isolated lineages defined by molecular studies (Moritz, 1994b), and sometimes phenotypic differentiation as well (Dizon et al., 1992;

Vogler & DeSalle, 1994; Wainwright & Waples, 1998), although where the molecular and phenotypic data are congruent the distinct populations could be regarded as separate species or subspecies and protected accordingly. Much of the variation seen or implied between ESUs belongs to categories 1, 3, 5, 7 in Fig. 2.2, the essentially irreplaceable variation. Whether or not it was neutral in the past, this variation may currently be selectively neutral (category 1), but could also be of potential significance to current or future adaptation (categories 3, 5, 7). Loss of ESUs can be inferred from dramatic declines in species with limited dispersal and wide geographic ranges (e.g. declining frogs throughout the world, and ghost bats in Australia: Churchill & Helman, 1990; Worthington-Wilmer *et al.*, 1999) but there is little direct evidence because of lack of historical samples. Many papers discuss the importance of conservation of separate ESUs for long-term management (and indeed for short-term also, if long-term options are to remain open) (Dizon *et al.*, 1992; Vogler & DeSalle, 1994; Moritz, 1994*b*; Wainwright & Waples, 1998). This chapter will focus upon variation between populations within ESUs.

Within one ESU, genetic variation is often partitioned geographically, either as clines across a continuum (Smith *et al.*, 1997; Schneider *et al.*, 1999), or in discrete populations which may or may not be connected by some contemporary gene flow. Discrete populations which are distinguished from one another by significant differences of allele or haplotype frequency are called management units (MU: Moritz, 1994*b*). This differentiation indicates some degree of short-term demographic independence, which should be considered for the purposes of demographic management and monitoring. For example, if a particular MU becomes extinct, it is unlikely to be quickly repopulated by a neighbouring MU, but populations that are part of a continuum may do so. Identifying the level of connectivity between populations requires care, because artificial gene flow can eliminate differentiation between populations that are essentially demographically independent, as in the case of koalas in south-eastern Australia (Houlden *et al.*, 1996; Sherwin *et al.*, 2000). Conversely, arbitrary divisions of a continuum can be misinterpreted as separate populations, which could have very adverse outcomes for population management (Taylor, 1997; Taylor *et al.*, 1997). Also, it is difficult to specify what degree of isolation is necessary to define separate MUs, since there is debate on the amount of gene flow necessary to prevent differentiation of populations by drift (Varvio *et al.*, 1986; Mills & Allendorf, 1996).

If the habitat varies, some of the variation between MUs, or between the extremes of a continuous cline, may be adaptive. For example, alleles conferring stress resistance may be more common in ecologically marginal populations than in optimal habitats where they tend to be selected against (Hoffmann & Parsons, 1991). Variation between populations is especially likely to be adaptive in cases where there is high gene flow between phenotypically distinct populations. As stated before, some of this type of adaptive variation could probably be relatively quickly re-created if lost due to threatening processes or management actions (categories 4 and 6 in Fig. 2.2). However, where there are differentiated populations in varied habitat, managers will maximise current viability, and the chances of continued adaptation, if they maintain this geographic variation.

A number of management actions or threatening processes could result in reduction of the genetic diversity embodied in the suite of populations which comprise a single ESU. One obvious problem would be loss of entire MUs, especially if this reduces the range of conditions experienced by populations in the ESU, which we call the 'ecological amplitude'. Accidental or deliberate mixing of stocks could also jeopardise the conservation of adaptive genetic diversity. It has been suggested that populations from the same ESU may be subject to deliberate exchange of migrants for conservation purposes such as demographic or genetic support, and that a low level of managed migration between differently adapted MUs is not expected to disrupt local adaptation (Moritz, 1994b, 1999b; Hedrick, 1996). However, this approach needs to be managed carefully to avoid swamping of local adaptation and consequent reduction in population viability. For example, koalas appear to have only a single ESU (Houlden et al., 1996, 1999), but the low diversity in southern populations is most likely the result of extensive restocking from a genetically depauperate stock, which has largely swamped any remaining local gene pools (Sherwin et al., 2000). Allendorf and others have debated whether introgression is reducing genetic variation between Pacific salmon stocks (Allendorf et al., 1997; Currens et al., 1998; Wainwright & Waples, 1998). Introgression into wild populations can even occur from introductions of a congeneric species (Abernethy, 1994; Rhymer & Simberloff, 1996). Hedrick (Chapter 7, this volume) discusses the possibility of introgression from congeners into wolf populations.

How likely is loss of populations or introgression to seriously reduce the adaptive component of genetic diversity? The expected distribution of (additive) genetic diversity between populations is well established for neutral traits, and depends on mutation, dispersal and effective population sizes (Hartl & Clark, 1997); the pattern is expected to be similar for single-locus and multilocus traits (Rogers & Harpending, 1983). In one of the few studies relevant to this point, Lynch *et al.* (1999) showed that quantitative and molecular genetic variation between populations was strongly correlated in *Daphnia*, although it is unclear whether the morphological markers were under selection. For adaptive traits, single-locus genetic models (Endler, 1977) and data (McKenzie & Parsons, 1974) suggest that a balance of selection differentials and dispersal can maintain differentiation between localities despite substantial levels of gene flow. For single-locus adaptive variation, these models could be used to predict the effects of altered gene flow. However, for quantitative traits, it is not so easy to make such predictions. Certainly many of these traits show substantial genetic variation between populations (e.g. wing length and bristle count in *Drosophila*: Coyne & Beecham, 1987). However, for a single species, some characters may show substantial among-site variation, while others do not (Mackay, 1981).

How then can we best conserve adaptive genetic variation between populations? Despite the paucity of direct experimental evidence pertaining to the geographic distribution of diversity underlying quantitative traits, the existing theory is sufficient to suggest that at least some components of adaptive diversity can be retained by maintaining viable populations across the full range of environments occupied by a species or ESU, preferably within connected and heterogeneous landscapes. Further to this, Lesica & Allendorf (1995) suggested that conservation of peripheral populations is particularly important for maintenance of evolutionary potential. This landscape approach to genetic conservation is consistent with ecological approaches to maintaining other components of biodiversity which also emphasise landscape-scale approaches. It should also be remembered that the adaptive component of quantitative variation within and between populations may be maintained by mechanisms other than spatial variation (e.g. temporal variation: Mackay, 1981), and also may be regenerated by mutation much faster than molecular variation (Lynch *et al.*, 1999).

There remains the question of what would be the best form of monitoring, so that in at least a proportion of management programmes, we can check whether our management is achieving its genetic aims. Ideally some measure of adaptive differentiation is needed, based on either molecular markers or estimates of quantitative genetic variation from experiments such as transfers between localities and 'common garden' experiments (Lynch & Walsh, 1998; Lynch *et al.*, 1999). Given the operational difficulties of reciprocal transplants, even in abundant species, we must consider the appropriateness of molecular markers. Lynch *et al.* (1999) showed correlation of quantitative and molecular genetic variation between populations, but the level of adaptive divergence is not necessarily correlated with the level of differentiation of neutral markers (Legge *et al.*, 1996; Moritz, 1999*a*). We are therefore forced to conclude that relying exclusively on neutral molecular markers is inadequate, and it is important that measures of genetic diversity between populations include adaptive traits. In the absence of data on spatially structured adaptive diversity, a useful surrogate may be 'ecological amplitude', the diversity of environments occupied by a single ESU.

SUMMARY AND RECOMMENDATIONS FOR MANAGEMENT AND RESEARCH

Fragmentation, decline or disturbance of a species can lead to genetic changes. Since genetic variation can affect fitness and adaptive potential, one important goal in precautionary conservation is to avert or remedy erosion of genetic variation within ESUs. Since adaptive genetic variation is difficult to analyse directly, our management and monitoring strategies must usually encompass surrogates (Brown *et al.*, 1997). Methods of using surrogates for direct manipulation of genetic variation in management include:

- minimising reduction of effective population size
- minimising change of natural levels of gene flow (this may sometimes require artificial relocations)
- minimising loss of separate management units
- minimising loss of peripheral populations, especially where clines are evident
- maintenance of the ecological amplitude of the species through retention of populations in different environments (this will require careful attention to choice of ecological factors to be measured)
- maintenance of normal temporal fluctuations.

Under these circumstances, any adapted phenotypes that are lost would be expected to be relatively rapidly reconstituted by the action of natural selection on the remaining stock.

This ecological approach to genetic conservation should be coupled with monitoring of genetic indicators, to allow us to assess whether the programme is having the desired genetic effect. Ideally these indicators will not be limited to neutral markers, and will have maximal information content and minimum cost (Brown *et al.*, 1997). There is an expanding range of genetic analyses that allows us to target specific aspects of population biology, and further research is required to identify which of these are most closely associated with fitness, adaptation and population viability, while being easily analysed in small populations. For each indicator, there must be careful consideration to the bias inherent in its derivation, such as the method of identifying loci under selection.

3

Inbreeding and outbreeding depression in fragmented populations

MICHELE R. DUDASH & CHARLES B. FENSTER

ABSTRACT

The goal of this chapter is to review inbreeding and outbreeding depression in the context of habitat fragmentation and to show how smaller, fewer populations of any organism separated by distance may exasperate the effects of these two genetic phenomena. We review the genetic basis of each, provide examples, and discuss specific empirical issues that need to be addressed in future research. We conclude with an illustrative case study of how both genetic phenomena can act simultaneously in a single species.

INTRODUCTION

Most rare and endangered species exist as small, isolated populations (Holsinger & Gottlieb, 1989). Unfortunately this seems to be the fate of even common species as natural populations are becoming increasingly fragmented. Fragmentation reduces the number of breeding individuals within a population while reducing gene flow between populations. Consequently, mating between individuals in fragmented populations is more likely to represent selfing (if genetically feasible) and/or biparental inbreeding (matings between related individuals) resulting in inbred offspring. The deleterious consequences of inbreeding are manifold. Inbred progeny may suffer from inbreeding depression, i.e. a decline in fitness, where the relative performance of the resulting inbred progeny is lower compared to progeny produced from matings between unrelated individuals within a population (Falconer & Mackay, 1996). Continued inbreeding associated with small populations also results in the loss of within-population genetic diversity (e.g. Schoen & Brown, 1991). Genetic diversity may influence the colonising ability and persistence of a population (Barrett & Kohn, 1991; Lande, 1994). Decreased genetic diversity may also be associated with increased susceptibility to pathogens and pests (Frankham, 1995b). Furthermore, as deleterious mutations are introduced to populations at a relatively high rate (Lynch, 1988) their accumulation and fixation are much more likely in small populations (Lande, 1994; Lynch et al., 1995a). In sum, fragmented populations may have reduced population mean fitness and suffer increased extinction rates because of increased expression of inbreeding depression, decreased levels of genetic diversity and higher probabilities of fixing deleterious mutations, relative to pre-fragmentation population structure.

A positive relationship between population size and genetic diversity is often observed (e.g. van Treuren et al., 1991; Leberg, 1993; Sun, 1996). Isolated populations frequently possess limited neutral molecular variation relative to less isolated populations (Brussard, 1984; Bayer, 1991; Holderegger & Schneller, 1994; Siikamaki & Lammi, 1998), providing empirical circumstantial evidence that isolation can also lead to the loss of genetic diversity and perhaps increase the expression of inbreeding depression. Manipulation of population size demonstrates that the fitness of a population will decrease following a bottleneck (Polans & Allard, 1989; Newman & Pilson, 1997). In a few cases, a direct relationship between reduced local population levels of heterozygosity (implying inbreeding) and the probability of local population extinction has been noted in natural populations of Granville fritillary butterfly (Saccheri et al., 1998) and the greater prairie chicken (Bouzat et al., 1998a).

Ecological considerations resulting from fragmentation may include difficulty in obtaining mates (i.e. Allee effect: Groom, 1998) and a lack of genetically compatible genotypes. For example, DeMauro's (1993) study of the self-incompatible lakeside daisy (*Hymenoxys acaulis* var. *glabra*) in Illinois, USA, demonstrates the potential role of chance/genetic drift on the population genetics of an endangered organism. The self-incompatibility system played a key role in obtaining successful seed production once the last native population had become so small that natural recovery was impossible. DeMauro found that only a handful of individuals remained and all possessed the same self-incompatibility type. Hand cross-pollinations between plants from Ohio and Illinois successfully produced viable progeny that were initially maintained in glasshouses. Following seedling establishment they were transplanted into protected nature preserves to re-establish this species in Illinois.

Clearly, if endangered populations are small and inbred then consideration has to be given to manipulation of the remaining populations to counteract the erosion of their fitness. Two methods to restore a population's vigour whose decline is due to inbreeding are (1) purge the population of its mutational load and (2) seed populations with progeny from interpopulation crosses (e.g. Fenster & Dudash, 1994). The success of both methods depends on the genetic basis of variation within and among populations. Purging a population of its genetic load by selecting among high performance (most fit) progeny following intense inbreeding will only be successful if inbreeding depression is due to the expression of recessive or partly recessive deleterious alleles expressed in the homozygous state (Fenster & Dudash, 1994). However, because purging requires close inbreeding, it is likely that weakly deleterious alleles will be fixed during the purging process owing to an increase in the role of drift relative to selection, leading to an overall reduction of population vigour (Hedrick, 1994; Lynch et al., 1995a). An alternative method to restore heterosis is to make interpopulation crosses. However, this raises the issue of outbreeding depression, or the loss of vigour that may result from crossing individuals from different/and or distant populations (e.g. Fenster & Dudash, 1994; Frankham, 1995b).

The goal of this chapter is to review inbreeding and outbreeding depression, and their potential role in population restoration within fragmented landscapes. We will briefly review what theory and empirical data suggest at this time, and propose future directions for research and management strategies. The empirical emphasis will be on flowering plants; however, we will make comparisons wherever possible with animal systems.

INBREEDING DEPRESSION

Inbreeding depression is defined as the reduction in the mean phenotype of a population associated with increasing homozygosity which results from matings between relatives, i.e. biparental inbreeding and selfing (mating occurs within an individual) (Falconer & Mackay, 1996). Traits known to exhibit inbreeding depression include such components of fitness as seed production of the parent, germination, juvenile survival and growth/reproduction of the offspring (e.g. Charlesworth & Charlesworth, 1987; Husband & Schemske, 1996), pollen and ovule production (e.g. Dudash et al., 1997), plant physiological traits (Norman et al., 1995), sperm production (e.g. O'Brien et al., 1987), egg-hatching rates (e.g. Westemeier et al., 1998) and long-term survival (e.g. Jiménez et al., 1994). Individual fitness is ultimately measured by both the quantity and quality of progeny produced by an individual that contributes to the next generation (e.g. Dudash, 1990). Comparisons of progeny performance utilising a multiplicative fitness function, incorporating all measured components of fitness into a single value for each cross type, indicate that the magnitude of inbreeding depression experienced in populations is often great (e.g. Sakai *et al.*, 1989; Dudash, 1990; Fenster, 1991*b*; Carr & Dudash, 1996). However, the environment in which one chooses to examine inbreeding depression can influence the magnitude of detection in both plants (e.g. Schemske, 1983; Dudash, 1990) and animals (e.g. Miller, 1994; Pray *et al.*, 1994) and needs to be considered when assessing the state of any population. Thus concerns raised about the consequences of inbreeding in fragmented populations should be addressed in the appropriate environment, i.e. nature, if maintenance of natural populations is of primary importance.

Darwin (1876) documented inbreeding depression in both cultivated and native plant species. His results have been confirmed across the range of diversity of the plant kingdom including cultivated plants such as maize (e.g. Hallauer & Miranda, 1985), naturally occurring annual plants *Gilia achilleifolia* (Schoen, 1983), *Chamaecrista fasciculata* (Fenster, 1991*b*) and *Impatiens capensis* (McCall *et al.*, 1994), hermaphroditic obligate biennials *Sabatia angularis* (Dudash, 1990) and *Hydrophyllum appendiculatum* (Wolfe, 1993), gynodioecious shrubs *Hebe subalpina* (Delph & Lloyd, 1996) and *Schiedea* spp. (Sakai *et al.*, 1989, 1997; Culley *et al.*, 1999), herbaceous perennials *Costus* (Schemske, 1983) and *Lobelia* spp. (Johnston, 1992), ferns (e.g. Kirkpatrick *et al.*, 1990; Soltis & Soltis, 1990, 1992), gymnosperms (e.g. Bush *et al.*, 1987; Williams & Savolainen, 1996; Sorensen, 1999) and flowering trees (e.g. Eldridge & Griffin, 1983; Brown *et al.*, 1985). Investigations of the impact of inbreeding depression in natural animal populations include studies of prairie dogs (Hoogland, 1992), mice (Jiménez *et al.*, 1994) and song sparrows (Keller, 1998), as well as captive populations of various animals (e.g. Ralls & Ballou, 1983; Lacy, 1993*a*; Lacy *et al.*, 1993). Given its widespread occurrence it is likely that we need to factor inbreeding depression into the management of natural populations.

Plants especially display a range of breeding systems, from selfing to complete outcrossing. We would like to know whether our concerns about inbreeding depression should be applied equally to all taxa. Selfing is the most extreme form of inbreeding since each generation of selfing results in a 50% increase in homozygosity of the progeny or in other words a 50% decrease in heterozygosity each generation (Falconer, 1981). Biparental inbreeding refers to matings between two related individuals. In this situation, increase in homozygosity levels will be slower than from selfing, thus decreasing the immediate potential expression of inbreeding depression. Although the rate of increase in homozygosity levels differs among the

Table 3.1. Expected homozygosity levels for different modes of inbreeding

Generation	Mode of inbreeding		
	Half-sib	Full-sib	Selfing
1	0.125	0.250	0.500
2	0.219	0.375	0.750
3	0.305	0.500	0.875
4	0.381	0.594	0.938
5	0.448	0.672	0.969
10	0.692	0.886	0.992

various mating strategies, ultimately all familial lines could become homozygous to the same degree over time regardless of the mating system (Table 3.1). How rapidly one can generate inbred lines influences the balance between the role of selection in removing deleterious alleles and the random fixation of deleterious alleles. If inbreeding is intense, then the role of drift is predicted to be more important than if inbreeding is weakly enforced across many generations (Lynch, 1988; Ehiobu *et al.*, 1989).

With no gene flow between fragmented populations individuals will eventually become inbred within a population. In animals, biparental inbreeding is the most common form of inbreeding since dioecy is the norm (separate male and female individuals within a population). Some plants also exhibit dioecy where related individuals are mated via a pollen vector (biotic or abiotic) owing to proximity between two individuals. In some hermaphroditic plant species there exist mechanisms that prevent inbreeding which include self-incompatibility systems, and temporal and physical separation of male and female function within a flower (dichogamy and herkogamy, respectively) as well as monoecy where there are separate male and female flowers on the same individual (all above reviewed in Briggs & Walters, 1997). Selfing contributes substantially to mating patterns in plants (Schemske & Lande, 1985) and can occur both within a single hermaphroditic flower or between two flowers on the same individual, i.e. geitonogamy. Some animals also exhibit selfing including freshwater snails *Lymnaea peregra* (Jarne & Delay, 1990) and *Physa heterostropha* (Wethington & Dillon, 1997). Jarne & Charlesworth (1993) recently reviewed the presence of selfing and its potential evolutionary path in both hermaphroditic plants and animals. As we shall see below, the interaction between inbreeding depression and mating system largely depends on the genetic basis of inbreeding depression.

There are two genetic mechanisms thought to be responsible for inbreeding depression; however, they are not mutually exclusive, confounding empirical elucidation of their genetic basis (Charlesworth & Charlesworth, 1987). The first is dominance, in which loss of fitness is due to increased expression of recessive or partially recessive deleterious alleles as homozygosity accumulates (Wright, 1977; Falconer, 1981; also known as partial dominance in Charlesworth & Charlesworth, 1987). The second is overdominance or heterozygote advantage. Under this mechanism, heterozygosity is advantageous *per se*, and inbreeding depression results from a breakdown of this advantage as heterozygosity declines (Wright, 1977; Lande & Schemske, 1985). The genetic basis of inbreeding depression will directly affect the ability to purge the genetic load of a population. Utilising the simplest scenario, the dominance hypothesis of inbreeding depression predicts that the amount of inbreeding depression *decreases* with increasing self-fertilisation in the presence of selection. This occurs because inbreeding increases both homozygosity and the efficiency of selection in removing deleterious recessive alleles from the population. A population's genetic load is expected to be more difficult to purge if overdominance is responsible for the observed inbreeding depression (Lande & Schemske, 1985; Charlesworth *et al.*, 1990), since the most-fit heterozygous genotypes continue to re-generate the less-fit homozygous genotypes. The overdominance hypothesis predicts that the amount of inbreeding depression *increases* with increasing self-fertilisation unless selection on viability against homozygotes is asymmetrical (Charlesworth & Charlesworth, 1987, 1990; Ziehe & Roberds, 1989). Other factors, however, such as epistasis (e.g. Crow & Kimura, 1970; Bulmer, 1985; Lynch, 1991), linkage (or pseudo-overdominance: e.g. Comstock & Robinson, 1952; Wright, 1977 and references therein), selection and drift may all influence the subsequent expression of inbreeding depression and consequently affect the ability of a population to purge its genetic load as well.

Understanding the genetic basis of inbreeding depression is important in predicting the success of a purging programme. The genetic basis of inbreeding depression has historically been examined primarily in crop plants. Evidence exists for dominance-based inbreeding depression in alfalfa (El-Nahrway & Bingham, 1989) and maize (Moll *et al.*, 1965; Hallauer & Miranda, 1985). Overdominance-based inbreeding depression has been suggested from data on orchard grass (Aprion & Zohary, 1961), cherry (Williams & Brown, 1956), barley (Gustafson, 1950) and maize (Hallauer & Miranda, 1985). The relative importance of dominance-vs. overdominance-based inbreeding depression in natural populations is largely unknown. Studies of allozyme variation in pitch pine suggested overdominance as the mechanism to explain fitness differentials between self and outcross progeny (Bush *et al.*, 1987). However, a growing number of other studies points to dominance-based inbreeding depression. Indirect evidence of dominance-based inbreeding depression has been found in studies of *Eichhornia paniculata* (Barrett & Charlesworth, 1991). Furthermore, approximate levels of dominance in two largely self-fertilising *Amsinkia* species provide further evidence for the role of deleterious recessive alleles in natural populations (Johnston & Schoen, 1995). Finally, a quantitative genetic study of two *Mimulus* species with contrasting breeding systems primarily supports dominance-based inbreeding depression as well (Dudash & Carr, 1998).

Can genetic load be purged? In other words, can the genetic load responsible for inbreeding depression be purged following either a natural population bottleneck (i.e. dramatic reduction in population size owing to declining habitat and/or fragmentation) or through a controlled inbreeding programme? Examination of population bottlenecks in nature has provided controversial support for reduction of genetic load in Speke's gazelle (Templeton & Read, 1984; but see Lacy, 1997) and the European bison (Simberloff, 1988; Lacy *et al.*, 1993; Ballou, 1995).

To conduct a purging experiment one must simultaneously inbreed to increase levels of homozygosity while selecting for high-performance genotypes across generations. A purging experiment assumes that the genetic mechanism responsible for the reduced fitness is deleterious recessive alleles which in a homozygous state can be eliminated, which has been largely demonstrated (see above). Additionally, epistasis can either enhance ('reinforcing') or inhibit ('diminishing') inbreeding depression as a function of increasing inbreeding coefficients (F) (Wright, 1977) and is documented by the non-linear decline in fitness because the expression of partially deleterious mutations is not independent (Crow & Kimura, 1970). Ideally one would want to compare the performance of elite inbred lines purged of their genetic load with the performance of outcrossed progeny in a natural environment.

Numerous empirical investigations of natural plant populations demonstrate significant variation among maternal lines in expression of inbreeding depression (e.g. Sakai *et al.*, 1989, 1997; Dudash, 1990; Norman *et al.*, 1995; Carr & Dudash, 1997; Dudash *et al.*, 1997; Mutikainen & Delph, 1998; Culley *et al.*, 1999). This information suggests that purging by selecting among maternal lines can be accomplished. However, we are unaware of any ideal purging studies. We can gain some insight from a number of serial inbreeding depression studies that have tried to assess the likelihood of purging in their study systems. At the population level Barrett

& Charlesworth (1991) and McCall *et al.* (1994) suggest that purging has occurred in populations of *Eichhornia paniculata* and *Impatiens capensis*, respectively. For example, individuals from an outcrossing, tristylous population in Brazil of *E. paniculata* were self-pollinated for five generations and then matings were performed between the inbred lines. Barrett & Charlesworth (1991) suggest that purging has occurred owing to an increase in flower production following intercrossing the inbred lines when compared to flower production following random mating (prior to any artificial inbreeding) in the parental population. In contrast, population-level purging was not indicated in other serial inbreeding and outcrossing studies of *Mimulus guttatus* (Carr & Dudash, 1997; Dudash *et al.*, 1997) or *Tribolium* (Pray & Goodnight, 1995). However, in the studies of *M. guttatus* and *Tribolium* cited above purging in some maternal families was demonstrated because self progeny were produced that outperformed the outcrossed progeny from the same female. Other maternal lines exhibited a consistent performance advantage of outcrossed progeny compared to self progeny, while still others revealed no pattern in performance, illustrating the importance of drift, i.e. the random fixation of a trait. Furthermore, a laboratory serial inbreeding depression study (without selection) of three subspecies of *Peromyscus* suggested purging of the genetic load, no purging and an increase in the genetic load for the three different subspecies examined (Lacy & Ballou, 1998). Additionally, Ballou (1997) found little evidence of purging after examining pedigrees of 25 captive populations of mammals. These investigations provide further support for the need to conduct true purging studies as described above.

In an attempt to generate inbred lines and enhance purging of a population's genetic load numerous family lines may be lost (e.g. McCall *et al.*, 1994; Dudash *et al.*, 1997). Losses can range from 30% to 80% of the original families and this is an important issue when one is dealing with a threatened or endangered organism where every individual is valuable. Nonetheless, loss of lines in nature is quite common through death and unequal reproduction among individuals from mammals and birds (e.g. Gompper *et al.*, 1997) as well as plants (e.g. Dudash, 1990, 1991; Dudash & Fenster, 1997), which suggests that purging may not decrease overall population vigour to a great extent. However, the loss of maternal lines increases the likelihood of drift leading to the fixation of slightly deleterious alleles as well as an overall general loss of genetic diversity.

Inbreeding depression can be reduced as the mating system of a population evolves toward selfing (Lande & Schemske, 1985), supporting the efficacy of purging. Thus, if selfing species exhibit reduced inbreeding depression compared to a related outcrossing species this suggests an increased efficiency through inbreeding to reduce the genetic load. Theoretical expectations in favour of dominance/partial dominance predict that inbreeding depression should decrease with increased inbreeding as deleterious recessive alleles are expressed and purged via natural selection (e.g. Charlesworth & Charlesworth, 1987). A review of 54 species of vascular plants by Husband & Schemske (1996) demonstrated a significant reduction in inbreeding depression in selfing species compared to outcrossing species. Their review also revealed that outcrossing species exhibited greater inbreeding depression in early life-history stages while selfing species tended to exhibit greater inbreeding depression in later life-history stages. This trend suggests that inbreeding depression expressed during the early life-history stages is caused by mutations of major effect while inbreeding depression expressed in later life-history traits is caused by mutations of small effect. In addition they found a general lack of correlation in the expression of inbreeding depression between traits thought to be associated with an individual's fitness throughout the life history. This final point also suggests that numerous genes are responsible for inbreeding depression throughout the life history. One study examined critical reproductive traits (pollen and ovule production), and demonstrated that natural selection had been effective in removing the genetic load associated with these fitness traits in the selfing *Mimulus micranthus* when compared to its mixed-mating congener species, *M. guttatus* (Carr & Dudash, 1996). However, given that some traits still exhibit inbreeding depression in selfing species also indicates that selection may be unable to purge the complete genetic load because (1) overdominance may be acting at some loci, (2) mutations introducing new deleterious alleles are fixed by drift and (3) the load of some traits may be polygenic, i.e. composed of alleles at many loci of slight deleterious effect making purging difficult. Additionally, research on *Drosophila* indicates that about one-half of inbreeding depression is due to lethals and the other half of inbreeding depression is due to detrimentals of smaller effect (Simmons & Crow, 1977).

The ability to purge genetic load may also be dependent on ploidy levels (e.g. Ronfort, 1999). This is an important issue in plants since upwards of 50% of all plant species are polyploids (Lewis, 1979; Grant, 1981; Masterson, 1994). Polyploidy is relatively rare in animals, being confined to self-compatible hermaphrodites (e.g. flatworms, earthworms, and freshwater snails) and parthenogens that are capable of producing offspring without fertilisation, (e.g. some species of shrimp, goldfish and salamanders) (Lewis, 1979). Polyploidy, i.e. having more than two sets of homologous chromosomes, can occur from independent doubling (autopolyploidy) or through matings between two individuals of different species that subsequently results in a doubling of genetic material (allopolyploidy). Allopolyploidy, the predominant mode of polyploidy, results in fixed heterozygosity which in part may explain the success of allopolyploids compared to their diploid ancestors. Fixed heterozygosity implies that allopolyploids will be less affected by inbreeding depression than their diploid ancestors. However, this would seem to be dependent on the age of the polyploid and the rate of accumulation of mutations. Whether allopolyploids actually express less inbreeding depression after long-term inbreeding than their diploid ancestors is an important empirical question. The case is certainly far more complex for autopolyploids which are now recognised to be much more evolutionarily common (Soltis & Soltis, 1989). The assumption of partial dominance/dominance predicts a reduction (or slowing) in the expression of inbreeding depression since homozygosity increases following selfing by only 17%–20% in an autopolyploid compared to 50% in a diploid (e.g. Haldane, 1930; Mayo, 1987). Thus purging may be slower in autopolyploid species compared to diploids. However, this is conditional on the degree to which deleterious mutations are recessive and how closely linked they are to the centromere, and is also related to the interactions among the mutations in the various heterozygous states (fitness of $Aaaa$ vs. $AAaa$ vs. $AAAa$, where a is the deleterious recessive allele) and homozygous ($aaaa$) states (Ronfort, 1999). Overdominance-based inbreeding depression predicts an increase in inbreeding depression in polyploids compared to diploids, owing to a decrease in potential higher-order heterotic allelic interactions following selfing (Bever & Felber, 1992). Empirical data on natural populations of flowering plants are inconclusive. Overdominance-based inbreeding depression is supported by increased expression of inbreeding depression in polyploid populations compared to diploid populations of *Amsinckia* (Johnston & Schoen, 1996). However, dominance-based inbreeding depression is supported by the observation of decreased inbreeding depression in polyploid populations compared to diploid populations of *Epilobium angustifolium* (Husband & Schemske, 1997). Even if we assume that dominance-based inbreeding depression is the basis for inbreeding depression we still have no empirical understanding of the fitness of the various heterozygous classes relative to one another. Thus we can make little prediction as to the success of purging in polyploids at this point. Clearly we need comparative studies that assess the ability to purge the genetic load of related diploid and polyploid congeners.

An alternative approach to purging is to conduct crosses among populations to combat inbreeding due to fragmentation. However, this latter approach raises the issue of outbreeding depression, which is discussed below.

OUTBREEDING DEPRESSION

Following the definition for inbreeding depression it seems reasonable to define outbreeding depression as the phenomenon where outcrossed offspring have lower relative performance or fitness than the parents (Lynch & Walsh, 1998). The term has been used to reflect the loss of fitness following crosses among individuals within a single population (Parker, 1991), between individuals in nearby and distant populations (Fenster, 1991b; Waser, 1993) and the product of interspecific hybridisation (Orr, 1995). Recall that fitness is a relative measure of performance of genotypes (Crow & Kimura, 1970). In most studies (reviewed in Waser, 1993), the term outbreeding depression is used slightly differently than as defined above with the difference in usage focusing on the word 'relative'. Thus in many studies, outbreeding depression has also been used to describe loss of fitness relative to some optimum crossing distance, which itself may reflect a cross not normally observed in nature, e.g. an interpopulation or higher cross. There is no inherent reason to restrict usage of the term outbreeding depression but none the less it is important to be clear on what is actually being described when measuring outbreeding depression. Using the definition of Lynch & Walsh (1998) above seems most useful for conservation purposes where interest focuses on whether to keep the genotypes of parental populations intact or to introduce hybrids in order to combat the cumulative effects of inbreeding depression (see discussion above).

There are further problems with the Lynch & Walsh (1998) definition of outbreeding depression because outbreeding depression may reflect complex interactions among ecological, evolutionary and genetic processes and the nature of these interactions may change with the scale of cross between the parents. The inherent confusion arising from using the term outbreeding depression is perhaps more clearly illustrated by a more detailed description of how outbreeding depression arises. Firstly, dilution of genes associated with local adaptation may lead to loss of fitness of the hybrid offspring. For example, if each parental population represents an ecotype where a certain number of loci confer local adaptation then where these loci are fixed for alternative alleles between populations, hybrids will have on average only half the genes of either parent. Consequently, hybrids may be less fit than either parent in either parental environment. Secondly, hybridisation may also result in the disruption of coadapted gene complexes (Fenster & Dudash, 1994). Genetic coadaptation reflects epistatic gene action, i.e. the interaction among loci that enhance fitness (Falconer & Mackay, 1996). Thus, if the selective advantage of a particular allele depends on alleles present at other loci, and if each population represents a unique mixture of alleles across loci, then mixing gene pools may lead to the disruption of well-integrated genotypes.

The problems associated with using the term outbreeding depression arise when we realise that both hybrid vigour and hybrid breakdown can be simultaneously expressed in hybrid populations, due to the simultaneous masking of recessive deleterious alleles, dilution of genes that confer local adaptation and disruption of coadapted gene interactions (Lynch, 1991). Furthermore, the interaction among these genetic effects is likely to change depending on which particular hybrid generation one is comparing with the parents. Thus we expect the F_1 generation to express the most heterosis, because all individuals are heterozygous at the maximum number of loci. With continued random mating heterozygosity is reduced by one-half (as a consequence of Mendelian $1:2:1$ segregation); thus the expression of heterosis in F_2 and later segregating generations is expected to be one-half that of the F_1. Compounding the loss of heterozygosity in these later segregating hybrid generations is the continued action of recombination which will result in gene combinations among the populations at smaller and smaller intervals of the chromosome. If coadapted gene complexes reflect linked genes, as predicted by theory (reviewed in Fenster *et al.*, 1997), then the loss of fitness due to the disruption of coadapted gene complexes is expected to be greater with each passing generation. Since populations exist on an ecological time-scale and gene flow is often restricted, populations may be inherently limited in their response to selection pressures because what represents an adaptive genotype in any given population may not reflect the testing of all combinations of genes found in a species. Thus by chance we might expect that a small proportion of the gene combinations that arise via interpopulation crosses may actually confer higher fitness than either parent. This latter scenario has been invoked in several recently described cases of recombinational speciation among a number of plant taxa (e.g. Rieseberg, 1997). The spread of these successful hybrid combinations will then depend on how often these chance positive associations arise and the selective advantage they confer (Kruuk *et al.*, 1999).

The environmental context may also increase the complexity of the interpretation of what is and is not outbreeding depression. For example, is it outbreeding depression if the hybrid performs poorly relative to the home parent, but better than the transplanted parent? The answer depends on the interest of the investigator. For a conservation biologist or land manager interested in maintaining the highest-performing population, lower hybrid performance relative to any one parent would constitute outbreeding depression. Thus the obvious strategy would be to keep population genotypes intact and not allow hybridisation. To summarise, when measuring outbreeding depression, we must be cognisant of the particular hybrid generation we are measuring relative to the parents and in what environment we are making the comparisons.

Mating system and ploidy levels should theoretically have a large influence on the expression of outbreeding depression. As the degree of sexuality decreases either through increasing asexuality or selfing, it is more and more likely that adaptation to very local environmental conditions will occur (e.g. Antonovics & Bradshaw, 1970; Endler, 1977; Schmitt & Gamble, 1990). Thus, gene dilution effects are likely to be greater in more highly selfing organisms following artificial production of hybrids. Furthermore, as discussed above, there is expected to be decreased expression of heterosis because of purging of the genetic load, and the loss of fitness through the disruption of coadapted gene complexes is likely to be greater. Any mating system which effectively reduces recombination will also in turn promote the evolution of coadapted gene complexes (Fenster et al., 1997), thus crosses among populations with reduced tendencies of sex and recombination will be more likely to increase the chances of disrupting coadapted gene complexes. The evolution of gene combinations may also be more important in polyploid organisms (Breese & Mather, 1960; Honne, 1982), because there are more genes and hence a greater opportunity for their interaction to contribute to genetic variation. Thus crosses among polyploid populations may also suffer the increased chance of disrupting coadapted gene complexes. We also expect an optimal outcrossing distance where the maximum amount of heterosis might be conferred on the progeny while minimising dilution of gene effects and disruption of coadapted gene complexes.

For comprehensive reviews on outbreeding depression see Waser (1993), Fenster & Dudash (1994), Frankham (1995b), Whitlock et al. (1995) and Fenster et al. (1997). The loss of fitness of hybrid generations relative to their parents first received prominence with the experiments of the early *Drosophila* geneticists who demonstrated the role of coadaptation in population divergence (Wallace, 1953; Brnic, 1954; Wallace & Vetukhiv, 1955;

Anderson, 1968) (but see McFarquhar & Robertson, 1963). Recent studies using marker-assisted techniques have demonstrated the contribution of coadaptation among genes to the adaptation of experimental populations of cultivated barley (Clegg et al., 1978), to reproductive isolation among sibling species of Drosophila (Palapoli & Wu, 1994), to introgressive hybridisation in Helianthus (Rieseberg et al., 1995), to genetic differentiation among cultivars for traits correlated to yield (Doebley et al., 1995; Lark et al., 1995; Li et al., 1997) and to differences among mouse strains for body weight (Routman & Cheverud, 1997). Furthermore, the role of coadaptation in adaptive evolution has been investigated by quantifying epistasis in laboratory environments which simulate natural conditions and testing the contribution of epistasis to the divergence of adaptive characters. Examples include the evolution of alcohol tolerance in laboratory populations of Drosophila melanogaster (Cavner & Clegg, 1981), population differentiation of photoperiod requirements for diapause in the pitcher-plant mosquito (Hard et al., 1992, 1993; Armbruster et al., 1997), osmoregulation of the tidepool copepod Tigriopus californicus (Burton, 1987, 1990), gill-raker length differences between sympatric species of stickleback (Hatfield, 1997) and others (reviewed in Whitlock et al., 1995; Fenster et al., 1997).

Outbreeding depression can also occur at a very local scale. In some studies loss of fitness occurs in crosses among asexual or inbred lines within the same population (Templeton et al., 1976; Parker, 1991; Deng & Lynch, 1996) and in others (Burton, 1987, 1990) disruption of coadapted gene complexes occurs among populations as close as 10 km apart. In several other studies (notably Price & Waser, 1979; Waser & Price, 1989, 1994) outbreeding depression in the F_I of crosses among plants only tens of metres apart have been quantified, but in this case breakdown may reflect a loss of fitness relative to other outbred progeny and not necessarily to the original parental population. The breakdown of presumably adaptive characters in the progeny of interpopulation crosses also provides examples of outbreeding depression. For example, between-population crosses result in larger retina size in the blind cave fish (Wilkens 1971), breakdown of pesticide resistance in houseflies (King, 1955) and the recovery of the wild-type breeding system in Eichhornia paniculata (Fenster & Barrett, 1994).

While an impressive number of studies indicates the presence of outbreeding depression as defined in a number of different ways, we still need studies that address the following issues. Can heterosis offset hybrid breakdown? At what spatial scale can crosses be conducted where the beneficial consequences of heterosis are stronger than hybrid breakdown? And what are the relative magnitudes of heterosis and hybrid breakdown when compared to the performance of the original parental populations? These remain important unanswered questions. Thus, the long-term consequences of mixing populations of endangered or threatened species has not been adequately documented (Whitlock *et al.*, 1995; Fenster *et al.*, 1997).

CASE STUDY: SIMULTANEOUS INVESTIGATION OF INBREEDING AND OUTBREEDING DEPRESSION IN *CHAMAECRISTA FASCICULATA*

A long-term project by Fenster and colleagues investigated the questions posed above by examining genetic structure and the role of inbreeding and outbreeding depression in population differentiation of the native North American, highly outcrossing legume, *Chamaecrista fasciculata*. The species is discretely distributed across eastern North America so it represents a model system to examine the consequences of fragmentation in terms of inbreeding and the consequences of crossing among populations to combat inbreeding effects. To maintain brevity we refer the reader to other publications where a detailed description of methods and results and much deeper discussion is presented (Fenster, 1991*a*, *b*, *c*; Fenster & Galloway, in press *a*, *b*, *c*; Galloway & Fenster, 1999, 2000).

Chamaecrista fasciculata's breeding unit (i.e. deme or neighbourhood) is small (radius <5 m) and corresponds to a density of approximately 100 individuals (Fenster, 1991*a*, *b*). Proximity-dependent inbreeding depression (Fenster, 1991*b*), suggests that most mating events within the neighbourhood are between close relatives. Heterosis in the progeny increased rapidly with interparent distance and then plateaued (by 10–40 m), with little increase of fitness following crosses between parents in different neighbourhoods.

To further evaluate how far crosses can be conducted and still obtain heterosis in the offspring, populations were crossed over six spatial distances, ranging from 100 m to 2000 km. For each population pair, three generations of hybrids (F_1, F_2, F_3) were created and within population crosses were used to produce parental seed under greenhouse conditions. The design was replicated across three regions of the United States (Maryland, Illinois and Kansas). Field-study plots were established at each location in each region and hybrid and parental seed for all crosses were planted in them for two years. Each generation and crossing distance included at least 360 planted seeds each year for a total of at least 8000 seeds per site per year. Fitness was estimated as the number of fruit produced per seed planted.

Because of limited genetic effects expressed in Illinois we present the results from only the Maryland and Kansas sites. Progeny from crosses between populations from 100 m to 2000 km apart express F_1 heterosis. Often the F_1 outperformed both parents, even for longest-distance crosses (Fig. 3.1). Since the expression of heterosis in the progeny depends on genetic differentiation between the parents (Falconer, 1981), these results further demonstrate population differentiation in *C. fasciculata*. The spatial pattern of genetic variation, determined by an extensive isozyme survey (Fenster, 1988; Fenster & Dudash, 1994: Fig. 2.3; L. F. Galloway & C. B. Fenster, unpublished data), support this conclusion.

The F_2 generation overall performed intermediately between the parents and F_1, conforming to a model of heterosis but no coadaptation (data not shown). However, the F_3 suffered a drop-off in fitness that reflects both the loss of heterozygosity as well as the disruption of coadapted gene complexes (Fig. 3.2). The F_3 performance was still often equal to those of the parents, suggesting that heterosis outweighs the loss of coadaptation except for the longest-distance crosses. However, at the longer distances, greater than or equal to 1000 km, interpopulation crosses often resulted in F_3 performance less than that of either parent, and indicated hybrid breakdown and true outbreeding depression.

For *C. fasciculata*, it appears that crossing populations originating from distances of up to intermediate length (100s km) has a short-term beneficial effect (owing to proximity-based inbreeding depression) on progeny performance through the F_1, and that longer-term effects are not necessarily disruptive of fitness, at least relative to parental performance for crosses less than or equal to 100 km. Because negative epistasis is mostly observed in the F_3 and less consistently in the F_2 generation, an additional round of recombination appears to be necessary to disrupt putatively linked epistatically interacting genes. These results conform to theoretical predictions that the evolution of interacting gene systems is facilitated where recombination is limited either through restricted gene flow or linkage as discussed above. The results suggest that deleterious alleles are likely to become fixed in populations through drift and that crossing populations results in the recovery of heterosis. The lack of an expected scale effect here may simply indicate that populations of *C. fasciculata* become so differentiated from one another so quickly with distance that we cannot detect increased heterosis with increased crossing distance. It is still unclear what the ultimate performance of the hybrids will be since further recombination may lead to even greater disruption of fitness, while chance might also bring together favourable gene combinations.

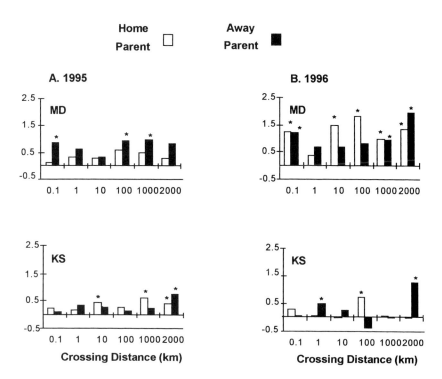

Fig. 3.1. Fitness of the F_I hybrids between populations of *Chamaecrista fasciculata* 0.1–2000 km apart relative to each of the parental populations contributing to the cross. Results are presented for the replicate crosses conducted for Maryland (MD) and Kansas (KS) in the years 1995 and 1996. Analysis of covariance was used to evaluate the performance of the F_I relative to the parents (details in Fenster & Galloway, in press, *a, b*). We quantified the performance of F_I hybrids by constructing a priori contrasts of the F_I with the parents contributing to the cross. The contrasts took the following form: F_I fitness − Home Parent fitness; F_I fitness − Away Parent fitness and were conducted for each crossing distance for each site and year. Square-root transformed means adjusted for initial seed size are presented. Comparisons that differ at $P < 0.05$ following sequential Bonferroni correction are indicated by *. Open bar: F_I fitness − Home Parent fitness; dark bar: F_I fitness − Away Parent fitness. If bar < 0 (horizontal line), then performance of $F_I <$ parent. If bar > 0, then performance of $F_I >$ parent.

RECOMMENDATIONS

In order to determine the consequences of fragmentation on population viability, researchers planning investigations of inbreeding depression need to quantify its effects at as many life-history stages as feasible at both the population and family level, and in the field if possible. Knowledge of

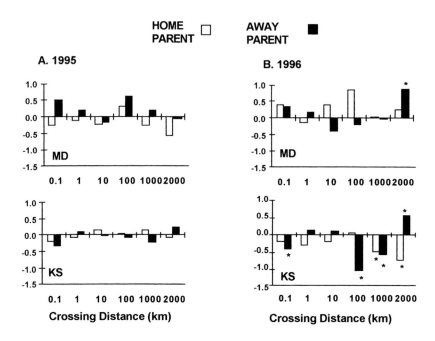

Fig. 3.2. Fitness of the F₃ hybrids between populations of *Chamaecrista fasciculata* 0.1–2000 km apart relative to each of the parental populations contributing to the cross. Open bar: F₃ fitness − Home Parent fitness; dark bar: F₃ fitness − Away Parent fitness. If bar <0 (horizontal line), then performance of F₃ < parent. If bar >0, then performance of F₃ > parent. See Fig. 3.1 for details.

the mating system, basic reproductive biology and the genetics underlying the expression of inbreeding depression is also desirable. The purging of genetic load through enforced inbreeding and selection of superior genotypes may be successful in some maternal lines or populations but the random fixation of traits whether favourable or not and the loss of maternal lines in the inbreeding process should be weighed carefully in light of any potential benefits. Ideal purging studies are needed to gain insights into artificial breeding programmes and their likelihood of success. Given the lack of detail that we now have on the genetic basis of inbreeding depression, it seems prudent to encourage further research on these questions using model organisms.

Whether one can successfully combine distant gene pools to produce viable persistent populations is still in great need of further empirical work. We would like to see investigations that quantify outbreeding depression in natural study systems which vary in mating system, ploidy level and degree of fragmentation, etc. If possible both inbreeding and outbreeding depression should be jointly investigated to understand how a fragmented landscape may magnify their effects and influence long-term persistence of the organism or population in question.

ACKNOWLEDGMENTS

The authors thank R. Frankham, A. Young and two anonymous reviewers for comments on a prior version of the manuscript. Many of the ideas and results cited here reflect fruitful collaboration with D. Carr and L. Galloway. The National Science Foundation supported the research conducted by M. Dudash (DEB-9220906) and C. Fenster (DEB-9312067 and DEB-9815780).

4

Demography and extinction in small populations

KENT E. HOLSINGER

ABSTRACT

Human-caused changes in the environment are by far the greatest threat to persistence of many species. Populations subject to long-term deterministic declines are certain to go extinct within a relatively short period, and large populations are expected to persist only a little longer than those that are small. Even populations that tend to grow from one year to the next, however, may persist for only a short time. If the year-to-year variability in population growth rate exceeds about twice the average annual growth rate, the population will decline to extinction over the long term, and large population size provides only a little extra protection against extinction. Persistence times of populations increase greatly with increasing population size only if there is relatively little year-to-year variability in population growth rate. Isolated populations are unlikely to persist unless either their annual growth rates are high relative to the variability in growth rate or they are carefully monitored and managed.

INTRODUCTION

Humans have an enormous influence on the earth, its ecosystems and its plant and animal populations, and it is our activities that are the primary cause of most species declines. Human activities are now responsible for more nitrogen fixation than all other biological and non-biological sources (Vitousek, 1994), they capture 54% of the accessible runoff in terrestrial ecosystems (Postel *et al.*, 1996) and they have transformed between one-third and one-half of the earth's land surface (Vitousek *et al.*, 1997). In light of these enormous impacts, it should come as little surprise that over 80% of species listed as endangered or threatened under the Endangered Species Act of 1974 in the United States are imperilled, in part, because of habitat loss and that nearly half are threatened through competition with introduced exotic species (Wilcove *et al.*, 1998). In the face of such overwhelming impacts it is clear that efforts to conserve a significant portion of the world's biodiversity can succeed only if we can find ways to limit our impact on the global environment. The techniques for doing so will come more from economics, psychology and sociology than from biology. None the less, biology has an important role to play that goes well beyond identifying the species and ecosystems particularly in need of conservation attention.

It has been known for more than a century and a half that small populations of plants and animals face particular risks of extinction (e.g. Darwin, 1859: 109; Ford, 1945: 143; Andrewartha & Birch, 1954: 664). Genetic factors may sometimes contribute to the failure of small populations, whether from a lack of appropriate genetic diversity or from the accumulation of deleterious mutations that lower average reproductive rate in populations (Barrett & Kohn, 1991; Dudash & Fenster, Chapter 3, this volume; Sherwin & Moritz, Chapter 2, this volume). Recent work has suggested that even populations of the order of 500–1000 individuals can accumulate deleterious mutations that lower population fitness sufficiently to cause a 'mutational meltdown' (Gabriel *et al.*, 1993; Lynch *et al.*, 1995*b*). None the less, a lack of genetic diversity in small populations is more likely to be a symptom of endangerment than its cause (Holsinger & Vitt, 1997). More importantly, no population can survive if fewer offspring are born in each generation than in the one before, and even populations that increase in most generations may suffer a relatively rapid extinction if their growth rate varies substantially from one year to the next. Demographic processes determine whether populations are likely to persist, and understanding those processes is fundamental to designing conservation programmes to protect them.

In this chapter I describe some of what is known about demographic threats to persistence, i.e. those that derive from changes in the number of individuals present from one generation to the next without respect to any associated changes in their genetic composition. Although most species, even those that are rare and threatened with extinction, occur in multiple populations, I also focus on the dynamics of isolated populations that neither receive immigrants from nor send emigrants to other populations. While persistence of a metapopulation system is possible when all component populations are declining if recolonisation is frequent enough, even a metapopulation system is more likely to persist for a long time if each of its component populations is relatively resistant to extinction. Understanding the properties of isolated populations that make them vulnerable to extinction is an important step in designing conservation programmes that will lessen the risk to imperilled species.

THREATS TO PERSISTENCE

Demography is a numbers game. Its basic variable is the size of the population at a given time. Its parameters are related to aspects of life history, reproduction and survival that determine whether the population size will increase or decrease from one generation to the next. Even for an isolated population in which immigration and emigration can be neglected, the number of factors that cause changes in population size can be enormous, and a complete description of its dynamics would require a mathematical model with an equal number of parameters. In spite of this enormous complexity, it is useful to recall that for any single time period even the most complex dynamics can be summarised by this simple equation:

$$N_{t+1} = (1 + R_t)N_t, \tag{4.1}$$

where N_t is the size of the population at time t and R_t is the growth rate of the population at time t.[1] Notice that R_t in equation 4.1 is the *actual* population growth rate. The actual growth rate will usually differ somewhat from the expected growth rate simply because of the vagaries of which individuals happened to reproduce or to die and how many offspring they left.

Two classes of demographic threats are easily distinguished using this simple framework. Deterministic threats are those that cause the mean of R_t to be negative, i.e. those that cause the population to decline from generation to generation on average. Clearly if deterministic threats to persistence are not reversed the population is doomed to extinction regardless of its current size. Stochastic threats are those that arise from variability in R_t, because even populations in which the mean of R_t is positive are not immune from extinction. If R_t varies at all, a series of unlucky years in which R_t is negative can lead to extinction (cf. Ludwig, 1975).

Most deterministic threats are unrelated to population size, although the Allee effect (see p. 59) is an obvious exception. The magnitude of stochastic threats, however, directly depends on population size. Thus, increasing the size of populations may provide some protection against

---

[1] This equation applies even in age- or stage-structured populations, though $1 + R_t$ is equivalent to the leading eigenvalue of the corresponding Leslie or Lefkovitch matrix only asymptotically. I use a discrete-time population model for the results derived in this chapter, but analogous results apply for continuous-time population models.

stochastic threats, while it affords little protection against most deterministic threats. Within the field of conservation biology what Caughley (1994) referred to as the 'declining-population paradigm' has focused on reversing or mitigating deterministic threats to populations while what he called the 'small-population paradigm' has focused on managing stochastic threats.

Deterministic threats

Some deterministic threats are obvious and pervasive. Species declines have been caused primarily by habitat destruction and conversion, overexploitation, species translocations and introductions, and pollution (Lande, 1998a). Others are more subtle and idiosyncratic. The conventional story of the dodo is that it was hunted to extinction. Caughley & Gunne (1996) point out, however, that the forests of the Mascarene Islands were still largely intact when the dodo went extinct in 1662. They suggest instead that when 20 domestic pigs were introduced to the islands in 1606–7, competition for food and destruction of nests caused the dodo to decline to extinction. Whether obvious or subtle, however, all deterministic threats have this in common: unless reversed they ensure that the population will be driven to extinction, often in a relatively short period of time. One of the first tasks for any manager of a threatened population must therefore be to identify deterministic threats to persistence and to mitigate or reverse them.

Habitat destruction is both the most obvious cause of species declines and the one most difficult to reverse. It is, none the less, vital to remember that many of the species about which we are concerned are imperilled, in large part, because the habitat on which they depend has been converted to other uses. Red-cockaded woodpeckers, for example, were once widely distributed in open pine savannas throughout the south-eastern United States. Today only about 30 isolated populations remain on about 1% of the original species range (Conner & Rudolph, 1991). The remainder of the habitat the red-cockaded woodpecker once used has been converted for agriculture and forestry.

When habitat is destroyed it is also usually fragmented. Not only does this fragmentation eliminate many populations, it often increases the isolation of remaining populations, and populations that do remain may face new threats. Physical edge effects in tropical and temperate forests – increased sun, decreased soil moisture, proliferation of disturbance-adapted plants, for example – commonly extend 10–15 metres from the forest edge and may, in some circumstances, extend as much as half a kilometre (Ranney et al., 1981; Laurance 1991b). As a result, small habitat fragments may not contain habitat suitable for forest interior species. Askins *et al.* (1987), for example, found that five especially area-sensitive forest interior birds required unfragmented forest areas of between 400 and 800 hectares to persist in Connecticut.

Although the enormous impact of non-indigenous invasive species on natural systems has been widely noted (e.g. Ruesink *et al.*, 1995; Vitousek *et al.*, 1996), it is less widely appreciated that they may also pose a significant threat to species threatened by habitat loss or fragmentation. Wilcove *et al.* (1998) point out, however, that unfavourable interactions with non-indigenous species are the second most frequently cited threat to endangered and threatened species in the United States. In at least one case, well-meaning biologists were responsible for introduction of a non-indigenous species that may now pose a threat to already imperilled species of plants. A flowerhead weevil *(Rhinocyllus conicus)* was released in several different parts of the United States in the late 1960s and early 1970s. The intent was to control Eurasian thistles *(Carduus* spp.) that had become extensively naturalised. Unfortunately, this beetle is now found in association with several North American thistles *(Cirsium* spp.), including one *(C. canescens)* that is sparsely distributed and geographically restricted in addition to being the putative progenitor of another species that is already regarded as threatened (Louda *et al.*, 1997).

As if these threats were not enough, populations that become sufficiently small face an additional threat: the inability to find and attract mates. Although this effect was first described nearly half a century ago in animal populations (Allee, 1951), the same phenomenon can occur in plant populations when small populations are insufficiently attractive to animal pollinators. Groom (1998) recently showed, for example, that patches of *Clarkia concinna* suffer reproductive failure if they become too small or isolated. Similarly, Robertson *et al.* (1999) suggest that declines in two species of New Zealand mistletoes *(Peraxilla colensoi* and *P. tetrapetala)* are attributable, in part, to a paucity of visits from bird pollinators on which they depend for sexual reproduction (see also Kelly *et al.*, Chapter 14, this volume). Of the deterministic threats to populations, only the Allee effect can be regarded as primarily intrinsic to the dynamics of the population. If the size of a population can be increased sufficiently, it should escape the threat posed by an Allee effect. But any population, no matter how large, is doomed to extinction if its habitat continues to be destroyed or fragmented or if it continues to suffer from its interactions with non-indigenous species.

Stochastic threats

Stochastic threats arise from the simple fact that it is always possible for a population to decline over a series of generations even if it tends to increase from generation to generation on average. This is reflected in the fact that the long-term growth rate of a population is determined by the geometric mean of $1 + R_t$ rather than the usual arithmetic mean. Because the geometric mean is always less than the arithmetic mean, the long-term growth rate of a population may be negative even if the arithmetic mean growth rate is positive. In other words, a population may decline over the long term even if it tends, on average, to increase in size from one year to the next. Throughout the rest of this chapter I will refer to R_t as the 'population growth rate', to the arithmetic mean of R_t as the 'average population growth rate', and to one subtracted from the geometric mean of $(1 + R_t)$ as the 'long-term population growth rate'. For example, the average growth rate in a population with density-independent growth will be positive when the arithmetic mean of R_t is positive. The long-term population growth rate of that same population will, however, be negative whenever the variance in R_t is more than about twice its mean (see Appendix 4.1 for details). Clearly, to understand the magnitude of stochastic threats to populations it is vital to understand both the causes of variation in population growth rate and their magnitude.

There are two sources of variation in population growth rate: demographic stochasticity and environmental stochasticity (Shaffer, 1981).[2] Demographic stochasticity refers to the variance in population growth rate arising from vagaries of which individuals happen to survive or die and how many offspring those that survive happen to produce. More precisely, demographic stochasticity refers to the variance in population growth rate that would occur even if both the probability distribution describing the number of offspring any individual has and the probability that any individual dies were identical for all individuals in the population and did not change over time. Environmental stochasticity refers to the variance in population growth rate arising from generation-to-generation differences in either the probability distribution describing offspring number or in the probability of death or both.

Because demographic stochasticity arises from individual-to-individual variation and reflects variability associated with independent sampling from a single probability distribution, its magnitude is inversely propor-

[2] See the section entitled 'Environmental stochasticity and natural catastrophes' for a discussion of the relationship between environmental stochasticity and catastrophic reductions in population size.

tional to population size. Specifically, if σ_d^2 is the variance in population growth rate that would be seen in a population with only one individual, then σ_d^2/N is the variance in population growth rate due to demographic stochasticity in a population of size N (cf. Keiding, 1975; Leigh, 1981; Goodman, 1987b). Environmental stochasticity, on the other hand, arises from variation extrinsic to the population, and we would expect it to be independent of population size.

Unfortunately, the relative contributions of demographic and environmental stochasticity to variation in population growth rate cannot be directly measured. If we understood the relationship between environmental variables and population growth rate well enough to predict accurately what population sizes should be, we could estimate the contribution of demographic stochasticity to variability in population growth rates from the difference between observed and predicted values. Similarly, we could estimate the contribution of environmental stochasticity as variation among predicted values. There are few, if any, populations for which such an approach is possible. None the less, we can use available data on population size changes from several animal species (great tit: Fig. 4.1, heron, Laysan finch, palila and grizzly bear: Fig. 4.2) to illustrate an alternative approach to determining whether demographic stochasticity alone is likely to account for the observed variation in population growth rates. If we assume that the number of births and deaths is approximately Poisson distributed, we can use the observation that the demographic variance will then be equal to $(1 + \tilde{R})/N$ to calculate the population size (labelled 'Equivalent N' in Table 4.1) that would produce a variance in population growth rate equivalent to that observed.

In both the Laysan finch and the palila, populations two to three orders of magnitude smaller than those observed would be necessary to account for the observed variance in population growth rate, while in the grizzly bear the observed variance is consistent with the observed population size. In the great tit and heron populations would have to be between one-third and one-thirtieth of those observed to account for the observed variance in population growth rate. Thus, in large populations, like the Laysan finch and the palila, and in some small populations, like the great tit and heron, environmental stochasticity is a much more important contributor to variation in population growth rate than is demographic stochasticity. In some small populations like the grizzly bear, however, demographic stochasticity may contribute importantly to variation in population growth rate.

It can be shown that the long-term growth rate of a population subject

Table 4.1. Demographic versus environmental stochasticity

Species	$\bar{R}^a$	Variance	Equivalent N	Observed N
Great tit[b]	0.20 (−0.045, 0.44)	0.4151	3	20–95
Heron[b]	0.025 (−0.035, 0.084)	0.01438	71	274–484
Laysan finch[c]	0.89 (0.33, 1.44)	0.0802	24	5000–21000[d]
Palila[c]	0.52 (−0.66, 1.70)	0.3625	5	2000–6400[e]
Palila[f]	0.0017 (−1.0, 1.0)	0.3931	3	2000–6400[e]
Grizzly bear[g]	−0.0018 (−0.036, 0.032)	0.006444	155	33–47

[a]Numbers in parentheses are the lower and upper 95% confidence limits.
[b]Direct estimates from population sizes reported in Lack (1954).
[c]Estimates from Ludwig (1999). $\bar{R}$ and variance calculated assuming interval is approximately normal.
[d]Cited in Ludwig (1999).
[e]Scott et al. (1984).
[f]Direct estimates from Table 1 of Scott et al. (1984).
[g]Direct estimates from three-year summed totals in Table 3 of Eberhardt et al. (1986). Reproductive females only.

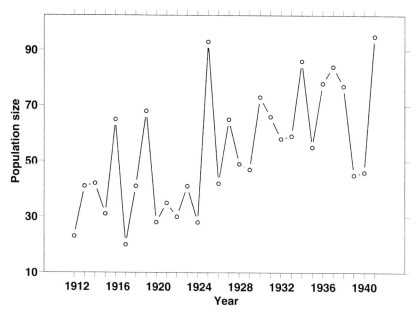

Fig. 4.1. Estimated population size of great tit at Oranje Nassau's Oord from 1912 to 1941.

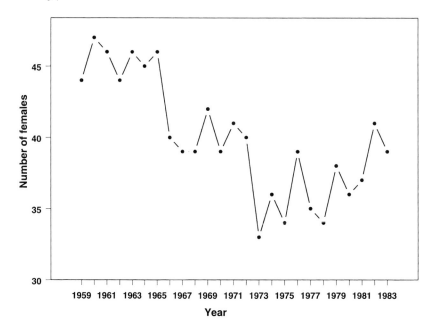

Fig. 4.2. Estimated number of adult female grizzly bears in Yellowstone National Park, Wyoming, USA from 1959 to 1983.

only to demographic stochasticity will be negative in the absence of environmental stochasticity if

$$2\rho < \frac{\sigma_d^2}{N},$$ (4.2)

where ρ is the mean population growth rate. Given that ρ_d^2 is likely to be 1 or less (Leigh, 1981; Engen et al., 1998; Sæther et al., 1998), it is evident from equation (4.2) that demographic stochasticity alone cannot cause a long-term population decline in populations with a mean growth rate of more than 1% in populations with more than 50 reproductive individuals. In short, environmental stochasticity is likely to be the predominant source of variation in population growth rates, except in populations that are already very small. As a result, environmental stochasticity is likely to play a far more important role in determining the probability that a population persists than is demographic stochasticity.

PERSISTENCE TIMES

The long-term fate of a population subject to a deterministic decline is obvious. Unless the causes of that decline are reversed the population will ultimately go extinct. It is less obvious but no less true that expanding populations, i.e. those with a positive average growth rate, may also go extinct. Consider, for example, a population growing exponentially and without bound. If it is subject only to demographic stochasticity with a per-capita birth rate of b and a per-capita death rate of d with $b > d$ and a current size of N, it will go extinct with probability $(d/b)^N$ even though both its average and long-term growth rates are positive (Karlin & Taylor, 1975: 147). More to the point, because no population can increase in size indefinitely, all isolated populations are guaranteed to go extinct given enough time if there is any stochastic variation in birth and death rates. In some cases 'enough time' may be thousands or tens of thousands of generations, but in others it may be less than one hundred. As a result, it is important to understand both the deterministic and the stochastic factors that influence how long an isolated population is likely to persist if left unmanipulated.

Besides the usual lack of demographic data that faces most conservation biologists, however, there are two other factors that complicate investigations of persistence times. Firstly, persistence times are approximately exponentially distributed (Mangel & Tier, 1994), meaning that most populations will go extinct before their mathematically expected time of extinction. If $T(N_o)$ is the average number of generations a population will

Table 4.2. Probability of extinction

Species	Years[a]	Quasi-extinction probability[b]
Laysan finch	25	2.7×10^{-5} (0, 0.035)
Snow goose	18	1.8×10^{-4} (0, 0.99)
Palila	9	0.031 (7.4×10^{-6}, 1.0)

[a]Number of years of demographic data on which extinction estimates are based.
[b]Probability that the population reaches 10% of carrying capacity within 100 years starting at 20% of carrying capacity. Least-squares estimate and lower and upper 95% confidence bounds. From Table 1 of Ludwig (1999).

persist given that it started with a size of N_o, for example, then there is a 50% chance that it will have gone extinct in $(\ln 2)T(N_o) \approx 0.69 T(N_o)$ generations and the chance that it will persist $T(N_o)$ generations or longer is only about 37%. Thus, average persistence times may significantly underestimate the risk of extinction over time-frames of interest to conservation biologists. Fortunately, the entire distribution of persistence times changes in approximately the same way as $T(N_o)$, so one can investigate the relative influence of different factors on $T(N_o)$ without concern that important parts of the extinction dynamics are neglected (contra Ludwig, 1996).

Secondly, because demographic parameters vary dramatically from one generation to the next in most populations, there is also enormous uncertainty about the probability of extinction within any particular time period, even when a point estimate for this probability is quite small (Table 4.2). As a result, estimates of persistence time are also associated with enormous uncertainty. Fortunately, it is possible to make reliable general statements about the relationship between population size, persistence times and persistence probabilities.

Deterministic and stochastic decline
It is straightforward to show that a population declining deterministically with logistic dynamics will reach a population size of one individual in

$$T(N_o) = \frac{\ln\left[\dfrac{K - N_o}{(K-1)N_o}\right]}{r}$$

$$= -\ln N_o / r + \ln\left(\frac{K - N_o}{K - 1}\right) / r,$$

where $N_o < K$ is the initial population size. If $N_o \ll K$, this reduces to

$T(N_o) \approx -\ln N_o/r$. Lande (1993) shows that this is a general property of population declines whenever the long-term population growth rate is negative (see also MacArthur & Wilson, 1967; Ludwig, 1976; Brockwell, 1985). Whether the growth rate is negative because deterministic factors cause a decline every generation on average or because the variability in population growth rate exceeds twice the average growth rate, the mean time to population extinction is proportional to the logarithm of initial population size.

The lesson for conservation biologists is clear. Not only may isolated populations with a negative long-term growth rate require constant management and occasional supplementation to prevent their extinction, but they are likely to need supplementation very frequently. Doubling population size from 100 to 200 individuals, for example, increases the expected time to extinction by only about 15%. Even increasing population size by a factor of 10 would increase the expected time to extinction by only 50%. Only if the factors that cause the long-term growth rate of a population to be negative are reversed can persistence times be significantly extended.

Demographic stochasticity

Chance events in survival and reproduction are likely to play an important role only in the persistence of populations that are already quite small. As pointed out earlier, even a small population is unlikely to show a negative long-term growth rate unless its average growth rate is very small. More importantly, when the long-term growth rate of a population subject only to demographic stochasticity is positive ($2\rho > \sigma_d^2/N$),

$$T(N_o) \approx \frac{\exp\ [(2\rho/\sigma_d^2)K - 1] - 1}{(2\rho/\sigma_d^2)K} \left[1 + \frac{1}{(2\rho/\sigma_d^2)K} \right],$$

where K is the population carrying capacity and $N_o \approx K$ (Lande, 1993). As a result, persistence times increase very rapidly, almost exponentially, as the carrying capacity increases. If demographic stochasticity were the only source of variation in population growth rate in the Laysan finch, for example, we can calculate from the data in Table 4.1 that a population with a carrying capacity of 10 individuals would persist for about 530 generations, while one with a carrying capacity of 20 individuals would persist for over 3 million generations.[3]

---

[3] This calculation assumes that σ_d^2 is equal to the observed $\bar{R}$ (1.89). If the demographic variance were twice as large as this, the mean persistence times would change to 20 and 1000 generations for populations with carrying capacities of 10 and 20 individuals respectively. Remember also that the 95% confidence interval for ρ overlaps zero, meaning

Notice, however, that persistence times also depend heavily on ρ. For realistic values of σ_d^2, populations with a small average growth rate have a shorter expected time to extinction than those with a larger growth rate. Using the smaller of the two values calculated for $\tilde{R}$ in palila (0.0017) gives a persistence time of 55 generations if the carrying capacity is near 2000 and a persistence time of over 48 million generations if the carrying capacity is near 6400. In short, even large populations can be threatened by demographic stochasticity if their average growth rates is very small (much less than 1%), but demographic stochasticity poses a significant threat to populations only if they are already quite small or if they are barely replacing themselves.

None the less, it is important to note that in a population with appreciable demographic stochasticity there may be a threshold size below which most population trajectories tend to decline even if their expected long-term growth rate is positive. This phenomenon might be called a stochastic Allee effect, because it implies that the population is likely to go extinct if it drops below this size. Specifically, Lande (1998b) shows that most population trajectories will decline if

$$ N_o < \frac{\sigma_d^2/4}{\rho - \sigma_e^2/2}, $$

where σ_d^2 is the individual demographic variance and σ_e^2 is the environmental variance. Given that σ_d^2 is likely to be of the order of 1 (Leigh, 1981; Engen et al., 1998; Sæther et al., 1998), N_o will be very small unless the environmental variance in population growth rate is almost exactly twice as large as the average population growth rate. If it is much smaller, N_o will also be small. If it is larger, the long-term growth rate of the population is negative and all population trajectories will eventually decline.

Environmental stochasticity and natural catastrophes

Demographic stochasticity is an intrinsic source of variation in population growth rates, i.e. it is an inescapable part of the probabilistic birth–death process which produces changes in population size from one generation to the next. Shaffer (1981) suggested that we recognise two extrinsic sources of variation in population growth rates: environmental stochasticity, which produces changes in population size as a result of 'changes in weather, food

---

that we cannot exclude the possibility that the long-term population growth rate is negative.

supply, and the populations of competitors, predators, parasites, etc.', and natural catastrophes, which produce changes in population size as a result of floods, fires and droughts (cf. Shaffer, 1987). It is clearly important for conservation biologists to recognise that rare events can reduce population sizes dramatically. Indeed, these rare events may pose greater threats to population persistence than the year-to-year variation in population growth rates that is more easily measured.

Consider, for example, a population growing exponentially with a constant per-capita birth rate of α and a per-capita death rate of β (implying $\rho = \alpha - \beta$). If this population is subject to catastrophes that remove a proportion p of the individuals in it with probability γ in any particular generation, the long-term growth rate of the population is negative if

$$\alpha \leqslant \beta - \gamma \ln (1 - p)$$
$$\rho \leqslant -\gamma \ln (1 - p) \tag{4.3}$$

(Brockwell, 1986; Ewens et al., 1987). The sobering message of Fig. 4.3 is that even very rare events can doom a population to extinction, if they are severe enough. A catastrophe that eliminated 80% of a population growing exponentially at 7% per year, for example, would only have to occur about once every 23 years to ensure its extinction. As the intensity of catastrophes increases, the interval between them that will cause negative long-term growth rates grows very rapidly. Not surprisingly, several authors (e.g. Goodman, 1987b; Ludwig, 1999) have pointed out that conservation biologists may seriously underestimate the hazards faced by small populations if they neglect these low-frequency events. Unfortunately, those with the greatest long-term impact may also be those that occur least frequently.

If year-to-year variation in birth and death rates is uncorrelated with the occurrence of catastrophes and catastrophes are rare, then the combined effects of environmental stochasticity and catastrophes will cause the long-term growth rate to be negative if

$$\rho \leqslant (1/2)\sigma_e^2 - \gamma \ln (1 - p), \tag{4.4}$$

where σ_e^2 is the environmental variance in growth rate (excluding the effect of catastrophes; see Appendix 4.2 for details). By making the assumption that observed variance in population growth rates in the heron, Laysan finch and palila is entirely environmental, we can use equation 4.4 to calculate catastrophe frequency necessary to cause long-term growth rates of these species to be negative (Table 4.3). Only if catastrophes occur more frequently than indicated in this table will long-term population growth rates be negative.

Table 4.3. Catastrophe frequency needed to cause extinction

Species	Fraction of population eliminated	
	$p = 0.5$	$p = 0.9$
Heron	38	129
Laysan finch	0[a]	2
Palila[b]	2	6

[a]If a 50% catastrophe occurs once a generation or less frequently, the expected long-term growth rate is positive.
[b]Using the estimate of $\bar{r}$ and variance derived from Ludwig (1999)

While it might appear from these figures that both the Laysan finch and the palila are quite secure, it should be remembered that the palila may actually be undergoing a deterministic decline (Scott *et al.*, 1984; Ludwig, 1999). The 95% confidence interval for population growth rate extends down to -0.66. More importantly, the heron results illustrate that a population with an average growth rate of a few percent may be threatened by rare catastrophes even if it has never been known to consist of fewer than several hundred individuals. If a catastrophe eliminating half of the heron population were to occur only once every 38 years, it would be frequent enough to ensure that the long-term population growth rate is negative. A more severe catastrophe that occurred only once each century would have the same effect.

When the long-term growth rate of a population is positive, it will still decline to extinction given enough time, unless it is able to grow without bound. The time to extinction increases proportionally to K^{ϕ} ($\phi > 1$), however, implying that populations with a positive long-term growth rate are likely to persist for a very long time – provided that they continue to have enough habitat to sustain a high carrying capacity. In the presence of environmental stochasticity alone, for example, the average time to extinction for a population near its carrying capacity is

$$T(N_0) \approx 2N_0^{(2\rho/\sigma_e^2)-1} / \{\sigma_e^2 [2\rho/\sigma_e^2] - 1]^2\}$$

(Lande, 1993). If we ignore the tremendous uncertainty in parameter estimates for the palila, this would suggest an expected time to extinction of more than 2.3×10^6 years with a carrying capacity of 2000. Assuming an exponential distribution of extinction times, this would correspond to a probability of extinction within 100 years of $1 - \exp [100/(2.3 \times 10^6)]$ $\approx 4.2 \times 10^{-5}$.

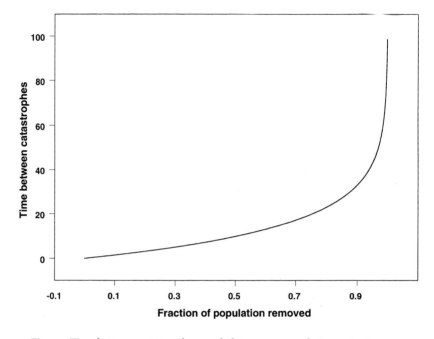

Fig. 4.3. Time between catastrophes needed to ensure population extinction assuming exponential growth between catastrophes with $\rho = 0.07$.

The relationships between population size and time to extinction summarised in this section are derived from specific, simple models of population dynamics. None the less, there is reason to think that the general properties they exhibit are quite general. Chesson (1982) points out that the conditions under which infinite-population models predict population persistence correspond with the finite-population models described here in which persistence times are proportional to K^{ϕ} ($\phi > 1$). The conditions under which infinite-population models predict population extinction correspond with the finite-population models described here in which persistence times are proportional to log K. In short, it is likely that persistence times increase logarithmically with population size if the variance in population growth rate exceeds twice its mean and that persistence times increase as a power (> 1) of population size when the variance is smaller.

CONCLUSIONS

The lesson for conservation biologists is clear. Populations with a negative long-term growth rate, whether from deterministic or stochastic effects, will require constant management and frequent supplementation to prevent their extinction. Populations with a positive long-term growth rate, however, are much more resilient and are likely to require far less attention – provided that they can be given enough habitat to support reasonably large populations. Caughley (1994) argued that the focus on threats to small populations characteristic of the small-population paradigm was a misdirection of effort and that conservation biologists could use their time more effectively by identifying deterministic threats to persistence, as exemplified by work in the declining-population paradigm. The results reviewed in this chapter, however, make it clear that these paradigms should be viewed as complementary rather than competing (see also Caughley & Gunne, 1996).

Populations faced with deterministic forces causing decline every year are doomed to extinction, and increasing their size without removing the causes of this decline will do little to enhance their persistence. If variation in population growth rates, whether the result of environmental variation or rare catastrophic events, is great enough, however, even populations that are stable or increasing on average face decline with no less certainty. More importantly, whether the cause of the inevitable decline associated with negative long-term population growth rates is primarily deterministic or stochastic, the ability of the population to persist is relatively insensitive to its current size. Only if a way can be found to make the long-term population growth rate positive is a population likely to persist more than a few tens of generations without intervention.

The lesson from the declining-population paradigm is that deterministic threats to population persistence must be reversed if populations are to have a long-term future. The small-population paradigm adds the caution that even large populations that grow from year to year, on average, may be in danger if their growth rates are highly variable. In spite of their differences in emphasis, the lessons for conservation biologists are similar. Deterministic threats to population persistence must be reversed, populations must be provided with ample habitat, small populations must be helped to grow rapidly, and the variance in population growth rates must be minimised if long-term persistence of populations is to be assured (cf. Holsinger & Vitt, 1997). Accomplishing these tasks in populations of many species may be extraordinarily difficult, but we have the satisfaction of knowing that if we are successful, those populations are likely to persist long after we are gone.

ACKNOWLEDGMENTS

I am indebeted to Peter Turchin and an anonymous reviewer for comments on an earlier version of this chapter. My research has been supported by the University of Connecticut Research Foundation and by the National Science Foundation (DEB-9509006).

APPENDIX 4.1

Long-term growth rate of a population

Equation 4.1 implies that

$$N_{t+T} = \prod_{k=0}^{T-1} (1 + R_{t+k}) N_t$$

$$= (1 + \tilde{R}_t)^T N_t$$

where $(1 + R_t) = \exp\left[(1/T) \sum_{k=0}^{T-1} \ln(1 + R_{t+k})\right]$ is the geometric mean of $1 + R_t$. If the R_t values are independently and identically distributed, as might be appropriate for a population not subject to density regulation, the strong law of large numbers (Feller, 1968: 258) guarantees that for T sufficiently large

$$(1/T) \sum_{k=0}^{T-1} \ln(1 + R_{t+k}) \approx E \ln(1 + R_t), \qquad (4.1.1)$$

where $E(\cdot)$ refers to the mathematical expectation operator.

Using the definition of $E(\cdot)$ and expanding the right-hand side of equation 4.1.1 in a Taylor series around the mean of R_t reveals that

$$E \ln(1 + R_t) \approx \ln(1 + \rho) - \frac{1}{2}\left(\frac{\sigma_r}{1+\rho}\right)^2,$$

where ρ is the expectation of R_t and σ_r^2 is the variance of R_t. Thus, the long-term growth rate of a population, $\tilde{R}$, is given by

$$1 + \tilde{R} \approx \exp\left[\ln(1 + \rho) - \frac{1}{2}\left(\frac{\sigma_r}{1+\rho}\right)^2\right]$$

and $\tilde{R} < 0$, i.e. the population will tend to decline, if

$$\ln(1 + \rho) < \frac{1}{2}\left(\frac{\sigma_r}{1+\rho}\right)^2,$$

which will occur if, approximately, $2\rho < \sigma_r^2$ (see also Tuljapurkar, 1982; Lande & Orzack, 1988).

APPENDIX 4.2
Long-term growth rate with catastrophes

For a population growing exponentially with catastrophes, the population size in generation T just before the next catastrophe is

$$N_{t+T} = \sum_{k=0}^{T-1} (1 + R_{t+k}) N_t \tag{4.2.1}$$
$$= (1 + \tilde{R}_t)^T N_t,$$

where $(1 + \tilde{R}_t)$ is the geometric mean of $1 + R_t$ over this time interval. If a fraction p of individuals dies in each catastrophe, then N_{t+T+1} is a binomial random variable with parameters N_{t+T} and $1 - p$. If catastrophes occur independently with probability γ in each generation, then T is a geometric random variable and

$$E\left[\ln (N_{t+T+1})\right] = \sum_{T=0}^{\infty} \gamma (1 - \gamma)^T \{T \ln [(1 + R_t)] + \ln N_t + \ln (1 - p)\} \tag{4.2.2}$$
$$\approx \ln N_t + \ln (1 - p) + \left[\ln (1 + \rho) - \frac{1}{2}\left(\frac{\sigma_r}{1 + \rho}\right)^2\right] / \gamma,$$

assuming γ is small enough that we can approximate $(1 + R_t)$ with its expectation. Because catastrophes are assumed to occur independently of one another, the long-term growth rate will be negative if and only if $E\left[\ln (N_{t+T+1})\right] < \ln N_t$, or equivalently

$$\ln (1 + \rho) < \frac{1}{2}\left(\frac{\sigma_e}{1 + \rho}\right)^2 - \gamma \ln (1 - p)$$

which will occur if, approximately,

$$\rho < (1/2)\sigma_e^2 - \gamma \ln (1 - p),$$

where σ_e^2 is the variance in ρ that occurs in intervals between catastrophes.

5

The metapopulation paradigm: a fragmented view of conservation biology

PETER H. THRALL, JEREMY J. BURDON & BRAD R. MURRAY

ABSTRACT

In the past, single-population approaches have dominated ecology and evolutionary biology. However, populations are not isolated either in time or space, but are connected by among-population processes such as migration and gene flow. While this concept is not new, until recently, there have been relatively few studies that have explicitly investigated the effects of spatial structure on demographic and genetic processes in the context of conservation. The metapopulation framework explicitly recognises and provides a conceptual tool for dealing with the interactions of within- (e.g. birth, death, competition) and among-population processes (e.g. dispersal, gene flow, colonisation and extinction). The ever-growing diversity of empirical and theoretical studies that demonstrate the importance of spatial structure in determining ecological and evolutionary trajectories also indicates that long-term conservation programmes need to focus on regional rather than local within-population persistence. In this regard, it is important to realise that ultimately all populations are ephemeral, and therefore colonisation processes must also be preserved. Clearly, not all species whose populations have undergone fragmentation fit the definition of a metapopulation. Nevertheless, a metapopulation approach to conservation biology is likely to provide a useful tool for developing management strategies as it addresses genetic, species and community effects of fragmentation in a single framework, thereby making explicit questions regarding extinction, population connectedness, species behavioural patterns and the survival of coevolved systems. In essence, a metapopulation perspective ensures a process-oriented, scale-appropriate approach to conservation that focuses attention on among-population processes critical for persistence of many natural systems.

THE METAPOPULATION PARADIGM: AN INTRODUCTION

Traditionally, ecological studies of natural systems have been dominated by a focus on single populations, or several populations sampled at a single point in time. Reflecting this, much of the theory associated with population dynamics, genetics and interspecific interactions (e.g. competition, predation) has been framed in the context of single-population deterministic models, in part because of the usefulness of tractable mathematical formulations for stability analysis. However, notwithstanding their narrow focus, these single-population approaches have been invaluable in disentangling interactions between various within-population demographic and genetic processes, and their impact on ecological and evolutionary dynamics.

Despite the practical emphasis on single populations, ecologists and evolutionary biologists have long been intrigued with understanding the processes that lead to patterns of abundance and distribution at a range of spatial scales (e.g. Andrewartha & Birch, 1954), beginning with the early work of Wallace, and including the development of the influential theory of island biogeography (MacArthur & Wilson, 1967; see Quammen, 1996 for a recent discussion and review). The recognition that spatial structure might influence species persistence in interspecific interactions has also resulted in a range of studies, including Huffaker's classic experiments on mites (Huffaker, 1958), and theoretical work on spatial models of predator–prey interactions (Hassell & May, 1988; Taylor, 1988, 1990; Hassell *et al.*, 1991; Wilson & Hassell, 1997) and competition (Comins & Hassell, 1987). In contrast, population geneticists and evolutionary biologists have shown particular interest in spatial structure with respect to the way in which among-population processes of gene flow and migration might influence genetic change and evolution (e.g. shifting balance theory: Wright, 1940, 1943; island model: Wright, 1951). More recently, the importance of colonisation and extinction processes for regional persistence and population genetic structure have also come under investigation (Slatkin, 1977; Wade & McCauley, 1988; Whitlock & McCauley, 1990; Gilpin, 1991; Thrall *et al.*, 1998). This has led to the development of a whole range of biologically realistic spatiotemporal modelling approaches (Kareiva, 1990; Czárán & Bartha, 1992; Durrett & Levin, 1994; Gilpin, 1996; Tilman & Kareiva, 1997; Bascompte & Solé, 1998).

Indeed, it is now widely recognised not only that most species are patchily distributed in nature, but that in many cases, populations occupying these patches are ephemeral (e.g. Andrewartha & Birch, 1954; Antonovics *et al.*, 1997; McCauley, 1997*a*). Following the work of Richard Levins (1969, 1970), these ideas have been broadly embraced by a single conceptual framework (metapopulations), incorporating the idea that local populations (or communities) are connected by among-population processes, and these processes can potentially affect, and be affected by, within-population demographic and genetic processes. Hanski (1991) defined a metapopulation as a system of local populations (the scale at which individuals move and interact with each other on a regular basis) connected by dispersing individuals (in plant metapopulations, this would also include gene flow through pollen or seeds). A crucial component of the definition of a metapopulation *sensu stricto* is its emphasis on colonisation/extinction processes whereby all local populations have a significant probability of extinction (Moilanen & Hanski, 1998). However, population size is an important determinant of extinction probabilities, and therefore differential probabilities of extinction (e.g. island–mainland situations) are likely to be more generally the case. Correlated environmental effects may also be important in some systems, leading to similar extinction probabilities at some spatial scales greater than that of local patches (Harrison & Quinn, 1989). As a consequence of these caveats, the metapopulation concept should perhaps be broadened to include any system where colonisation/extinction dynamics play a significant role in the dynamics and persistence of a species or community (Hanski, 1998). The metapopulation approach has been widely publicised in two recent edited volumes (Gilpin & Hanski, 1991; Hanski & Gilpin, 1997).

There is some debate about what types of genetic structures might be found in metapopulations, especially where population turnover is an important feature of the dynamics (Harrison & Hastings, 1996). In general, theory suggests that among-population genetic structure will depend on the nature of among-population dispersal (Whitlock & McCauley, 1990; McCauley, 1991), with genetic differentiation of populations being magnified in proportion to the probability that colonists are drawn from a single source. Empirical studies explicitly relating these colonisation/extinction processes to genetic variation are scarce. However, McCauley and co-workers (McCauley, 1994; McCauley *et al.*, 1995) found that the genetic structure of local populations of the plant *Silene alba* was enhanced by colonisation and extinction processes, with relatively little mixing of individuals from multiple source populations and greater divergence in younger populations. Other work on the metapopulation dynamics of *S. alba* has also demonstrated significant rates of population turnover in this system (Antonovics *et al.*, 1994; Thrall & Antonovics, 1995). Similarly, studies on a metapopulation of *S. dioica* indicated that colonisation played an important role in determining the genetic structure of populations, although colonising propagules were derived from many, rather than few source populations (Giles & Goudet, 1997*a*, *b*). In contrast, counter to classical metapopulation model predictions, Dybdal (1994) found that even though colonists in tidepool populations of copepods were likely to be drawn from only one or a few sources, younger populations were less differentiated than older ones. He postulated that this might be due to the fact that colonists were actually less likely to be drawn from older populations. It is still largely an open question as to how genetic outcomes should depend on the type of metapopulation structure and life-history features of the organisms involved (Thrall & Burdon, 1997).

The extent to which any natural systems depend on a balance between colonisation and extinction processes for persistence has also been questioned (Harrison, 1991, 1994). However, when broadly defined as the interaction between within-population processes and among-population movement, metapopulation dynamics are inevitably a feature of all natural systems (Antonovics *et al.*, 1994; Husband & Barrett, 1996). Indeed, there is now a wide range of empirical studies that have demonstrated the importance of among-population demographic processes, ranging from insects and other invertebrates (Addicott, 1978; Bengtsson, 1989; Hanski *et al.*, 1994) through to amphibians (Gill, 1978; Sjögren, 1991*a*; Driscoll, 1998), mammals (Paillat & Butet, 1996; Moilanen *et al.*, 1998) and plants (Menges, 1990; Ouborg, 1993*a*; McCauley, 1994; Overton, 1994; McCauley *et al.*, 1995; Husband & Barrett, 1996), as well as host–pathogen interactions (Antonovics *et al.*, 1994; Burdon *et al.*, 1995; Thrall & Antonovics, 1995; Grenfell & Harwood, 1997; Burdon & Thrall, 1999). Eriksson (1996), however, suggested with respect to plants that long-lived species with clonal propagation or extensive seed banks may be much less dependent on the balance between colonisation and extinction required for regional persistence in a metapopulation. Certainly, metapopulation dynamics are likely to be much more apparent in annual plants or other short-lived species, but it is inevitable that population turnover will be a feature of most systems at some spatial or temporal scale. In fact, studies on a wide range of organisms, including predator–prey, host–parasitoid and plant–herbivore systems, provide solid evidence that local population extinction is not infrequent in nature (Sabelis & Diekmann, 1988; Fahrig & Merriam, 1992; Merriam & Wegner, 1992).

ECOLOGICAL AND GENETIC NEIGHBOURHOODS: THE IMPORTANCE OF REGIONAL SPATIAL STRUCTURE

Landscapes at many spatial scales are becoming increasingly fragmented as a result of human intervention – an activity that further emphasises the importance of understanding spatial processes. In this context, it should also be noted that 'metapopulation-like' situations can be created by habitat degradation as well as outright fragmentation. Such processes will be a permanent feature of many (if not all) ecosystems for the foreseeable future, and are rapidly becoming of significant concern among conservation biologists (Fig. 5.1), in large part because of the greater risk of extinction faced by small isolated populations. Given the enormous impact that fragmentation can have on genetic structure, ecological and evolutionary dynamics, and therefore population persistence, it is of paramount importance that strategies for conservation management also acknowledge this fundamental process. Moreover, stochastic and historical effects operating at local scales will often produce misleading (and in some cases erroneous) conclusions; one very real effect of fragmentation is to increase the likelihood that local populations will be in a non-equilibrium state (Olivieri *et al.*, 1990). Thus, understanding the causes and consequences of genetic and demographic processes at the within-population level requires that neighbouring demes also be taken into account because of gene flow and migration. Such considerations may have practical as well as theoretical implications, e.g. with respect to establishment of artificial populations for conservation purposes. A cogent demonstration of this was recently provided by a study showing that gene flow from diploid populations of an endangered daisy (*Rutidosis leptorrhynchoides*) into nearby re-established tetraploid populations was causing high levels of chromosomal abnormalities and reduced seed set (Young & Murray, in press).

For these reasons, a compelling argument can be made that viable long-term conservation programmes must place greater emphasis on regional persistence than has been done in the past. In this regard, it is crucial to realise that ultimately all populations are ephemeral, and hence for any species to have a long-term future, colonisation processes must also be preserved. In other words, this means setting aside areas of suitable habitat where species are not currently found, as well as maintaining existing populations. This is the core of the relevance of the metapopulation concept for conservation, where there is a continuum from large essentially 'global' systems through to an opposite extreme of many single, unconnected populations or habitat fragments (Fig. 5.2).

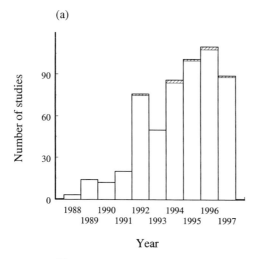

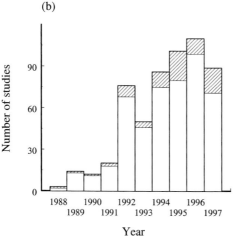

Fig. 5.1. Year-by-year account of the total number of conservation biology studies (empty bars) and the total number of studies incorporating both conservation biology and (a) metapopulation theory (hatched bars) and (b) fragmentation (hatched bars). Data obtained from searches of *Current Contents* (CAB – Abstracts) for 1988 to 1997, using terms 'conservation biology', 'conservation biology and metapopulation' and 'conservation biology and fragmentation'. Note that the 1997 search is up to and including October.

Identifying fragmented systems that have long-term viability

Perhaps the most important conservation issue that the metapopulation approach can usefully address is to identify conditions in which frag-

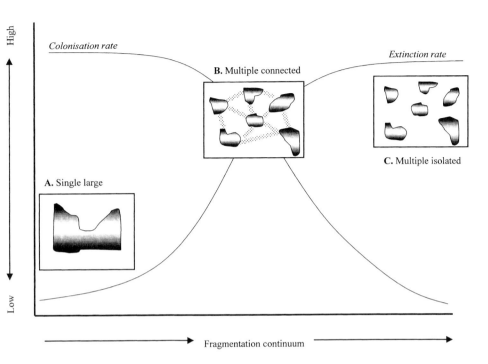

Fig. 5.2. Schematic diagram of the relationship between the continuum of species and community fragmentation and the relative importance of among-population processes for the dynamics. Where individual species or communities fit on the metapopulation axis is determined by a complex of factors including (but not limited to) life-history features that influence dispersal, colonisation and extinction phenomena. Note that even large natural areas (inset A) are rarely homogeneous, and thus among-patch movement will still play an important role in the dynamics. Inset B depicts the level of fragmentation at which the balance between colonisation and extinction processes can potentially result in long-term regional persistence. Conservation efforts centred on preserving such processes may thus provide the best chance of maintaining biodiversity, especially given that fragmentation is becoming a dominant feature of most landscapes.

mented systems can persist as 'habitat area is lost and the remaining habitat becomes ever more fragmented' (Hanski, 1998). In order to integrate recent work on spatial structure and metapopulation dynamics into conservation practices, we need to become better informed as to how demographic and ecological processes are affected by fragmentation in natural systems (there are several excellent case studies in the present volume that address some of these issues). For example, how can we utilise knowledge about the impact of fragmentation to manage biodiversity effectively? Specific concerns that relate to spatial structure and fragmentation include the loss of genetic diversity through genetic drift and inbreeding effects which may lead to increased extinction rates, and even system collapse (Gilpin & Soulé, 1986; Lynch *et al.*, 1995*a, b*; Thrall *et al.*, 1998). As a case in point, increased risk of extinction due to inbreeding effects has recently been demonstrated in a fritillary butterfly metapopulation (Saccheri *et al.*, 1998). In general, the importance of genetic and demographic stochasticity will depend on the type of metapopulation structure (i.e. stochastic effects will be relatively more important in systems composed primarily of small populations; Holsinger, Chapter 4, this volume).

Other issues relate to how various kinds of fragmentation processes might influence the relative persistence of species with different life histories. This will be a particularly important consideration in situations where the focus of conservation efforts is on species or communities that can function in a metapopulation situation. Some authors have also suggested that consideration should be given to how rates of spread of diseases depend on the degree of connectedness of patches across the landscape (Simberloff, 1988; Hess, 1994, 1996*a, b*). Whether pathogen or parasite movement should be taken into account when designing networks of habitat patches has not, to our knowledge, been critically addressed, although several theoretical studies have focused on how among-population connectedness influences pathogen dynamics and persistence (Thrall & Antonovics, 1995; Grenfell & Harwood, 1997; Thrall & Burdon, 1999).

The metapopulation paradigm provides a useful conceptual framework within which to address such questions, as it explicitly acknowledges that local populations are not isolated either in time or space. Furthermore, while fragmentation has certainly contributed to increased extinction rates in small populations, metapopulation theory tells us that even in fragmented systems, many species may have the ability to persist if individual patches are connected by dispersal (Moilanen & Hanski, 1998). While real-world metapopulations are unlikely to match several of the assumptions of the original Levins model (e.g. homogeneity of habitat quality and patch size, equal probability of dispersal to any site), increasing numbers of researchers are employing a range of descriptive, experimental, theoretical and computer-simulation approaches to ask how ecological and evolutionary processes are influenced by the incorporation of spatial structure in natural systems. These studies range from empirical work on small mammals inhabiting archipelagos to biologically realistic computer simulations of host–pathogen interactions that are integrated with empirical studies. In fact, much of the recent theoretical work on metapopulations has centred on exploring the consequences of departures from the classical Levins models. In the context of conservation, the most important of these are considerations of the interactions that occur between within- (e.g. birth, death, competition) and among-population (e.g. dispersal, gene flow, colonisation and extinction) processes, and how altering these may impact on species persistence.

Carrying out metapopulation studies in a conservation context

While all metapopulations (by definition) consist of spatially structured sets of habitat patches, not all fragmented ecosystems show the minimum requirement of stochastic extinction–recolonisation dynamics that fit the definition of a metapopulation (Hanski, 1994a). Therefore, it also follows that not all species in recently fragmented systems will have evolved to persist in a metapopulation situation. Nevertheless, a metapopulation approach to conservation biology is likely to provide one of the most useful tools for developing management strategies that optimise among-population processes critical for persistence of many natural systems. For example, Hanski (1994b) outlines methods for estimating the risk of metapopulation extinction and stochastic steady state. This approach can be used to investigate the likelihood of different outcomes when a fraction of patches is removed from the system (a possibility that is all too pertinent to conservation biology). Hanski's approach is based on simple metapopulation models which allow for estimation of colonisation and extinction probabilities from empirical patch occupancy data (often the very data that is most readily available from field studies). Hanski (1994b) illustrates this with an example from a metapopulation of fritillary butterflies (*Melitaea cinxia*) that assesses the 'relative importance of each patch for persistence of the metapopulation'. Similarly, Day & Possingham (1995) used a stochastic metapopulation approach to examine the spatial structure of populations of malleefowl, *Leipoa ocellata*, in southern Australia. With their model, they were able to quantify how further habitat loss (and consequent alterations in patch arrangement) would influence metapopulation extinction probabilities.

How do we approach the seemingly daunting task of carrying out metapopulation-type studies? Both practical issues associated with sampling multiple sites across time and also aspects of the species biology that determine relevant spatial scales for within- vs. among-population interactions must be taken into account (Antonovics *et al.*, 1994). As noted by Husband & Barrett (1996), one needs to consider the distribution of a species within its geographic range, the discreteness of habitats occupied and dispersal capability – these all play a role in determining connectivity and therefore what constitutes a local population. Demographic and genetic information may provide other criteria for defining local populations and metapopulations (Husband & Barrett, 1996); in this regard the concept of ecological and genetic neighbourhoods (Antonovics & Levin, 1980) is likely to provide a useful starting point. It is also possible to study metapopulation dynamics experimentally, using 'model' species that are easily manipulable in laboratory situations, such as bacteria and protozoa (Burkey, 1997), or where artificial populations can be readily established and maintained in the field (Thrall & Antonovics, 1995). However, in many cases, especially where endangered species are involved, experimental manipulation may be excluded and inferring metapopulation processes such as colonisation/extinction rates from pattern data becomes essential (Hanski, 1994*a*).

Both implicit and explicit modelling approaches have been developed to investigate spatial issues (Hanski, 1994*b*); the former include analytically tractable but often somewhat unrealistic theoretical models based on Levins's original concept, as well as more complex structured metapopulation models (e.g. Verboom & Metz, 1991), while the latter are often centred around biologically realistic, but analytically less friendly computer simulations. Both theoretical and computer-simulation studies have a special role to play as they also allow the investigation of many questions related to spatial scale that are not tractable empirically. Thus, these approaches have focused particularly on local vs. regional dynamics and their longer-term coevolutionary effects. Spatially explicit simulations have also proved useful as predictive tools for investigating the consequences of varying patch number and arrangement within reserves, and thus may be of assistance in identifying, not only how many, but which patches should be kept in a system. It is encouraging that in many cases, useful information may be derived from models based entirely on patch area and isolation (Moilanen & Hanski, 1998), both factors for which information is readily available.

LESSONS FOR CONSERVATION FROM METAPOPULATION STUDIES

Genetic effects

There are several general problems that have been raised with respect to increasing fragmentation and the consequent loss of connectivity among local populations. Notable among these are issues relating to genetic structure, for example, concerns over the loss of genetic diversity and inbreeding effects (Lacy, 1987; Gilpin, 1991; Thrall *et al.*, 1998). While a subject of intense interest among conservation biologists, empirical studies explicitly relating metapopulation dynamics (i.e. population turnover) to genetic fitness are almost non-existent, and non-conclusive. However, Richards (1997, and Chapter 16, this volume) has clearly demonstrated that smaller, more isolated populations of the plant *Silene alba* have higher levels of inbreeding, manifested as reductions in seed germination rates. Importantly, experimental work showed that fitness in these populations could be significantly enhanced by increased gene flow from other populations (Richards, 1997) – one of the criteria suggested by Harrison (1994) as a justification for using metapopulation models. Similarly, Young *et al.* (1999, and Chapter 19, this volume) found in the daisy *Rutidosis leptorrhynchoides* that the degree of correlated paternity was negatively related to population size and positively correlated with isolation, indicating the possibility of biparental inbreeding effects. Stochastic population simulations based on several years of demographic data for *R. leptorrhynchoides* also hint at a negative relationship between persistence time and the level of correlated paternity (Young *et al.*, Chapter 20). In contrast, work on a metapopulation of pool frogs (*Rana lessonae*) in northern Sweden indicated that while higher levels of population turnover resulted in reduced heterozygosity, this did not translate into lower fitness (Sjögren, 1991*a*, *b*). On several grounds, it has been argued that extinction of small populations is likely to occur for demographic or environmental reasons long before inbreeding effects will become important (Lande, 1988; Holsinger, Chapter 4, this volume). Clearly, this is an issue that is still far from resolved. However, in species with strong self-incompatibility systems, genetic effects may be manifest at a far earlier stage, effectively reducing population size very substantially (Young *et al.*, Chapter 20).

Species and community effects

Other important concerns relate to the maintenance of diverse species assemblages and communities. Under what conditions can fragmented systems maintain their integrity with respect to species composition? Much has been written about one single aspect of this issue, and that is with respect to fragment size in relation to effects on community structure [i.e. single large or several small (SLOSS); see Quammen, 1996 for a general review of this debate]. For example, long-term large-scale experiments in the Amazon basin (Lovejoy & Bierregaard, 1990) were explicitly focused on addressing this question. A great many researchers have argued both for and against particular sizes and numbers of reserves on economic, political and biological grounds, and it may well be that metapopulation theory can aid in determining optimal spatial structuring of patches. However, while fragment size is clearly important, it does not take into account issues of connectedness and the need to preserve colonisation processes at the among-population level for both ecological and genetic reasons.

Metapopulation viability

A wide range of recent studies that explicitly use the metapopulation framework has already indicated a number of important features of fragmented systems that are crucial to conservation biology. Firstly, various empirical studies have provided evidence that both population size and isolation can affect demographic and genetic parameters (Barrett & Kohn, 1991). For example, small populations of *Eucalyptus albens* (white box) have been shown to have less genetic diversity than large ones, this effect being more pronounced in more isolated sites (Prober & Brown, 1994). Similar results have been obtained in studies of the roadside weed *Silene alba* (Richards, 1997, and Chapter 16, this volume). Secondly, whether due to demographic, genetic or environmental causes, local patches often go extinct, which is a major reason for the simple rule that we should be preserving as much habitat as possible. In this regard, one of the most important lessons to be learned from metapopulation models is that 'currently unoccupied habitat fragments may be critical for long-term persistence' (Hanski, 1998). At the same time, both empirical and theoretical studies of metapopulations indicate that even though local extinction is common in a system, regional persistence may still be possible.

Paradoxically and deceptively, simulation studies have demonstrated the possibility that in many cases, metapopulations below the colonisation/extinction balance (and hence on the path to annihilation) can still have a high percentage of occupied sites at any particular point in time (Hanski, 1998). This may occur during the lag phase before past habitat destruction becomes manifest as declining numbers of populations in remaining areas, and begs the critical question of how we will know when we have passed a threshold. This highlights the importance of identifying 'indicator variables' in relation to species and community characteristics that can provide clues to the current and potential future state of fragmented systems (Brown et al., 1997; Thrall et al., 1998). It also shows quite clearly the potentially misleading nature of 'snapshot' studies taken at a single point in time or restricted to one or a few populations.

As noted by Hanski (1998), based on simple models, several important general conclusions can be drawn about how metapopulations respond to processes of fragmentation. Firstly, because among-population connectivity is lost in a non-linear fashion, gradual loss of habitat can result in a sudden shift from a positive equilibrium to regional extinction. Secondly, metapopulation decline in response to fragmentation occurs with a time-lag, sometimes referred to as the 'extinction debt' (Tilman *et al.*, 1994), which basically means that current extinctions are happening because of past habitat destruction. Thirdly, the equilibrium number of suitable but unoccupied patches existing in a metapopulation before fragmentation will be equal to the colonisation/extinction threshold. All of these factors high-light the difficulty of determining precisely when a system has passed the threshold for collapse, and moreover emphasise the fact that, in all likeli-hood, this point will be reached well before a decline in population numb-ers is obvious (Gutiérrez & Harrison, 1996; Thrall *et al.*, 1998).

On the positive side, the corollary to these observations is that meta-population theory also suggests several obvious ways in which the likeli-hood of regional persistence can be enhanced: these include increasing the number of habitat patches, increasing the degree of clumping of these patches, increasing variance in patch size (e.g. moving towards a core–satellite structure) and increasing connectivity (Harrison & Fahrig, 1995; discussed in Wiens, 1996).

PRACTICAL APPLICATIONS OF METAPOPULATION THEORY

In the previous sections, we have argued for the general importance of understanding spatial structure, and taking it into account when imple-menting conservation strategies. How can we use knowledge about the effects of fragmentation to manage biodiversity? In practical terms this may translate into understanding how metapopulation extinction is a function of the number, size, quality and connectivity of habitat patches in the sys-tem (Drechsler & Wissel, 1998). Are there situations where such ap-proaches have been implemented?

One clear example of how metapopulation thinking has been useful in a practical sense is with respect to the northern spotted owl in the north-western United States (Gutiérrez & Harrison, 1996; Noon & McKelvey, 1996). As has been widely publicised, old-growth forests in this region have been heavily logged, leaving only discrete fragments in which spotted owls can persist. The application of metapopulation modelling to the conserva-tion of spotted owls has led to a management approach that emphasises patch size and spacing consistent with persistence in a metapopulation (Wiens, 1996). Other examples include the Florida scrub jay (Stith *et al.*, 1996), kangaroo rats (Price & Gilpin, 1996), greater gliders in Australia (Possingham *et al.*, 1994) and pool frogs in Sweden (Sjögren-Gulve & Ray,

1996). In contrast, there appear to be relatively few examples where meta-population thinking has been explicitly applied to conservation of plant species (but see Menges, 1990, 1991b with respect to Furbish's lousewort *Pedicularis furbishae*) although a wide range of studies has focused on genetic and demographic consequences of fragmentation in plant populations (e.g. Richards, 1996; Young & Brown, 1998).

Will we ultimately be reduced to artificially dispersing individuals among reserves? It has been suggested that knowledge of metapopulation dynamics may allow us to perform the functions of dispersal and re-colonisation of locally extinct habitat patches, and that the metapopulation approach may also be able to guide conservation efforts in terms of optimal numbers and placement of translocated plants and animals, as well as genetic sources (McCullough, 1996). This may be of particular importance if we accept the argument that as suitable habitat areas become ever more isolated, and intervening areas increasingly hostile due to further development, most species will require help to move among patches (Simberloff, 1988).

Another area where metapopulation theory may inform conservation biology is in restoration ecology, where the realisable goal is often to regenerate a semi-natural system (in the sense of containing a subset of the original community). In a practical sense, it may well pay to identify and focus on species that have at least some potential to persist in a meta-population-like situation. Biologically realistic metapopulation simulations should prove useful in identifying general characteristics that such species are likely to possess.

SPATIAL AND BEHAVIOURAL INTERACTIONS TO CONSIDER

Extinction thresholds

A major and increasingly urgent problem for conservation biologists is understanding to what extent systems can be fragmented, and the number of individual populations lost, before collapse occurs (whether for genetic, demographic or environmental reasons). This issue clearly relates to the concept of minimum viable metapopulations, defined as the 'minimum number of interacting local populations necessary for the long-term persistence of the metapopulation' (Hanski et al., 1996a). Understanding factors that influence extinction risks in endangered metapopulations is an important component of management (Drechsler & Wissel, 1998). In this regard, it is important to develop a conceptual framework that allows some predictive power with respect to how spatial structure (e.g. core–satellite vs.

many small populations) interacts with biological factors to determine persistence thresholds. Theoretical work on metapopulations certainly shows that thresholds for persistence exist, and at the most basic level this translates into a balance between extinction and colonisation processes (Levins, 1969, 1970). However, exactly how any threshold depends on various life-history features (e.g. dispersal capability, life span, the existence of seed banks in plant metapopulations) is not well understood. Indeed for most species we have very little knowledge of either their dynamics or their dispersal ability (Wiens, 1996). However, there are some theoretical studies that relate dynamics in metapopulations to dispersal (e.g. Herben & Söderström, 1992; Olivieri et al., 1995). In the context of island biogeography, studies of colonisation ability in relation to life history suggest that the best colonisers are those that are generalists or are self-fertile; with respect to mammals, size appears to be positively correlated with colonisation ability, but the opposite has been shown to be true for insects (see Ebenhard, 1991 for a review). There is no reason to believe these general patterns will not be applicable in fragmented environments. However, ultimately at some spatial scales most organisms have limited dispersal, and therefore beyond that distance, increasing isolation brings a decreasing probability of colonisation. This provides an upper limit to the extent of fragmentation (with its almost inevitable dual effects of increased isolation and habitat loss) that is possible while yet maintaining the metapopulation.

Ascertaining what factors determine the conditions under which extinction rates exceed colonisation rates is essential to developing a true appreciation of the vulnerability of individual species (Harrison & Quinn, 1989; Frankel et al., 1995). When integrated across multiple species this has direct links to the concept of conservation at the community level (i.e. meta-communities: Hanski & Gilpin, 1991a; Wilson, 1992). Within a single community type, species may have very different disturbance requirements or dispersal distances (e.g. consider the range of fruit-dispersal mechanisms in tropical forest tree species) leading to different extinction thresholds. While it may be possible to determine extinction thresholds for a range of single-species metapopulations, this may grossly underestimate the threshold required to maintain the intact community. Identifying extinction criteria for those organisms that are highly mobile may provide a conservative estimate of the total amount of habitat necessary to preserve particular community types (Frankel et al., 1995), although it will still be necessary to understand how loss of connectivity within conserved regions impacts less vagile species (or those with more specialised requirements). A major challenge for the future is then to develop the concept of minimum community thresholds – otherwise, fragmentation may lead to qualitatively different communities, with drastically altered coevolutionary interactions and characteristics (e.g. with respect to ecosystem function).

A further complexity is that distance-independent extinction (correlated environmental variation) could lead to correlated extinction probabilities for conspecific populations (Harrison & Quinn, 1989). Such a possibility effectively reduces the number of demes in a metapopulation, as well as global persistence time, yet most metapopulation models still assume that local populations experience completely independent extinction (Harrison & Quinn, 1989). The very real potential for correlated environmental variation at spatial scales greater than single-habitat patches clearly indicates the need for management strategies that conserve as many habitat/environment types as possible, so as to maximise the probability of regional persistence.

Movement of pests, pathogens, parasites and predators

While, in general, metapopulation modelling has indicated that increased connectivity will lead to increased persistence at the regional scale (e.g. Harrison & Fahrig, 1995), some authors have suggested that there may also be negative consequences, at least in highly connected networks. For example, increased ability to move among populations may also facilitate the spread of infectious diseases and parasites (Dobson & May, 1986; Simberloff, 1988; Hobbs, 1992; Hess, 1994, 1996a, b). In fact, Simberloff (1988) has gone so far as to suggest that fragmentation may actually have the positive effect of reducing disease. Indeed, positive correlations have been found between disease incidence and the disease status of neighbouring populations in two natural plant–pathogen systems (Burdon et al., 1995; Ericson et al., 1999). Whether or not this is a real issue is an open question as there may well be trade-offs between the positive effects of increased ability of the host to move among patches and the negative effects of disease. Indeed, using simple Levins-type models, Hess (1994, 1996a, b) showed that there may be intermediate levels of connectedness that are optimal. However, these results do not take into account the fact that increased movement will also result in greater movement of resistance genes, and may also aid in disease avoidance, nor do they incorporate other aspects of coevolutionary interactions between host and pathogens. For example, recent empirical studies of a protozoan parasite (*Ophryocystis elektroscirrha*) of monarch butterflies with different patterns of migration suggest that life history and population structure may interact to determine disease prevalence, and the evolution of virulence (S. Altizer, unpublished data); parasites in non-migratory host populations were generally more virulent and disease prevalence was higher than in migratory populations. Overall, we would argue that disease is generally unlikely to provide a novel threat for species that have been artificially fragmented, as increasing connectivity is not likely to lead to the introduction of diseases that were not present in pre-fragmented times.

With respect to predation, increased connectivity may assist the movement of predators from one prey population to another. A classic illustration of this is given by the loss of native species on oceanic islands when predators such as rats and cats have been introduced (Simberloff, 1988). In a metapopulation context, a particularly good example is provided by the decline in frogs (*Rana muscosa*) as a consequence of the spread of introduced predatory fish in streams and drainage ditches that connect populations (Bradford *et al.*, 1993). Furthermore, where corridors are narrow it has been argued that they may often act as 'demographic sinks' doing more harm than good (Soulé & Gilpin, 1991). Thus, the effective degree of connectivity with respect to species persistence may be quite different from the perceived level if factors such as predation and behavioural complexities (see below) are not taken into account.

Behavioural issues

It has been suggested that increasing connectivity could lead to loss of individuals because of movement into suboptimal habitats (Hobbs, 1992). As a cautionary note, patterns of species behaviour may also alter model predictions and thresholds for extinction. For example, in a study of Columbian ground squirrels (*Spermophilus columbianus*), Weddell (1991) showed that squirrels did not disperse to new patches but settled near other individuals. Thus, social attraction can lead to reduced probabilities of patch colonisation, making the system more sensitive to fragmentation than would be otherwise expected. It seems likely that this will be a general issue for social species. Of critical importance for many species will be distinguishing between habitat patches used for different purposes (e.g. breeding vs. feeding). Moreover, movement among patches may often be non-random with respect to identity – e.g. young males leaving home territories to find mates. The extent to which such types of behaviour alter predictions for metapopulation dynamics is unknown. Overall, metapopulation management models for animals need to consider behavioural complexities (e.g. requirements for periodic migration: Hobbs, 1992) that would reduce the effective number of habitat patches, or change regional persistence thresholds. Other ecological features that apply to plants as well as animals, such as competitive interactions among species, may also be important factors to consider (Tilman *et al.*, 1994; Wiens, 1996).

Conservation of coevolutionary processes

Coevolutionary interactions between different organisms ranging from symbiotic (e.g. mycorrhiza fungal – plant and rhizobia bacterial – plant associations) to totally parasitic (e.g. fungal and viral diseases of plants and animals) are being increasingly recognised as both sources of biodiversity in their own right and as fundamental ecosystem processes that affect both population and community structure and function. However, conservation of these associations may be particularly complicated as the survival of one or both players is often inextricably tied to the continued existence of their partner(s). Examples of this phenomenon are myriad, including very many micro-organisms and invertebrates associated with plants.

Following the spatial pattern of their major component, coevolutionary associations typically occur as a series of individual, co-occurring populations distributed patchily across the landscape. As Thompson (1994, 1997) has pointed out, across this range of populations that represents a species' presence in an area (metapopulation) there are likely to be 'hotspots' of coevolutionary activity as well as others in which change is occurring far less rapidly. These hotspots may reflect a range of phenomena – both biological (e.g. enhanced recombination in the genome of either player through hybridisation with other ecotypes or species) and physical (e.g. sites that particularly favour the development of parasites) – that act as sources of variation in the metapopulation as a whole, and which consequently provide the driving force for continuing evolutionary change. A priori identification of such sites of high coevolutionary activity is very difficult if not impossible. Yet they must be maintained if such interactions are to continue to develop and change.

Increasingly, empirical evidence from coevolutionary studies is highlighting the diversity of responses achieved as coevolutionary partners respond to each other in different demes of single metapopulations. Thus in the interaction between *Linum marginale* (native Australian flax) and *Melampsora lini* (flax rust), the size of *Linum* and *Melampsora* populations, their ephemerality and their genetic structure often vary markedly between closely adjacent demes of a single metapopulation (Burdon & Jarosz, 1991; Burdon & Thrall, 1999). This interpopulational diversity provides a long-term 'insurance' for the continued existence of the association as individual local populations flicker in and out of existence. Equally importantly, though, the unpredictable changes in both the direction and intensity of selective interactions that occur as a result of these changes mean that we cannot predict the long-term trajectory of such associations nor can we expect to maintain them simply by preserving one or two populations. Indeed, recent increasingly sophisticated computer-simulation modelling clearly indicates the lack of clear predictive correlations between the behaviour of interactions at the level of the individual deme and that at the metapopulation level (Thrall & Burdon, 1999).

Failure to conserve such interactions within metapopulation arenas will inevitably lead to their eventual collapse. In many instances, this may result in the extinction of the antagonist. When this is a viral or fungal pathogen such a loss may raise little public concern (Dobson & May, 1986). However, what if the antagonist driven to extinction is a moth or spectacular butterfly? Or even a specialised variety of bird? Moreover, because of the unpredictable nature of interactions, we cannot say a priori what might be the long-term consequences of the extinction of individual players in co-evolutionary plays. Micro-organisms are frequently ignored by ecologists and conservation biologists, but their potential as powerful selective forces shaping the demography and genetics of their hosts is well documented in their use as biological control agents.

CONCLUSIONS AND FUTURE DIRECTIONS

As noted at the beginning of this chapter, much of population ecology has focused on studies of single populations, and has further assumed that dynamics in local populations are deterministic, with local stability a viable outcome. It is of considerable concern that this view may have influenced the development of conservation management practices centred around preserving existing populations and/or patterns across the landscape. However the metapopulation perspective has highlighted the fact that local populations are, in fact, ephemeral, thereby emphasising the need to ensure that long-term conservation planning focuses on spatial scales larger than the single population. Moreover, we would argue that, generally speaking, there needs to be a frameshift towards conserving process rather than patterns (patterns are also ephemeral). In the context of studies on the ecology of mosses, Herben & Söderström (1992) specifically note that the 'primary aim in the protection of species living on temporary, patchy habitats is to identify and preserve the conditions that enable the dynamic process to operate'. We concur wholeheartedly and argue that this applies to all species, although the relevant spatial and temporal scales will vary in different situations. In fact, this argument could be extended to preserving co-evolutionary processes rather than species themselves – ultimately, nothing is permanent except change! The importance of emphasising regional persistence applies equally to many different levels of organisation, ranging from metapopulations of a single species (or coevolutionary interaction) through to entire communities and ecosystems.

The bottom line is that fragmentation results in two fundamental changes across the landscape: firstly, it leads to increasing isolation of habitat patches, and secondly, it results in a decrease in the total amount of available habitat (Simberloff, 1988). The metapopulation paradigm explicitly recognises and provides a conceptual tool for dealing with the interactions of within- (e.g. birth, death, competition) and among-population processes (e.g. dispersal, gene flow, colonisation/extinction). The ever-growing diversity of studies that demonstrate the importance of spatial structure in determining ecological and evolutionary trajectories also makes clear that conservation biology must focus on regional persistence and not local within-population persistence. Simply put, it is just not good enough to put fences around existing populations as all local sites are ephemeral at some temporal scale. Rather, we must explicitly recognise the importance of colonisation/extinction processes. Unfortunately, while the importance of spatial structure is now generally recognised, it is still only rarely incorporated into population viability analyses (Harrison, 1994).

But, given the fact that most reserves are already far too small to support intact communities, and in the context of the ever-increasing destruction of habitat with the simultaneous decrease in the number and size of larger reserves, does a metapopulation perspective really provide any greater hope than any other approach? It would be relatively easy to advocate the rather pessimistic perspective that since (1) maintaining many small patches is both more time-consuming and expensive than single large reserves, and (2) small patches will go extinct anyhow, we should simply let the small reserves go and focus on the largest remaining areas. Simberloff (1988) has suggested that the metapopulation paradigm does provide some encouragement in that it has demonstrated that a partially connected system of many small patches can indeed have long-term viability. Thus, if it is to have value for conservation, the clear challenge for metapopulation theory in the immediate (and foreseeable) future is to enable managers (1) to determine when there is sufficient between-patch migration for persistence under realistic conditions (and if not, how this could be achieved), and (2) to guide efforts in maintaining appropriate numbers and placement of currently unoccupied patches in a state that they could become occupied again and not just overrun by weeds. Integration of management efforts within a regional metapopulation framework is invaluable in that it forces a process-oriented, scale-appropriate strategy to conservation.

6

Population viability analysis for conservation: the good, the bad and the undescribed

MARK BURGMAN & HUGH POSSINGHAM

ABSTRACT

Population viability analysis (PVA) is used in a variety of ways to solve conservation problems. These uses are defined in part by data availability and theoretical and biological understanding, and in part by social, regulatory and political context. The use of a PVA in one context does not preclude or invalidate its use in another. In this paper we attempt to discuss objectively the role of PVA in population management and conservation planning. We emphasise its role in organising information, engaging stake-holders, and making decisions. To be successful, some traditional views about population models and decision-making need to be suspended and reviewed.

INTRODUCTION

The use of population models in conservation biology is at a crossroads. In the 1980s many believed that population viability analysis (PVA) would provide a comprehensive framework for threatened-species management. Early enthusiasm has been tempered by problems with lack of data, lack of validation, and several studies that have demonstrated the sensitivity of results to uncertainty in the data (Taylor, 1995; Ruckelshaus *et al.*, 1997). More and more conservation ecologists are expressing disappointment at the inability of PVA to provide verifiable answers, to be impregnable to misinterpretation or to make the work of an academic conservation ecologist any easier (Doak & Mills, 1994; Beissinger & Westphal, 1998). Others who view PVA from a more applied perspective and have tried to use it to help with environmental legislation, land-use planning and making decisions about the management of populations have also been disappointed (Hamilton & Moller, 1995; Harcourt, 1995). The role of population modelling in conservation is suffering a backlash from early over-enthusiasm and an unrealistic expectation that PVA would solve all single-species conservation problems. The role of population modelling in conservation biology, and in population management in general, requires reassessment (Shea & NCEAS Working Group on Population Management, 1998).

We begin this paper by characterising the different roles PVA can play, roles that apply fairly generally to the application of population modelling tools in conservation. We move to a discussion of what we think a good PVA should be and do. We end with three aspects of population modelling for conservation that require further attention – the role of genetics in PVA, spatial population modelling for management and the use of decision theory with PVA.

THE VARIOUS ROLES OF PVA

PVAs are constructed at numerous spatial and temporal scales, and with different underlying theoretical bases. Individual-based models can create very detailed forecasts. They have particular utility when details of individual behaviour are important, when genetic processes are to be included in the model and when the modellers wish to explore the fine detail of the spatial arrangement of individuals and their interactions with environmental variables and management objectives (DeAngelis & Gross, 1992; Judson, 1994; McCarthy, 1996; Letcher et al., 1998). At the other end of the spectrum, incidence functions and related patch-based models ignore the details of population dynamics completely (Hanski et al., 1996b). People thinking about applying models to solve a particular problem are confronted with this very broad range of possibilities, and a range of apparently conflicting advice on what can and cannot be done. Below we summarise the different roles of PVA and their associated concerns.

The intellectual curio

PVAs have been considered to be of theoretical interest but little practical value. Caughley (1994) argued that there are two paradigms in conservation biology. The first is the small-population paradigm, driven by theoretical interest in the stochastic processes governing small populations, arising from the mathematics of birth and death processes. Here the concern is with chance extinctions in highly restricted and/or fragmented environments. The second is the declining-population paradigm where species have a long-term negative population growth rate throughout their range. In stark contrast to the small-population paradigm, this requires immediate pragmatic strategies and population modelling is of little value. Caughley (1994) suggested that the principal contribution of the small-population paradigm was to provide an answer to a trivial question: how long will a population persist if nothing unusual happens? The theory bears 'tenuous relevance to the problem of aiding a species in trouble' and any empirical generalisation constructed directly from simple theory is suspect (Harcourt, 1995) – it is an 'intellectual curio'.

The criticisms are certainly valid. However, one can argue that theoretical literature does not search for empirical generalisations. The search is for mathematical generalisations upon which solutions to empirical problems may be formulated. Much of the early PVA literature was motivated by theoretical interest (e.g. Goel & Richter-Dyn, 1974; Ginzburg et al., 1982; Lande & Orzack, 1988; Tuljapurkar, 1989). As Caughley (1994) noted, it served to make the scientific foundations upon which pragmatic PVA is built. Robust general principles permeate later applications.

Caughley (1994) and others (e.g. Harcourt, 1995) have not understood that the distinction between deterministic and stochastic issues is artificial (Boyce, 1992). Their narrow view of PVA as a simple, demographic tool that ignores 'external' deterministic processes and the ecology of the species is an exaggeration. Any PVA with these characteristics is just badly constructed. The ecology of species and the role of management should be, in the words of Boyce (1992), the nuts and bolts of modelling exercises.

The loaded gun

The 'loaded gun' view contends that PVA is dangerous in the hands of the untrained. This view argues that PVAs built with too few data are suspect and should be handled with care or discarded completely. For example, Harcourt (1995) evaluates PVA models of the Virunga gorilla (*Gorilla gorilla*) and concludes that demography dominates the models at the expense of ecology. This conclusion is echoed by Beissinger & Westphal (1998) who suggest that the advent of easy-to-use computer software makes it 'too easy' to construct a model that can be passed off as a sound and useful PVA. Similarly, Reed et al. (1998) argue that when demographic data are absent, effort should be given to more data collection or alternatives to PVA rather than the PVA itself. These arguments rest on the proposition that a lack of data and a poor level of understanding of the ecology of a species may precipitate a degree of uncertainty in models so that the results provide no credible management suggestions (Beissinger & Westphal, 1998).

The view that if data are insufficient then there is no point developing a PVA at all reappears from time to time, sometimes motivated by a published result that does not sit well with someone in the scientific or regulatory community. This problem plagues all models, not just PVAs, and it has consequences for model interpretation, sensitivity analysis and the generation of error bounds for stochastic and deterministic expectations (Shaffer, 1987; Goodman, 1987a; Boyce, 1992; Burgman *et al.*, 1993). However, it does not follow that all PVAs must be based on a full set of credible data. The antidote is not to urge people to desist from building models under some (vaguely specified) circumstances. It is to engage in building a better model that includes the features that the other model has disregarded.

Given we will never have a 'complete' data set for any species, we take the view that there are never insufficient data to construct a useful model. Incomplete information does not mean that meaningful results are impossible to obtain because there is very significant value in building a model, for its own sake (Akçakaya *et al.*, 1997). It clarifies assumptions, integrates knowledge from all available sources, and forces biologists to be explicit and rigorous in their reasoning. It allows us to identify, through sensitivity analyses, which model structures and parameters matter (in that they make a difference to the outcome), and which do not. In this sense it is a guide to further data collection (Possingham *et al.*, 1993). It results in a set of logical statements that are internally consistent, and it allows us to explore the consequences of what we believe to be true, even in the absence of all relevant data.

The magnifying glass

Under close scrutiny we will almost certainly find that other models will perform equally well – which one do we choose? This concern about PVA comes from thinking like an 'ecological detective' (Hilborn & Mangel, 1997). There will always be more than one plausible model for each natural population, as long as there is more than one ecologist thinking about it. Each of the alternatives represents a different idea of how the world works. It is unwise to discard any plausible hypothesis. Rather, we should treat each one on its merits, reflected in its associated likelihood. For example, Pascual *et al.* (1997) explore the expectations of a series of structurally different models in determining the best way to manage populations of wildebeest (*Connochaetes taurinus*). In this role, PVAs can be used iteratively to encapsulate the things we know currently, and to provide a benchmark against which we measure the improvement in the status of our knowledge. However, PVAs typically have many parameters, and given the uncertainty around these parameters, there are many ways of getting a good fit from population data to any model.

Models of biological systems are no different from any other scientific tool. If there are not 'enough' data to answer the questions at hand, or if the model builder has not taken care to represent ecological ideas and deterministic processes faithfully in the model's equations, PVAs may lead to wrong decisions. There are also many ways to abuse standard statistical tests; the scientific consensus is that the benefits of the availability of powerful computer programs for mathematical and statistical analysis outweigh their disadvantages (Akçakaya & Burgman, 1995). The suggestion that analyses may have been made 'too' easy (Beissinger & Westphal, 1998) emphasises the need for better training and for the specification of standards for a good PVA (see below).

The idea that all plausible models should be explored and their predictions weighted explicitly by the likelihood of the model is reminiscent of a Bayesian view of science (Hilborn & Mangel, 1997). Engineering and human-health-risk analysts distinguish structural uncertainty (different models), natural stochasticity (environmental and demographic stochasticity) and parameter uncertainty (measurement error) (e.g. Cohen *et al.*, 1996; Frey & Rhodes, 1996). Some elements of structural uncertainty may be examined using Monte Carlo simulation (Frey & Rhodes, 1996). If the model structure is incorrect or inappropriate for the species in question, serious errors in prediction are likely. Errors, together with uncertainties, are magnified into the future with each time step, so usually only a few time intervals can be predicted with any certainty. The omission of an important process, such as loss of habitat, competition or predation from introduced species, impacts of disease or parasites, or the impacts of rare catastrophic events, may substantially affect what it is best to do to manage a population to avoid extinction (Akçakaya *et al.*, 1997).

The facilitator

Often, a PVA is instrumental in easing the assembly and discussion of a problem. Very many PVAs have been conducted with the primary aim of getting people together to discuss a problem and to find a consensus approach to management. For example, Seal (1992) reported conducting 30 'Population and habitat viability' workshops involving 800 people in eight countries. PVAs in this context provide a structured framework that make assumptions plain, and provide a focus for discussions about data, ecological processes and management alternatives. Of course PVA is not essential for this process (Beissinger & Westphal, 1998) and one could argue the model plays a subsidiary, and possibly confusing, role to the process of mediation. The confusion can arise when the people concerned become too obsessed with the results of the PVA rather than the process of assembling and integrating information.

The stockbroker

One of the original objectives of developing stochastic models for population dynamics was to weigh expected gains and costs against stochastic catastrophes and uncertain benefits. Risk assessment is often most effective in the role of comparing risks and evaluating the best choice among a range of alternative management options. Managing risks for natural populations results in economic trade-offs within the decision-making framework (Possingham et al., 1993). Decisions made under uncertainty may depend on the attitude of the decision-maker to the prospective risks, on the costs of various alternatives (e.g. Akçakaya & Raphael, 1998), and on how these risks change as circumstances change (Possingham, 1997). Such applications make use of formal decision theory to solve problems (Shea & NCEAS Working Group on Population Management, 1998). PVA does not include a formal role for decision theory – although there is no reason why the results of a PVA cannot be included in a formal risk-assessment process (Maguire, 1986).

Risks must be managed. They are as much a part of the management of threatened species as they are of our personal choices about insurance, investment choices by companies and strategic defence alliances made by governments. Effective management may be achieved simply by finding a consensus of opinion among scientists, managers and other interest groups, facilitated by workshops.

Often, risk analysis suffers from the fact that it is at a distance from formal decision-making frameworks (Lackey, 1997). There is a well-developed literature on cost–benefit analysis and optimal decision-making under uncertainty that might be employed to improve decisions about the management of threatened species (Possingham, 1996). Our contention is that the most appropriate role for PVA is in comparing risks, rather than measuring risks (Possingham et al., 1993). It is the only available tool for deciding on the allocation of resources to solve problems where the outcomes are inherently unpredictable. It is the only tool available to us that allows assumptions to be stated unambiguously and transparently. That does not mean that these things are always done, only that they should be part of any 'good' risk assessment.

The money pit

The cost of carrying out a PVA is not negligible. The naïve manager will often initially view PVA as a tool that will deliver answers after an after-noon's playing on the computer. The idea of predicting the fate of a species for hundreds of years after guesstimating just a few parameters and then running a fast simulation is appealing. In circumstances where the PVA is used to facilitate discussion, this approach has value. Many species may be 'workshopped' in just a week or two with the intent of identifying key con-servation issues and defining the roles of people interested in conserving the species. We make mistakes in these contexts when we put faith in the absolute values derived from these analyses. Any results derived from a model developed over a few hours cannot be considered reliable, unless the model and the circumstances are so simple and so well understood that the exercise is almost trivial.

We believe that a good management-based PVA will take at least three months using existing software, and several more if one wishes to con-struct a tailored model. Despite the benefits of model building, there are potential costs. As Beissinger & Westphal (1998) point out, there are real risks of over-interpretation of results and inappropriate belief in model out-comes that are not justified by data. Modelling involves time and money that may be better spent elsewhere. Thus, the scale of the modelling exer-cise should match the benefits that are likely to result from the exercise.

The roulette wheel

A management decision (such as which patch to purchase first, or which of several populations to target for management) involves a gamble. The dy-namics of threatened species, and the management decisions that must be made for them, are acted out on a spatially complex landscape that itself can change randomly through time (Richards *et al.*, 1999*b*). In most cases, even if the process could be accurately described mathematically, the optimal management for such a spatially complex stochastic population would be impossible to find mathematically. Our management actions are no differ-ent from betting in a game of chance, but one in which we can try to inform ourselves about the different likelihoods of the range of possible events.

While the management of complex non linear spatial systems may seem daunting – and PVA a crude tool for tackling the problem – surely the tool can, at worst, aid the informed guess of the wildlife manager. In the absence of any other tool even a simple model is better than no model at all.

The correct role for a PVA

Of course, PVA plays all the roles listed above – they are not mutually exclusive. Each of the different roles may be entirely appropriate, depending on context and application. There are good and bad applications of PVAs. Attributes that can seem an advantage – such as model complexity (the 'intellectual curio' and the 'roulette wheel' above) – can at times be a burden (the 'loaded gun', the 'magnifying glass' and the 'money pit' above). These roles are not necessarily criticisms of PVA, but rather are views of a tool that is part of the maturing discipline of conservation biology.

Bearing all these roles in mind we advocate the use of PVA as a decision-support tool, rather than as a decision-making tool (Possingham *et al.*, 1993; Ralls & Starfield, 1995). Disquiet, if not distrust, develops when the model takes over the decision-making process. From our point of view, this unease is well deserved. PVA may play a valuable role in any of the lights in which it is seen. However, we must remember that, by definition 'All models are wrong but some are useful.' (Box, 1979). The only correct model is an entire reconstruction of the actual system – whereupon it ceases to be a model. The utility of a PVA is determined by several things, including the care taken to include all ecological intuition faithfully, the care taken to represent all views (hypotheses) as structural alternatives, the detail and transparency of statements about assumptions, and the role of the model within the decision-making framework.

One of the most important steps in establishing the credibility of a PVA is to communicate the uncertainties embedded in the model and its assumptions. These will involve natural variation in parameters (spatial, environmental and demographic uncertainty), uncertainty about the true parametric values of the parameters (means, variances, kinds of distributions, dependencies and correlations) and uncertainty about the structure of the model (representing uncertainty about ecological processes). Tools used in communication of these uncertainties include qualitative statements, sensitivity-test results and the use of a variety of graphical representations of the kinds of uncertainty and their consequences for model predictions. Modellers are responsible for communication of these details. Direct communication in workshops provides the least ambiguous form of communication and imparts a degree of ownership of the model to the participants.

THE ATTRIBUTES OF GOOD PVAS

The attributes that distinguish good risk assessment from poor are debated regularly by the wider community of environmental-risk analysts (Power & Adams, 1997; Warren-Hicks & Moore, 1998). Criticisms of risk-assessment models range from a lack of ecological realism (Karr, 1995) to objection to their use on the grounds that the analyses are distorted inevitably by the personal values and attitudes to risk of those who build the models (Funke, 1995). Furthermore, it may be argued that risk-assessment tools (such as PVA) can be used to support almost any predetermined policy position and provide for it a mantle of scientific acceptability (Merrell, 1995) when manipulated by a cunning user. These criticisms are, of course, made equally in any scientific arena. There is always a risk that performing a relatively sophisticated analysis will create the appearance of scientific rigour (Burmaster & Anderson, 1994).

Reporting requirements for Monte Carlo simulations should enable re-viewers to replicate the analyses. The requirements of a PVA should be such that it is appropriate for the questions at hand, and that it be transpar-ent and internally consistent. The value of any model depends on how clearly and thoroughly its limitations have been communicated. In the spirit of recognising the potential for misapplication of risk assessment, we suggest the following criteria be used to evaluate the quality of a PVA. These attributes are based on the description of protocols for best-practice quantitative uncertainty analysis by Burmaster & Anderson (1994), Oreskes *et al.* (1994), Shea *et al.* (1998), Ferson (1996a, b) and Warren-Hicks & Moore (1998).

The most important elements of a successful PVA

- A fundamental understanding of the species' ecology, including what constitutes suitable habitat, and the ability of the species to disperse between patches of habitat
- An understanding of the environmental disturbances that directly threaten (or indirectly threaten through habitat alteration) a species, their probabilities and their impacts on the population (or habitat); this should include the effects of both deterministic threats (e.g. timber harvesting, recognising that nothing is completely deterministic) and stochastic threats (e.g. hurricanes)
- An understanding of the response of the species to the threats (e.g. density dependence)
- An assessment of the current state of the population
- An assessment of the probabilities of future risks
- An evaluation of the habitat as well as the population itself, and
- The extent to which the PVA helps managers make decisions.

In explaining the motivation for, data behind, and implications of, a good PVA the conservation biologists needs to fully describe many things:

Describe the context

- Outline the motivation
- Explain the application of the results
- Identify any sources of contention or any resources that depend on the outcome
- State the position of the modellers in relation to these parameters
- Describe the management objectives, options and indicators of performance, and
- Describe constraints on decision variables and management options.

Describe the model

- Show all formulae
- Identify components based on scientific or professional consensus, professional judgement by an individual, policy
- Show alternative components and alternative formulae when no consensus exists
- Demonstrate that units make dimensional sense, and
- Identify critical assumptions.

Describe the data

- Provide an evaluation of the quality of the input data for all the means, variances, and dependencies
- Describe procedures used to calibrate parameters
- Discuss the ways in which the model captures structural uncertainty, natural variation and parameter uncertainty
- Provide sources and values for all parameters (sampling distributions, shape and position parameters, correlation structures), and
- Discuss potential consequences of extrapolations, alternative distributions, kinds of dependencies or correlations.

Describe the analysis

- Describe the ways in which dependencies are handled
- Demonstrate that relationships between variables are feasible and plausible (i.e. ensure the correlation structures are self-consistent and that sampled values fall within plausible ranges)
- Describe and evaluate the random-number generator used in the analyses

- Describe and rationalise the choice of summary statistics (i.e. state variables: interval extinction risk, terminal extinction risk, quasiextinction thresholds, quasiexplosion thresholds, time to extinction, bounds on extinction probabilities, mean population, median population, confidence intervals, interquantile ranges, harvest volumes, translocation numbers, occupancy rates, ranks of management options, costs, benefits)
- Rationalise any sources of ignorance or variability that may contribute to future surprises, and
- Describe approaches for verification and validation of model output.

Describe the output

- Provide complete information on each of the model's state variables
- Provide quantitative, deterministic sensitivity analysis
- Provide quantitative, stochastic sensitivity analysis
- Display central tendencies and tails of the output
- Describe the qualitative fit between model results, a priori expectations and empirical observations, and
- Describe the quantitative fit of the model to independent observations.

In judging the adequacy of a model in a given circumstance, there are several things that Monte Carlo simulation cannot do (Ferson, 1996a). They include:

- The propagation of non-statistical uncertainty: if uncertainty about parameters or model structures is qualitative, vague or linguistic, then the model cannot account for these uncertainties adequately
- Provision of reliable quantitative answers when statistical dependencies are unknown, when input distributions are unknown; and when the model structure is unknown; if the knowledge is not available through empirical studies, it must be present in the form of explicit assumptions, evaluation of alternatives and sensitivity analyses
- Provision of general bounds on uncertainty: if dependencies are unknown or uncertain, then the model will not provide bounds that are as broad as they could be if all possibilities were to be explored (Ferson & Burgman, 1995).

The above guidelines are intended to promote PVA quality assurance – something which we as much as others have failed to achieve. In addition, human errors can appear in the form of data transcription errors, incorrect model specification, logical and algorithmic errors, output errors and so on. In complex models, data and computational errors can be exceedingly difficult to detect and correct. Whenever PVAs are submitted for review, the source code or at least the compiled computer program used to produce them should be supplied so that models can be tested. This is perhaps the greatest strength of ready-made computer programs; they have the distinct advantage of being relatively free of computational mistakes. Because dealing with uncertainty is a special problem with PVA, both in the context of the input and the output, here we elaborate on some of these issues in more detail.

Dealing with uncertainty and parameter estimation

Models fall into disrepute when the people that build them fail to communicate fully the uncertainties in the model's construction, parameter estimation and output (Burgman et al., 1993). Most, if not all, PVAs constructed to date do not take full account of all sources of uncertainty. They underestimate the full extent of possible outcomes.

Uncertainty about stochasticity has several components. They include measurement error of the moments of the statistical distributions representing natural variation, uncertainty about the form of the statistical distributions from which to draw this variation, uncertainty about the form and strength of dependencies among random variables in the model and uncertainty about the structure of the equations representing growth, survival, reproduction and dispersal. Most models use a single kind of statistical distribution to represent uncertainty without any clear reasoning for the choice, most use one simple set of linear dependencies without exploring other possible scenarios, and most use a single model structure (McCarthy et al., 1995).

The most important weaknesses of most current PVAs are the failures to propagate correctly the stochasticity and structural uncertainties inherent in the model. These simplifications make the task of generating simulations easy, but they fail to communicate the full extent of our uncertainty in the outcomes. The choices we make about these seemingly arbitrary conditions may make a substantial difference to the model's predictions, particularly in the tails of the distributions (Haas, 1997).

There are methods for propagating uncertainty about statistical distributions and dependencies in risk-assessment models. Monte Carlo simulation of alternative distributions may be used in what is known as a two-stage analysis (Hoffman & Hammonds, 1994; Frey & Rhodes, 1996).

In these analyses, parameter uncertainty and structural uncertainty are treated in a factorial design in which a range of possible combinations of distributions and parameter settings is trialed, resulting in a cloud of possible trajectories. There have been developments in computations that use the principle of maximum entropy to select distributions to represent uncertainty (e.g. Lee & Wright, 1994). Other approaches use probability-bounds calculus (Ferson *et al.*, in press) to accommodate any level of knowledge about statistical uncertainty, and propagate these uncertainties faithfully through a chain of calculations. Detailed brute-force sensitivity studies should be able to accommodate most (but not all) of these issues, yet most sensitivity studies, when they are included at all, focus on a very limited set of possibilities that underplay important structural and statistical uncertainties.

Having defined how the PVA process should occur, there remain many issues that require further work as we strive for a better product. The resolution of these problems is a future challenge for us all. Three special issues are raised below.

Dealing with spatial complexity

Many recent PVAs include a spatial component which can be explicit (Lindenmayer & Possingham, 1996) or vague (Hanski *et al.*, 1996*b*). Spatially explicit models are being seen increasingly as an important decision-support tool when we need to make specific spatial decisions – such as where to put a reserve or corridor, and where to protect habitat from disturbance. Unfortunately the introduction of spatial heterogeneity brings with it an avalanche of increased data requirements such as detailed habitat maps, detailed information on movement and associated mortalities, spatial correlations of catastrophes and environmental variability, and dependence of population processes on habitat. Ruckelshaus *et al.* (1997) evaluated the effect on model predictions of misclassification of habitat suitability, errors in estimation of dispersal distances and errors in estimation of the mortality rate of dispersers. They found that parameter estimation errors concerning mortality during dispersal can have an important qualitative effect on the predictions of spatially explicit models. They suggest that models of animal movement that depend on GIS maps are sensitive to errors in the data used to construct the representation of habitat. Models may make robust predictions within sets of plausible conditions, and some models make reliable and unbiased predictions when based on sparse data (Groom & Pascual, 1997). Increasing model complexity results in realism only if the parameters may be estimated accurately. Estimates based on insufficient data represent additional assumptions.

Increased complexity can sometimes be valuable, even in the absence of requisite data (e.g. Letcher *et al.*, 1998), because it allows us to explore ideas and identify important parameters that we may not yet have measured, and it makes transparent the assumptions necessary to answer questions put by management. The important point is that the need for data does not vanish when we simplify a model. Rather, the required data are subsumed within the assumptions of the simpler model. The advantage of a complex model is that it forces us to be explicit about these assumptions. Of course, there is a trade-off between the time invested in modelling and the insight that the model provides. The art of model building is to find this balance.

How important is genetic information?

Since Lande (1988) there has been a minor backlash against the importance of genetic data and processes in making decisions about the management of threatened species. We can ask some of the questions alluded to above in a genetic context. How expensive are genetic data? How will it change our decision-making? What is the relationship between the genetic structure of a population and its demographic processes? Why have so many species appeared to recover from a seemingly fatal genetic bottleneck? The most apparent role of genetic information comes from molecular ecology, where the data are used to identify the object of any conservation effort (a population, metapopulation or species), and to provide estimates and plausible bounds for demographic attributes that are otherwise impossible to obtain (such as social structure, and dispersal rates) (Frankham, 1995*b*). But the caveats that apply to spatial structure also apply to the population-genetic dimension of management. If we do not deal with the genetic implications of population management explicitly, then they become buried in the assumptions of the decision-making process. The advantage of a model that accounts for genetic variation is that it forces us to be explicit about the effects of inbreeding, outbreeding depression, mutational meltdown and the range of plausible interactions between demographic attributes and genetic composition (Frankham, 1995*a*). Formal sensitivity analysis would allow an evaluation of the potential importance of these features. The fact that we may not know many of the parameters necessary to parameterise such a model does not mean we can subvert the need to know them by ignoring them in a simplified model structure. Experience may allow us to make assumptions and thereby make the process of modelling efficient, but a good model will rationalise these assumptions.

PVA and decision theory

Like much of the theory of conservation biology, PVA outside the context of decision theory is of little practical use (Possingham, 1997). For example, what is the value of finding the forest-management option that will minimise the extinction probability for an old-growth specialist when that option is unacceptable to the associated timber industry? There may be other solutions that are almost as efficient from the perspective of minimising added risks, but which are much less costly from a social perspective. Costs, constraints and objectives need to be built in to any decision-making process and PVA does not formally allow for these issues to be included.

If we recognise that a population-management model for a threatened species is inherently stochastic and decisions are likely to depend on current circumstances (e.g. the price of timber) then we must accept that stochastic dynamic programming represents the only way to find the precise optimal strategy (Millner-Gulland, 1997; Possingham & Tuck, 1998). Given the complexity of the state-space, dynamics and management options the full problem will almost always be insoluble. We believe that the only practical way forward is to develop some robust rules of thumb that work under common scenarios.

CONCLUSIONS

In any discipline, there is a time-lag between the development of theoretical ideas and their sensible application. The lag is created by the scientific community being duly cautious about developing a consensus on the ways in which a new method is best applied. Improvements in theory will result, ultimately, in better decisions. For the moment, we do the best with what we have. Currently we have tools available to us that we are not using. They include formal propagation of statistical uncertainty, exploration of structural alternatives and application of decision theory. Hence there remain challenges to theoreticians, empiricists and managers in the development of PVA as a tool. We reassert our final opinion, that when making a decision concerning a threatened species, it is better to make it having tried to build and run a population model, than otherwise. The absence of the requisite data is always going to be problematic, but it is not debilitating, and cannot be avoided by simplifying a model or by deciding to dispense with modelling altogether.

ACKNOWLEDGMENTS

This work was conducted as part of the Biological Diversity and Extinction Risk Working Groups supported by the National Center for Ecological Synthesis and Analysis, a center funded by the National Science Foundation (DEB-9421535), the University of California – Santa Barbara, and the State of California. We thank our numerous colleagues who have worked on PVA with us – especially Sandy Andelman, David Lindenmayer, Mick McCarthy, Martin Drechsler, Charles Todd, Helen Regan, Dick Frankham, Byron Lamont, Barry Brook and Resit Akçakaya.

7

Applications of population genetics and molecular techniques to conservation biology

PHILIP W. HEDRICK

ABSTRACT

Some endangered species have declined to very low numbers and a few populations. Particularly in these situations, genetic information is often critical for evaluation of the existing animals and for future management. Three examples are discussed here, winter-run chinook salmon and bonytail chub, both of which exist in only one natural population, and Mexican wolves, which have been extinct in the wild and have only recently been released into eastern Arizona. A supplementation programme for winter-run chinook salmon has been ongoing since the early 1990s. Using data from 11 microsatellite loci, it appears that returns from these hatchery-reared fish are consistent with random expectations over both female and male parents. The potential broodstock to use for captive rearing of bonytail chub is evaluated. However, the extent of genetic variation in the potential broodstock appears very limited and is descended from between 3.5 and 8.5 effective founders as determined from allozyme data and supported by mtDNA haplotype sequence information. As a result, it is critical to incorporate the contributions from other wild-caught individuals into the broodstock. Finally, there have been concerns that Mexican wolves are not well differentiated from northern gray wolves and may have some ancestry from either coyotes or dogs. Using 20 microsatellite loci, all three lineages of Mexican wolves cluster together, are different from other canid taxa, and do not show any ancestry from either coyotes or dogs.

INTRODUCTION

Although efforts to perceive and integrate the factors that influence the numbers and persistence of rare species are not just recent, conservation biology as a comprehensive approach to understanding problems in threatened and endangered species is only a few decades old. In addition to being a relatively young science, conservation biology is often viewed as applied science, i.e. using the principles from other biological disciplines to answer specific questions about the conservation of threatened and endangered species. While this in itself need not be negative, both because conservation biology is relatively new and also because it has relied on ideas from other disciplines, it has had growing pains.

As a result, when Caughley (1994) wrote a provocative review giving his perspective of the scientific approaches used in conservation biology just before he died, it caused some questioning of the success of conservation biology. In my opinion, Caughley developed a false dichotomy between the 'small-population paradigm' and the 'declining-population paradigm' and, in particular, had some misunderstandings about the application of ideas from the genetics of small populations to conservation biology (Hedrick *et al.*, 1996). For example, Caughley suggested that a major weakness of the small-population paradigm is that it has not significantly contributed to preventing extinctions and that 'no instance of extinction by genetic malfunction has been reported'. However, genetics does not operate in isolation but influences populations through its effects on disease resistance, viability, reproductive success, behaviour, physiology and other characteristics. Extinction may be influenced by these indirect genetic effects and an important issue for conservation biology is to determine under what conditions genetic factors are likely to influence population persistence (e.g. Nunney & Campbell, 1993; Mills & Smouse, 1994). Caughley did acknowledge that genetic considerations in avoiding inbreeding and maximising retention of genetic variation have played a major role in captive breeding.

Caughley ended his essay on a positive point and suggested that 'each paradigm has much to learn from the other and in combination they might enlarge our idea of what is possible'. It is in this spirit that I discuss the application of population genetics and molecular techniques to conservation of endangered species. It is in fact an exciting time when new molecular-genetic techniques promise detailed understanding of the genetics and evolution of endangered populations and species and may give us insight to problems that we have only guessed about in the past. On the other hand, while these advances are enticing, it should be kept in mind that many problems with endangered species cannot be solved with genetic techniques.

Below, I will summarize information from three situations that illustrate the usefulness of molecular techniques and give insights not possible with other approaches. Both winter-run chinook salmon and bonytail chub, the first two examples I will discuss, only have one extant population. Habitat modification has played a significant role in the decline of both species and the introduction of non-native fishes has eliminated recruitment in bonytail. Mexican wolves, the other example, appear extinct in the wild mainly because of concerted efforts over many decades to kill them, and now only exist as descendants of a few individuals captured in about 1980. Some of these animals have now been released into natural areas in eastern Arizona. In other words, all three species are severely endangered because of known human impacts. Perhaps partly because of the very low numbers in a single population, all three present particular problems that can be best addressed by genetic approaches.

EFFECTIVE POPULATION SIZE IN RETURNING WINTER-RUN CHINOOK SALMON

Knowledge of the effective population size of natural populations is funda-mental to understanding and predicting the extent of genetic variation in endangered species. However, most estimates of effective population size in natural populations are indirect, i.e. they are either based on demo-graphic information, such as the mean and variance in the number of progeny per parent, or on genetic data, such as temporal changes in allele frequency or the extent of linkage disequilibrium (Schwartz *et al.*, 1998). It is assumed that demographic and genetic estimates of the effective size will be concordant, an assumption generally supported by a review of the known estimates (Frankham, 1995c).

It is often difficult to acquire either the demographic or genetic infor-mation to adequately estimate the effective population size, N_e, but esti-mates of the number of adults in a population, N, are easier to obtain. If the relationship between N_e and N is known, then a general idea of the effective size is possible by estimating N. For example, in their recommendations for evaluating species for different categories of endangerment, Mace & Lande (1991) suggested that the ratio N_e/N is approximately 0.2 to 0.5. Nunney (1993) suggested that the ratio, over a broad array of mating and demographic situations, should be about 0.5 and not less than 0.25 within a generation. However, Frankham (1995c) found in a review of available data that the ratio over a number of species was only about 0.1. This low value appears to be explained in part by the fact that within-generation effects and between-generation effects were combined, and variation in effective popu-lation size over generations can greatly reduce the overall effective size (Vucetich *et al.*, 1997). In addition, Frankham used the harmonic mean to determine the mean effective population size and the arithmetic mean to determine the mean number of adults. Because the harmonic mean is always less that the arithmetic mean, and the difference increases with the amount of variation, this ratio may be much less than unity just because of different types of means used (S. Kalinowski, personal communication).

In organisms with high reproductive potential, such as many plants and invertebrates and a number of fish species, there is an opportunity for the effective population size to be small because even though the number of offspring is very large, the variance in contribution to the next generation may also be very large. This effect is particularly a concern with endangered fishes for which there may be hatchery programmes utilising a few adults to supplement the natural population with a large number of individuals (e.g. Hedrick et al., 1995, 2000a, b).

Winter-run chinook salmon, Oncorhynchus tshawytscha, from the Sacramento River, California drainage are federally listed as endangered. The annual estimated number of spawners dropped from over 100 000 in 1969 to approximately 200 in 1991. One of the major factors causing the initial reduction was the construction of Shasta Dam, which made the traditional spawning grounds for winter-run inaccessible. In addition to the lack of cool-water spawning grounds and adequate spawning gravel, other factors, such as pollution and impediments to upstream and downstream migration, have resulted in further stresses.

A supplementation programme for the winter-run chinook salmon at Coleman National Fish Hatchery on Battle Creek, a tributary of the Sacramento River, has been releasing pre-smolts (with known captured parents) since 1991. Hedrick et al. (1995) evaluated the impact of this programme on the effective population size in the first three years of the programme, 1991–3, and found that the supplementation did not appear greatly to influence the effective population size of the run, neither significantly reducing nor increasing it. Supplementation continued for 1994 and 1995 but was discontinued in 1996 and 1997 because some non-winter-run chinook were mistakenly used as spawners (in addition to the winter run, there are spring, fall and late-fall runs of chinook in the Sacramento River) and the returning spawners were found primarily in Battle Creek, above the site of the hatchery, and not in the normal spawning area in the mainstream Sacramento River (Hedrick et al., 2000b).

In estimating the effective population size of the released hatchery-raised fish and the overall population, Hedrick et al. (1995) assumed that survival and return migration were random. However, Geiger et al. (1997) have reported that survival may not be random with respect to family in pink salmon. The returns, primarily to Battle Creek in 1997 and 1998 of

Table 7.1. The number of winter-run chinook salmon progeny released and the number of returning spawners from the different females and males in the 1994 brood year and the ratio of the proportions of returns to releases for a given parent

Female	Releases	Returns	Ratio	Male	Releases	Returns	Ratio
3	3444	10	1.35	B	4433	9	0.95
4	3055	5	0.77	C	3152	9	1.33
5	2499	7	1.28	D	4360	16	1.61
6	2361	6	1.20	E	6013	8	0.62
7	2421	3	0.57	F	5223	15	1.34
8	2292	2	0.42	G	5098	6	0.51
9	2338	5	1.00	H	4432	11	1.06
11	2320	7	1.39	I	6353	14	1.17
12	2701	3	0.52	J	3012	4	0.46
13	3946	8	0.93	K	1270	1	0.38
14	1364	2	0.69				
15	3426	10	1.37				
16	2855	10	1.64				
17	2766	7	1.17				
18	3088	4	0.61				
19	2470	4	0.75				
Total	43335	93		Total	43335	93	

fish from 1994 and 1995 releases, allow us to determine whether the assumptions of random survival and return are true. This is possible because these returning fish are identified as releases from Coleman by clipped adipose fins and can be assigned to parents used in the hatchery in 1994 and 1995 with seven variable microsatellite markers (Banks *et al.*, 1996; D. Hedgecock, personal communication). From this information, we are able to estimate directly the effective population size of these released fish, evaluate our earlier predictions and give a direct measure of the N_e/N ratio.

Table 7.1 gives the number of progeny released from the different females and males in 1994 (for more details, see Hedrick *et al.*, 2000b). Of the 29 fish captured in 1994, 26 (16 females and 10 males) produced 43335 released progeny. Overall, the production was fairly equal over individuals with the proportions of progeny produced for females ranging from 0.032 to 0.091 (mean of 0.062) and for males from 0.029 to 0.147 (mean of 0.1). Also given in Table 7.1 are the 93 returning spawners that were assignable to families released in 1994 using the genetic analysis described above (Hedrick *et al.*, in press). Every 1994 male and female parent was represented in the returning spawners. Table 7.1 also gives the ratio of

Table 7.2. The predicted and observed effective population sizes of winter-run chinook salmon, and the ratio between them

Year	Returning spawners	Predicted N_e (95% confidence interval)	Observed N_e	N_e/N
1994	93	34.8 (28.1, 41.2)	31.5	1.21
1995	23	24.5 (16.1, 34.3)	18.0	0.50

the proportions of the returns to released fish for given females and males. These ratios are in a fairly narrow range from 0.38 to 1.64, suggesting no large differences related to either female or male parent.

Of the 47 adults captured in 1995, 36 (21 females and 15 males) produced progeny and a total of 51 273 progeny were released. The variation in contribution over females was much larger than in 1994 while the male variation was somewhat larger (Hedrick et al., 2000b). Many fewer spawners from the 1995 release returned and 26 were assignable to families. Both the higher variation in releases and the lower number of returns resulted in higher variation in the ratios of returns to release proportions (Hedrick et al., in press).

The predicted effective population sizes are given for the two years in Table 7.2 [the simulation approach described in Hedrick et al. (1995) was used]. The predicted value is larger for 1994, mainly because there were over four times as many returning spawners than in 1995 (see below). However, for both years the predicted N_e and observed N_e, based on the family representation of the returning spawners, were fairly close and were within the 95% confidence intervals both years. If we assume that there were 26 parents in 1994 and 36 parents in 1995 (the non-spawning adults were probably not winter-run and six captive-reared females only produced 1131 progeny total), then the N_e/N ratios for the two years were 1.21 and 0.50. However, N_e, as we have estimated using randomly returning individual spawners, is a function of the number of returning spawners (Hedrick et al., in press). For example, if a very large number of spawners returned, then the N_e would be that based on the proportions released from different families and approach $2N$ as the contributions from different families become more equal.

The overall finding is that it appears, using molecular genetic techniques to identify individual returning spawners to family, that the variance over families is close to that expected at random. As a result, the N_e/N ratio from fish reared in the hatchery and returning as spawners is relatively high and probably would not result in a large reduction in effective population size.

ESTABLISHING A BROODSTOCK FOR BONYTAIL CHUB

Even when threats to extinction of an endangered species are mitigated, long-term persistence depends on avoiding extinction from factors such as genetic deterioration resulting from inbreeding depression and loss of genetic variation. As a result, it is of utmost urgency to retain as much genetic variation in endangered species as possible so that the likelihood of recovery and the persistence of stable natural populations are maximised.

The bonytail chub, *Gila elegans,* is a federally listed endangered species and one of the most critically imperilled of North American freshwater fishes (Minckley *et al.,* 1989). This status is the result of destruction and alteration of natural habitat associated with water developments and introduction of non-native fishes (e.g. Williams *et al.,* 1985). The only wild bonytail chub known to remain are an unknown number of large, old adults persisting in Lake Mohave on the lower Colorado River and scattered individuals in the upper Colorado River basin. Little recruitment has occurred since closure in 1954 of Davis Dam, which forms Lake Mohave, and the subsequent establishment of non-native sport fishes in the lake.

In 1981, because of imperilment of the species and concern about apparent lack of recruitment in the wild, F_1 progeny were produced (Hamman, 1982) at Willow Beach National Fish Hatchery by the US Fish and Wildlife Service from wild adults captured in Lake Mohave (referred to as the original captive broodstock below). A sample of these F_1 progeny (about 2300 individuals) was retained at Dexter National Fish Hatchery and Technology Center as broodstock; approximately 350 still existed there in 1998. From these F_1 progeny, and potentially from any additional wild bonytail chub from Lake Mohave, US Fish and Wildlife Service now plans to establish and maintain a new broodstock, hereafter termed future captive broodstock. We evaluate here the status of the remaining original F_1 progeny produced from wild parents in 1981, and establish a protocol for how they, and any wild bonytail chub captured in the future, should be integrated into a future captive broodstock. The goal in initiating this future broodstock is to have as complete and representative a genetic sample from the remaining bonytail chub as possible. The broodstock itself could be used to re-establish natural populations and serve as a reservoir of genetic variation if the Lake Mohave population continues to decline.

Information about production of the F_1 progeny in 1981 is summarised below as background. Because there are no samples from the parental wild-caught fish for analysis, genetic data from samples of F_1 and F_2 individuals are the only information available from which to estimate actual genetic

Table 7.3. Reproductive performance of each of six pair-mated, wild-caught female bonytail chub

Female	Fecundity (number of eggs)	Fertilisation (proportion)	Fecundity × fertilisation	Proportion of fertilised eggs (p_i)
1	0	0.00	0	0.000
2	27 300	0.97	26 481	0.224
3	22 100	0.94	20 774	0.176
4	32 500	0.95	30 875	0.261
5	37 700	0.91	34 307	0.290
6	5 850	0.98	5 733	0.049
Mean (2–6)	25 090	0.95	23 634	0.200

contributions from the original matings. Therefore, we have summarised previously published genetic data from allozymes, and evaluated new allozyme and mitochondrial DNA (mtDNA) information (Hedrick *et al.*, 2000*a*). We present some of these data here to illustrate the potential contributions from the original matings.

The number of eggs produced and the proportion of fertilisation for each female from the wild fish are given in Table 7.3 (Hamman, 1982). Female 1 did not yield eggs and numbers of eggs from the other five females varied from 5850 to 37 700 with a mean of 25 090. The proportion of fertilisation was fairly uniform, from 0.91 to 0.98 with a mean of 0.95. Proportions of fertilised eggs (p_i) produced by the five females thus varied from 0.049 to 0.290 with a mean of 0.2. These values represent the largest number of matings (and individuals) that could have contributed progeny and a fairly equal distribution of progeny from the matings. Note that we assume that these fertilisation proportions represent the contributions to the surviving F_1, i.e. survival of F_1 fish to adult is independent of their parents. Below we assume these values result in a reasonable upper limit to the retention of genetic variation possible from the wild fish and therefore a maximum number of effective founders for the original broodstock. We hereafter refer to these proportions as the 'high' estimate.

A sample of 24 juvenile F_2 fish from the 1985 year-class was examined previously by Minckley *et al.* (1989) for six polymorphic allozyme loci. Further data are presented by Hedrick *et al.* (2000*a*) for a sample of 30 F_1 fish from Dexter (originally spawned at Willow Beach from the wild-caught individuals) for three polymorphic allozyme loci. There were six different three-locus genotypes for the polymorphic loci observed in the F_1 sample (Table 7.4). Genotype 1 was found in 22 of 30 individuals and was heterozygous for *Pgm-A*. The other five three-locus genotypes were present in one

Table 7.4. Three-locus genotypes observed in the 30 F_1 progeny of bonytail chub and the parental matings that could have produced them

N	sAat-A	Gpi-B	Pgm-A	Mating
22	bb	bb	ab	bb bb aa × bb bb bb
8	1 ab, 7 bb	1 aa, 4 ab, 3 bb	3 aa, 3 ab, 2 bb	ab ab ab × bb ab ab

to three individuals (total of eight individuals) and included heterozygotes at all three loci. The most likely explanation for the high frequency of genotype 1 is that 22 individuals (or most of them) were produced by a mating between two individuals of genotypes *bb bb aa* and *bb bb bb*. Assuming Hardy–Weinberg proportions in the wild population from which their parents were drawn, the probability that such a mating would occur is low (Hedrick *et al.*, 2000a) and it is likely that there was only one such mating in the five that produced progeny. Even though the other eight progeny sampled were spread over five different genotypes, it is possible that all were produced from a second mating, *ab ab ab × bb ab ab*. To be conservative, if we assume that the 22 individuals of genotype 1 are from one mating and that the rest are spread equally over two other matings, then $p_1 = 0.733$, $p_2 = 0.1335$ and $p_3 = 0.1335$ for these three hypothetical matings. These values represent a more unequal distribution from the matings than values given in Table 7.3 and consequently a lower retention of genetic variation and will be indicated hereafter as the 'low' estimate. In addition, three different mtDNA types in substantial frequencies were found in a sample of wild fish while only one was found in the captive fish (Hedrick *et al.*, 2000a) supporting the suggestion that a low number of matings was responsible for the surviving fish.

Given the proportions of eggs from different females in Table 7.3, the 'high' estimate from hatchery reproduction or the 'low' estimate based on genetic data (Table 7.4), the expected loss of genetic variation can be calculated compared to that in the population from which the parents were taken. Using the approach outlined in Hedrick *et al.* (2000a) and the proportions of surviving eggs from the 'high' estimate, then $N_e = 8.5$. On the other hand, if the proportional contribution is based on the 'low' estimate from genetic data, then $N_e = 3.5$. In this case, the effective number of original founders is very small, reflecting the facts that only three mating pairs contributed progeny and that most of the progeny were from one of the three mating pairs.

Inclusion of additional wild bonytail chub into the broodstock-development programme would be very desirable because even with the 'high' estimate, the effective number of founders with the original broodstock is at maximum approximately 8.5. Therefore, additional fish would be genetically important (Ballou & Lacy, 1995) and could greatly increase the effective number of founders. For example, one additional wild fish would increase the number of effective founders at a minimum using the 'high' estimate from approximately 8.5 to 9.5 (11.8%) or at a maximum using the 'low' estimate from approximately 3.5 to 4.5 (28.7%).

Avoiding further extinction of endangered species is one of the greatest challenges for contemporary biologists. Obviously, most critical to recovery of an endangered species is the elimination of threats that have reduced the numbers and/or distribution. In addition, other factors can influence the long-term persistence of an endangered species, including the extent of genetic deterioration in the form of loss of genetic variation. Evaluation of the genetic status and the development of a protocol for the establishment of a broodstock for the endangered bonytail chub is designed to reduce the possibility of genetic deterioration beyond what has potentially already occurred.

GENETIC EVALUATION OF MEXICAN WOLVES

The number of founders for a captive population for a species that has become extinct in the wild determines the amount of genetic variation in that species barring mutation, which may regenerate some variation over time. Some captive populations have been started with only a few founders, such as the black-footed ferret with six and Przewalski's horse with 13. In other words, to maximise the genetic variation it is important to include as many founders as possible in a captive population and to add more if at all possible, For example, the Speke's gazelle captive population was started with four individuals but three more wild-caught individuals were added in the early 1990s. Even though it is of great importance to expand the number of founders, it is more important to ensure that all the founders are of the species being preserved so that this closed population is not contaminated by genetic contribution from related species.

The gray wolf, *Canis lupus*, had a distribution throughout most of North America before eradication programmes, starting in the late nineteenth century, resulted in extirpation from most of its original range. The Mexican gray wolf, *Canis lupus baileyi*, listed as endangered in 1976, was probably extirpated from its range in the south-western United States by 1970, and it is thought to be extinct in the wild (there has been no confirmed evidence of wild Mexican wolves in Mexico for more than a decade). Wolves from Mexico were captured in the late 1970s and taken to the

Table 7.5. Frequencies of eight microsatellite alleles which illustrate the differentiation between the three lineages of Mexican wolves, gray wolves, coyotes or dogs

			Mexican wolf				
Gene	Allele	Certified	Ghost Ranch	Aragon	Gray wolf	Coyote	Dog
172	G	0.825	1.000	1.000	0.076	0.000	0.000
200	K	0.000	1.000	0.625	0.011	0.000	0.000
204	D	0.500	1.000	0.813	0.467	0.000	0.021
213	L	0.925	1.000	1.000	0.408	0.167	0.083
250	G	0.550	1.000	1.000	0.083	0.071	0.204
253	D	1.000	1.000	1.000	0.448	0.088	0.120
431	I	0.400	0.500	1.000	0.117	0.063	0.000
442	B	0.789	0.700	1.000	0.435	0.290	0.021

Source: Hedrick *et al.* (1997).

United States to start a captive breeding programme, which became known as the Certified lineage. Two other lineages, the Ghost Ranch and the Aragon, are both thought to be founded by Mexican wolves but there was some question about the founders and their ancestry in both these lineages based on morphology (Hedrick *et al.*, 1997). Therefore before including these lineages and their founder representation into the Certified lineage, an effort was made to determine whether these wolves are genetically Mexican wolves.

Earlier surveys of allozymes and mtDNA provided inconclusive results of the ancestry of the lineages so a survey of genetic variation for 20 microsatellite loci for samples of the three Mexican wolf lineages was undertaken. These data were compared to variation in gray wolf populations as well as samples of coyotes (*C. latrans*) and dogs (*C. familiaris*), two species that have been known to hybridise with gray wolves. Table 7.5 gives the frequencies of eight selected alleles at different loci that illustrate the differentiation found between the groups (Hedrick *et al.*, 1997). For example, the allele G at gene 172 was in high frequency or fixed in all three Mexican wolf lineages while it was at low frequencies in the northern gray wolf sample and missing in both the coyote and dog samples.

The data over all 20 loci can be summarized graphically as in Fig. 7.1 in a two-dimensional representation of allelic-frequency variation (Hedrick *et al.*, 1997). Here the Mexican wolf lineages cluster together to the right, away from the other taxa, indicating that they share the closest common ancestry and are distinct from the other taxa. The coyote populations from throughout the United States cluster to the left, the northern gray wolf populations

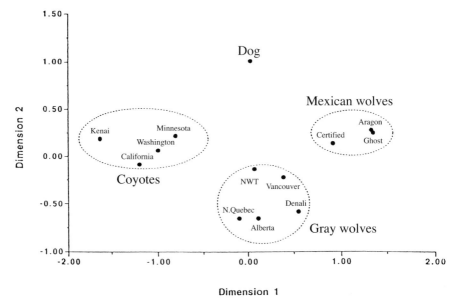

Fig. 7.1. A graphical representation of the allelic frequency variation at 20 microsatellite loci (Hedrick *et al.*, 1997). Coyote and gray wolf populations are indicated by state or province. Broken ellipses encompass all taxa of a given type, e.g. Mexican wolves, gray wolves or coyotes.

cluster to the lower centre, and the dog is in the upper centre. Further detailed analysis of the allelic frequencies suggests that there is no evidence of recent ancestry from either coyotes or dogs in any of the Mexican wolf lineages (Hedrick *et al.*, 1997).

As a result of these findings, a recommendation was made to incorporate the Ghost Ranch and Aragon lineages into the managed Certified lineage. Genetic analysis demonstrated that the Certified lineage was descended from three founders, one female and two males (Hedrick *et al.*, 1997). Both the Ghost Ranch and Aragon lineages are thought to be descended from only two founders each. As a result, the Mexican wolf now is descended from a total of seven founders, still a low number, but significant genetic variation has been added to the species because of the inclusion of the other two lineages of Mexican wolves.

CONCLUSIONS

We should use all the tools at hand to determine the status and recommendations for endangered species management. As shown in the instances discussed here with species of very low numbers, genetic analysis has provided insights that would not have been possible otherwise. This is not to say that genetics is going to save these or any other species from extinction but that this information will allow much more informed decisions about the management of these species than would have been possible without it.

PART II

Animal case studies

The conceptual framework underlying the predicted genetic and demographic consequences of habitat fragmentation has been founded broadly on the theory of island biogeography (MacArthur & Wilson, 1967; Soulé & Wilcox, 1980b). As a consequence, much of the early fragmentation literature revolved around the insular concepts of size and isolation and the balance between extinction and colonisation (e.g. Lovejoy et al., 1986; Wilcove et al., 1986). The majority of studies at this time focussed on patterns of species diversity (with an emphasis on birds) rather than the impacts on community structure, individual species or populations. As empirical data on species diversity accumulated and were found to generally support the island-biogeographic models of fragmentation, the principles began to be applied to individual species within fragmented landscapes. In addition, island-biogeography concepts formed an integral component of meta-population theory (Hanski & Gilpin, 1991a), reserve design (e.g. Diamond & May, 1976; Soulé & Simberloff, 1986) and wildlife corridors (Soulé & Gilpin, 1991; Hobbs, 1992).

Over the last decade the number of studies examining the effects of fragmentation on individual species has risen almost exponentially, again with an emphasis on the impacts of reduction in size and increased isolation of individual patches within a fragmented landscape. To date, most of these studies have dealt with animals rather than plants (see introduction to Part III, p. 237). Among the animals the primary focus has been on birds and mammals, with considerably less attention paid to fish, reptiles and amphibians and only a handful of studies on invertebrates (a symptom of most modern conservation).

It might be envisaged that it is virtually impossible to make generalisations about the impacts of fragmentation on such a heterogeneous group such as the 'animals', with such a diversity of life-history strategies. In terms of reproductive strategy alone, the 'animals' range from clonal reproduction, lifetime single-pair bonds, male- or female-dominated harems and incest, to species in which both males and females multiply mate with large numbers of partners. In addition, fecundities range from very few offspring over large time-spans to many thousands within a single year, average life spans range from decades to days, and vagility can vary from a few metres, or less, to many thousands of kilometres.

However, despite this incredible variability among taxa, there is an overwhelming amount of congruence among the results of fragmentation studies, although there are obviously many exceptions. In general terms the pattern is that increasing fragmentation leads to a reduction in population numbers, increased risk of extinction due to stochastic processes, reduced gene flow between populations, increased inbreeding and loss of genetic variability. These results are perhaps not surprising and in general are those predicted by island-biogeography, ecological and population-genetics theories.

The chapters in this section provide a representative sample of integrated studies using a range of approaches and methodologies to examine the genetic and demographic consequences of fragmentation for a variety of animal species. Chapters by Daniels, Priddy & Walters (Chapter 8) and Dietz, Baker & Ballou (Chapter 11) are focused primarily on the use of quantitative models to examine the effects of fragmentation on inbreeding and survival of red-cockaded woodpeckers (*Picoides borealis*) and golden lion tamarins (*Leontopithecus rosalia*) respectively. The chapters of Srikwan & Woodruff (Chapter 9) and Lindenmayer & Peakall (Chapter 10) investigate patterns of decrease in population numbers and concomitant genetic erosion within small-mammal populations following forest fragmentation in Thailand and Australia respectively. In Chapter 12, Clarke, consistent with the approach outlined by Hedrick in Chapter 7, utilises genetic data to infer some basic demographic and life-history parameters within an endangered species of moth (*Synemon plana*). Finally, Gutiérrez-Espeleta, Kalinowski & Hedrick (Chapter 13) use genetic data to estimate levels of population differentiation and patterns of gene flow, and to identify conservation management units, for bighorn sheep in Arizona. Unfortunately there are no chapters on fish, reptiles or amphibians; however the reader is referred to some excellent examples on these organisms (e.g. Vrijenhoek, 1996; Cunningham & Moritz, 1998; Driscoll, 1998, respectively).

8

Inbreeding in small populations of red-cockaded woodpeckers: insights from a spatially explicit individual-based model

SUSAN J. DANIELS, JEFFERY A. PRIDDY & JEFFREY R. WALTERS

ABSTRACT

Inbreeding depression can be a critical factor affecting viability of small populations and may be especially dangerous in combination with demographic and environmental stochasticity. For species with fragmented habitats and declining, isolated populations, estimating the threat of inbreeding depression is an urgent conservation need. In previous work, we documented inbreeding depression in a wild population of red-cockaded woodpeckers, an endangered, co-operatively breeding species that was once widespread but is now restricted to remnant patches of mature pine savannas in the south-eastern United States. Here, we investigated how rapidly inbreeding accumulates in red-cockaded woodpecker populations of varying size and with varying rates of immigration, using a spatially explicit individual-based simulation model of population dynamics. With this approach, we were able to assess the accumulation of inbreeding in the presence of environmental and demographic stochasticity, while also accounting for effects of the complex social system and extremely restricted dispersal of our study species. We found several meaningful results: (1) most populations were declining, (2) substantial inbreeding accumulated in small, declining populations with limited immigration, due mainly to high percentages of closely related pairs (numbering from 40% to 100% of all pairs after 50 years) and (3) moderately high levels of immigration (two or more migrants per year, equalling four effective migrants per generation) were required to stabilise small declining populations and obtain a mean inbreeding level under 0.10. We conclude that inbreeding depression is a very serious threat to the viability of the many small, isolated and declining populations of red-cockaded woodpeckers.

INTRODUCTION

Loss of genetic variation is a fundamental process of small isolated populations. Such loss may affect population viability through inbreeding depression and/or reduced evolutionary potential (e.g. Frankel & Soulé, 1981; Shaffer, 1981; Allendorf & Leary, 1986; Lande & Barrowclough, 1987; Lacy, 1997; Frankham, 1998). Although researchers have debated the relative importance of genetic, demographic and environmental effects on short-term population viability (e.g. Lande, 1988; Pimm et al., 1988; Schemske et al., 1994), inbreeding depression can be a critical factor in population extinction and may be especially dangerous in combination with demographic and environmental stochasticity (Gilpin & Soulé, 1986; Mills & Smouse, 1994; Frankham, 1995a, 1998; Lacy, 1997; Haig, 1998). Moreover, some authors have recently argued that genetic and non-genetic factors should not be viewed separately in viability assessment (Mills & Smouse, 1994; Lacy, 1997).

Evidence of inbreeding depression in wild populations is continually accumulating (e.g. Bensch et al., 1994; Kempenaers et al., 1996; Pusey & Wolf, 1996; Keller, 1998). In fact, costs of close inbreeding may be common or even universal. Lacy (1997) argued that for mammals, all studies that used appropriate statistical methods on adequate sample sizes have reported evidence of inbreeding depression. Similarly, Daniels & Walters (1999) held that, in studies of inbreeding depression in birds, reduced hatching rate was found in every species for which it was assessed. For species with fragmented habitats and declining, isolated populations, estimating the threat of inbreeding depression to population viability is an urgent conservation need. The development of new, holistic approaches – assessing genetic, non-genetic and interactive effects simultaneously – is especially critical given current landscape conditions.

Processes of population genetics are well understood in theory (e.g. Wright, 1931, 1943, 1951, 1978), but application of genetic theory to wild populations remains challenging. We need to better identify impacts of the loss of genetic variation on individual fitness and population viability. Documentation of inbreeding depression in wild populations is one small but difficult step towards this goal. Other valuable approaches to this problem include investigation of genetic impacts in extinction of island species (Frankham, 1998), theoretical models of accumulation of deleterious mutations (Lynch & Gabriel, 1990; Lande, 1994) and population-viability models incorporating genetic and non-genetic factors (e.g. Mills & Smouse, 1994). However, we also need to better understand how genetic variation is actually lost in wild populations. We know that loss of variation is a complex process affected by population size, demography, social and spatial structure, immigration rates and selection (Wright, 1951; Allendorf, 1983; Chepko-Sade & Halpin, 1987; Soulé, 1987; Lacy, 1997; and many others), but most theoretical and simulation models to date are unable to assess such factors simultaneously. Identifying rates of loss in natural populations can be enhanced by: (1) long-term field studies of marked individuals to collect appropriate and accurate demographic data, and (2) modelling techniques that incorporate complex social behaviour, spatial factors and demographic and environmental stochasticity.

Here we explore the potential effect of inbreeding on the population viability of the endangered, co-operatively breeding red-cockaded woodpecker (*Picoides borealis*). We first review our previous findings of inbreeding depression in this species. We then simulate the accumulation of inbreeding in small populations over time, using an individual-based, spatially explicit model of population dynamics developed recently by our research group (Letcher *et al.*, 1998). We explore effects of various immigration rates and initial population sizes on inbreeding levels. These simulations reflect inbreeding in natural populations as accurately as currently possible, because the model incorporates complex social behaviour, spatially restricted dispersal and subsequent non-random mating, and environmental and demographic stochasticity. Lastly, we compare simulated rates of inbreeding with those expected from random mating within the simulated populations, using Wright's (1931, 1951) mathematical inbreeding models and a previous estimate of the ratio of effective population size to census population size (Reed *et al.*, 1993) for this species. This work is an important contribution to the study of viability of small populations for several reasons. As emphasised, we assess inbreeding levels in the presence of demographic and environmental effects, while including complex social and spatial factors. Also, we use pedigree analysis, a rare but important approach to the study of genetic variation in natural populations. Finally, our work is based on a uniquely powerful data set collected over 15 years from a large, individually marked population (Walters *et al.*, 1988a; Letcher *et al.*, 1998). This study is only a first step, however, in identifying the relationships among inbreeding depression, environmental and demographic effects and population viability. Future work will incorporate declining survival and productivity due to inbreeding depression directly into the population simulations.

Red-cockaded woodpeckers

Red-cockaded woodpeckers are endemic to mature pine woodlands and savannas of the south-eastern United States. Historically, an estimated 3 million or more red-cockaded woodpeckers were distributed continuously throughout the south-east (Conner *et al.*, in press); today, there are roughly 9000 birds, many of which exist in small, isolated populations in remnant patches of natural pine ecosystems (James, 1995). Red-cockaded woodpeckers are permanent residents, restricted to mature pines because of their unusual habit of excavating nesting and roosting cavities in live, rather than dead or decaying, pines (Jackson & Jackson, 1986).

The presence of a critical resource – cavities – is considered the primary basis for co-operative breeding in red-cockaded woodpeckers (Walters *et al.*, 1992). In this co-operative breeding system, roughly half of male fledglings remain on their natal territory as helpers; these birds assist in raising young in subsequent years and commonly inherit a breeding position on their natal territory (Walters *et al.*, 1988a; Walters, 1990). Female fledglings rarely remain on their natal territory. Natal dispersal distances for both sexes are extremely short: male and female fledglings disperse an estimated median distance of one and two territories respectively (Daniels, 1997). Long-distance movements by females between populations have been documented (e.g. Walters *et al.*, 1988b), although this behaviour appears to be rare.

Inbreeding depression and avoidance in red-cockaded woodpeckers

Red-cockaded woodpeckers are one of the few species in which both inbreeding-avoidance behaviours and inbreeding depression have been documented. In our study population in south central North Carolina, we found inbreeding-avoidance behaviours exhibited by sub-adult and breeding females. Female fledglings rarely remain on their natal territory, but they remain less often if there are closely related breeding males present in the following year (Daniels & Walters, in press). Similarly, breeding females disperse in over 90% of cases in which their sons became breeders on their territories (Walters *et al.*, 1988a). In this same population, close inbreeding results in substantial fitness costs. Closely related pairs (kinship coefficient $\geqslant 0.125$) produced 44% fewer yearlings per year than did unrelated pairs (Daniels & Walters, in press). This difference in overall reproduction was the result of two separate effects: reduced hatching rates and lowered survival of fledglings. The substantial cost of close inbreeding, supported by evidence of inbreeding avoidance, prompts concern about the effect of inbreeding depression on the viability of the many small, isolated and declining populations of red-cockaded woodpeckers (Daniels & Walters, in press).

The individual-based, spatially explicit simulation model

Individual-based, spatially explicit models provide information about population dynamics by tracking the performance, fate and locations of individuals (DeAngelis & Gross, 1992; Judson, 1994; Dunning *et al.*, 1995). Individual behaviour is dictated by a set of rules that reflect the known biology of the species. Thus, these models are unique in their ability to incorporate complex social behaviour and spatially heterogeneous environments (DeAngelis & Gross, 1992). For red-cockaded woodpeckers, an individual-based, spatially explicit approach is able to simulate the delayed and spatially restricted dispersal behaviour of helpers, their failure to reproduce in the first and sometimes subsequent years, and the effects of extremely short dispersal distances of both sexes on demography and kinship structure. Letcher *et al.* (1998) developed such a model for red-cockaded woodpeckers and used it to assess effects of territory dispersion on the viability of populations of various sizes. In this study, we use the same model to examine inbreeding processes in the presence of stochastic, social and spatial factors.

METHODS

Study population and data collection

Demographic data used to construct the model were collected between 1980 and 1994 from a marked population of red-cockaded woodpeckers in south central North Carolina. By 1982 virtually all groups (roughly 220) within the 110 000-ha study area were being monitored. Individuals were banded with a unique colour combination and reproduction of all groups was monitored, from clutch size to number of fledglings. Most birds were banded as nestlings, and each breeding season all group members were identified and their status (breeder, helper, floater) determined based on behavioural observations and/or relative age. Further information on the study species, study area and methods of data collection is given elsewhere (Walters *et al.*, 1988a).

Model description

Here we briefly describe the individual-based, spatially explicit model of population dynamics in red-cockaded woodpeckers; further details, including input parameter values and illustrations of the spatial arrangement of territories, are available in Letcher *et al.* (1998). In each simulation, territories are fixed in space and separated by non-breeding space that the birds must cross. Territories are lost if unoccupied for more than five consecutive years, and new territories may be created by territorial budding (Hooper, 1983), a process of territory splitting for which each territory has a 1% probability annually. Both features mimic territory dynamics observed in the North Carolina study population. The number of territories and the level of clumping (the measure of dispersion) of these territories are determined prior to each model run. The size of the landscape is fixed, and therefore the density of territories within the landscape depends upon the number of territories.

Males and females behave according to separate sets of rules. Female fledglings disperse, and male fledglings either remain as helpers or disperse. Dispersing birds move in a random direction, and continue in that direction at a specified rate until they die, obtain a breeding position, or leave the population. Each time-step (three months), dispersing birds and helpers remaining on their natal territory compete for any breeding vacancy within 3 km. Helpers remain on their natal territory as helpers until they fill a breeding vacancy or die, and the oldest helper inherits the natal territory if a vacancy arises. Males compete for empty territories as well as breeding vacancies, and the closest male within the search radius wins. Females compete for breeding vacancies in territories that contain a male, and the oldest female within the search radius wins. Females do not pair with fathers or sons. Only pairs on territories produce offspring, because red-cockaded woodpeckers are monogamous (Haig *et al.*, 1993a, 1994). Reproduction consists of the probability of nesting successfully and the probable number of fledglings produced from successful nests; both are functions of breeder age and the number of helpers. Mortality rates are specific to each status class (breeder, helper, etc.), and birds die if a deviate from a random uniform distribution is less than the probability of mortality. Mean probabilities of mortality and fecundity change annually; these probabilities are drawn each year from a normal distribution with a mean and variance as observed in 15 years of data collected from the study population. Recent additions to the model (and therefore not described in Letcher *et al.*, 1998) include territorial budding and environmental stochasticity as described above.

In summary, model input parameters estimated from the study population include stage-based mortality and fecundity. Behaviours that are observed in nature and simulated in the model include avoidance of close inbreeding by females, delayed dispersal and reproduction of male helpers, the helper search radius, and monogamy, among others. Other behaviours simulated in the model, such as dispersal speed, dispersal direction and the search radius for dispersing birds, are approximations based on experience and judgment rather than directly supported by data. Emergent properties of the model include dispersal distance, recruitment of fledglings and helpers into the breeding population, population growth rate and net gain or loss of territories, among many others.

Model runs

We performed two sets of model runs to explore the effects of population size and immigration rate on the accumulation of inbreeding in this species. The first set of runs included three population sizes, referred to by initial number of territories rather than individuals: 25, 49 and 100 initial territories. Each of these three simulations (25, 49 and 100 initial territories) was continued for an arbitrary duration of 50 years and replicated 20 times. 50 years is equivalent to 12.5 generations, because estimated generation length in this species is roughly four years (Reed *et al.*, 1988). The second set of runs included two population sizes: 25 and 49 initial territories, and five levels of immigration: 0.125, 0.25, 0.5, 2 and 4 migrants per year. All migrants were female floaters, 1.75 years of age, and were placed in a random location on the edge of the smallest square that bounds all territories within the model landscape. Migrants had to compete for breeding positions and did not always reproduce; the effective number of migrants was roughly half (52%) of the absolute number of migrants.

Initial conditions were the same for all runs. We used a moderate level of territory clumping. Of all initial territories, 90% were inhabited by a breeding pair, and the remaining 10% contained an unpaired male. The initial number of helpers was equal to half the number of territories; roughly half of the territories with breeding pairs received one helper and a few received more than one helper. Ages of initial breeders were randomly assigned but were designed to reproduce the age distribution of the study population in 1991, a typical year (J. R. Walters *et al.*, unpublished data). No dispersing birds or fledglings were present initially, so that all dispersal and reproduction emerged during simulations.

Analyses of population persistence and demographic status

We described the persistence of populations of varying size and immigration rates by calculating the percentage of the 20 replicates in each treatment that had at least one successfully breeding pair remaining after 50 years (percentage replicates surviving). We described the demographic status (declining, stable or increasing) of these populations by calculating mean annual growth rate, mean percentage of initial territories that were occupied within the last five years (percentage territories surviving) and mean number of pairs successfully breeding in year 50 in surviving replicates.

Analyses of inbreeding

We constructed a pedigree file for each run containing all individuals and their parents. We then calculated coefficients of kinship for successfully breeding pairs using the SAS procedure, PROC INBREED (SAS, 1997). The coefficient of kinship between two individuals is the probability that two gametes taken at random, one from each, contain alleles that are identical by descent from a common ancestor (Falconer, 1989). By definition, the kinship coefficient of a breeding pair is equal to the inbreeding coefficient of its offspring. We averaged kinship coefficients for all successfully breeding pairs first by year within each replicate, then by year among the 20 replicates. Only successfully breeding pairs were included in these averages because the pedigree contained only individuals and their parents, not all known pairs. In this way, we estimated the average kinship coefficient between successfully breeding pairs. We refer to this measure as mean kinship of pairs; it should not be confused with mean kinship as defined by Lacy (1995) and others, which refers to the mean of kinship coefficients between a given individual and all others in the population. Other variables estimated by averaging among replicates were percentage of all successfully breeding pairs that were closely related (kinship coefficient $\geqslant 0.125$), moderately related (kinship coefficient $0-0.125$) and unrelated (kinship coefficient $= 0$).

We compared the accumulation of inbreeding estimated by pedigree analysis with that expected from random mating within the simulated populations. Expected inbreeding based on random mating was calculated using Wright's (1931, 1951) mathematical models for inbreeding over time,

$$F_t = 1/2N_e + (1 - 1/2N_e)F_{t-1} \qquad (8.1)$$

$$F_t = [1/2N_e + (1 - 1/2N_e)F_{t-1}] \times (1 - m)^2 \qquad (8.2)$$

where F is the inbreeding coefficient, t is the generation, N_e is the effective population size and m is the proportion of N_e that are migrants. Effective population size was computed every generation (4 years: Reed et al., 1988) by adjusting the simulated population size by 0.725, a ratio of effective to breeding population size (N_e/N) recommended by Reed *et al.* (1993) for this species. Thus, changes in simulated population sizes were incorporated into inbreeding estimated with Wright's models. Unpaired males and helpers were not included in the breeding population size nor in Reed *et al.*'s (1993) estimate of N_e/N. Effective migration, for Wright's island (migration) model, was estimated as half the absolute number of migrants in each generation. Migrants are assumed to be unrelated to each other and to the recipient population, in both Wright's island model and our simulated pedigrees.

RESULTS

Population status and persistence

Mean annual growth rates, mean percentage of territories surviving 50 years, percentage of replicates surviving 50 years and mean number of successfully breeding pairs in year 50 are presented in Table 8.1 for all population treatments. Closed populations of 25 and 49 initial territories declined steadily in size, with average growth rates of 0.962 and 0.956 respectively. For these populations, only 30% and 75% of replicates survived 50 years. In contrast, the closed population of 100 territories was stable, with a growth rate of 0.994 and 100% of replicates surviving. For simulations including immigration, populations of 25 and 49 initial territories showed declines at all immigration levels of less than two migrants per year. With two or four migrants per year, these populations exhibited relatively stable average growth rates ($\geqslant 0.990$), and 80% to 90% of replicates survived 50 years.

Inbreeding

For closed populations of 25, 49 and 100 initial territories, mean kinship of successfully breeding pairs increased somewhat linearly through the simulated 50 years, with final values of 0.33, 0.20 and 0.09 respectively (Fig. 8.1). Variation in mean kinship of pairs declined as population size increased: standard deviations in the final year for 25, 49 and 100 initial territories were 0.14, 0.10 and 0.02 respectively (Fig. 8.1).

For the closed population of 25 initial territories, mean kinship of pairs as estimated from simulated pedigrees was less than expected inbreeding based on Wright's model (1931, 1951) and an N_e/N of 0.725 (Reed *et al.*, 1993) (Fig. 8.1). This population was severely declining: mean number of successfully breeding pairs in remaining replicates was less than five (Table 8.1). For populations of 49 initial territories, pedigree analysis and Wright's

Table 8.1. Indicators of persistence and demographic status, as functions of initial number of territories and migration rate, for populations of red-cockaded woodpeckers simulated by the spatially explicit, individual-based model

Initial number of territories	Migration rate[a] (migrants per year)	Population growth rate ($\lambda \pm$ SD)	Percentage territories surviving (mean $\pm$ SD)	Percentage replicates surviving	Number of successfully breeding pairs remaining[b] (mean $\pm$ SD)
25	0	0.915 ± 0.043	12.6 ± 21.1	30	4.5 ± 4.3
25	0.125	0.904 ± 0.037	7.4 ± 15.7	15	3.7 ± 3.8
25	0.25	0.929 ± 0.037	14.6 ± 24.1	50	2.3 ± 3.1
25	0.5	0.946 ± 0.040	26.8 ± 26.3	60	4.8 ± 2.9
25	2	0.990 ± 0.014	70.8 ± 36.0	80	10.5 ± 5.4
25	4	0.999 ± 0.007	88.4 ± 29.0	80	14.1 ± 5.1
49	0	0.956 ± 0.038	34.3 ± 29.2	75	11.9 ± 8.5
49	0.125	0.959 ± 0.038	36.5 ± 29.7	80	11.9 ± 8.6
49	0.25	0.969 ± 0.031	44.5 ± 26.5	90	12.6 ± 6.7
49	0.5	0.966 ± 0.036	39.3 ± 26.3	95	9.9 ± 7.3
49	2	0.990 ± 0.009	67.7 ± 21.0	90	17.7 ± 7.3
49	4	0.993 ± 0.009	70.6 ± 24.0	90	22.0 ± 8.5
100	0	0.994 ± 0.008	83.6 ± 23.5	100	46.8 ± 16.8

[a]Migration rate is the absolute, not effective, number of migrants per year; all migrants are female.
[b]Number of successfully breeding pairs remaining is substantially lower than percentage territories surviving because the latter includes unoccupied territories not yet labelled abandoned, territories occupied by solitary males and territories occupied by unsuccessfully breeding pairs.

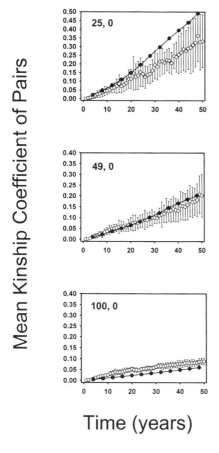

Fig. 8.1. Mean kinship coefficient between pairs of red-cockaded woodpeckers calculated from simulated pedigrees (open circles, ± SD) compared to inbreeding coefficients based on Wright's mathematical model (closed circles) for populations of 25, 49 and 100 initial territories and no immigration.

model gave similar results, and for 100 initial territories pedigree kinship of pairs was higher than Wright's estimate of inbreeding (Fig. 8.1).

Figure 8.2 illustrates the mean percentage of successfully breeding pairs that were closely related, moderately related and unrelated, for 25, 49 and 100 initial territories with no immigration. Mean percentage of closely related pairs increased in all treatments, but rose much less rapidly in the population of 100 territories. Final values for mean percentage of closely related pairs were 100, 70 and 19 for 25, 49 and 100 initial territories respectively. No unrelated pairs remained after 50 years at any population size. Moderately related pairs contributed more than closely related pairs to

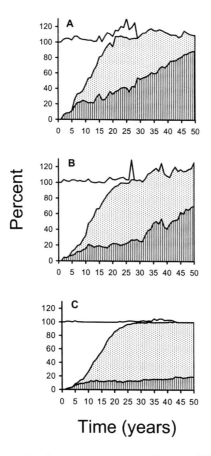

Fig. 8.2. Mean percentage of successfully breeding pairs of red-cockaded woodpeckers that were closely related (hatched), moderately related (dotted) and unrelated (open) in simulated closed populations of (A) 25, (B) 49 and (C) 100 initial territories.

the mean coefficient of kinship in simulated populations of 100 initial territories, but the reverse was true for simulated populations of 25 and 49 initial territories. Also, percentage of related pairs was much less variable across replicates of the largest population size, as expected.

Effects of immigration on the accumulation of inbreeding over time are presented in Fig. 8.3. Mean kinship of pairs, as estimated from simulated pedigrees, rises above 0.10 for all immigration rates of less than two migrants per year, and remains below 0.10 for immigration rates of two and four migrants per year for both population sizes. Variation among replicates of mean kinship between pairs decreased with increasing immigration.

Inbreeding estimated using Wright's models was higher than pedigree kinship of pairs for populations of 25 initial territories and low immigration rates (Fig. 8.3.) In these simulations, very few pairs remained at the end of the runs and immigration events sharply reduced mean kinship between pairs. Expected inbreeding under Wright's model is slightly lower than pedigree kinship of pairs in simulations of two and four migrants per year. However, Fig. 8.3 indicates that these populations receiving two and four migrants per year reach an equilibrium level of inbreeding fairly rapidly; that is, within 20 to 30 years. Equilibrium inbreeding is similar between populations of 25 and 49 initial territories, which is consistent with Wright's (1951) theory.

The effect of immigration on the mean percentage of closely related pairs is presented in Fig. 8.4. For both population sizes, there was a substantial reduction in the mean percentage of closely related pairs only when the number of migrants was two or more per year. For immigration rates of two or more migrants per year, mean percentage of closely related pairs was less than 20% in both population sizes. For immigration rates of one migrant every two years or less, mean percentage of closely related pairs was 60% or greater after 50 years in populations of 25 initial territories and roughly 40% after 50 years in populations of 49 initial territories.

DISCUSSION

Perhaps the major strength of this study is in the approach: we used a spatially explicit individual-based model, based on abundant demographic data, to simulate the accumulation of inbreeding in small populations. Thus, we were able to incorporate spatial structure and complex social behaviour as well as demographic and environmental variation into population models. This novel approach gave several meaningful results: (1) most populations were declining; (2) substantial inbreeding accumulated in small, declining populations with limited immigration, due mainly to high percentages of closely related pairs (numbering from 40% to 100% of all pairs after 50 years); (3) moderately high levels of immigration (two or more migrants per year, equalling four effective migrants per generation) were required to stabilise small declining populations and obtain a mean inbreeding level under 0.10; and (4) inbreeding predicted by population-genetics theory was fairly similar to that of our simulated populations, and cases in which the two approaches differed were consistent with known

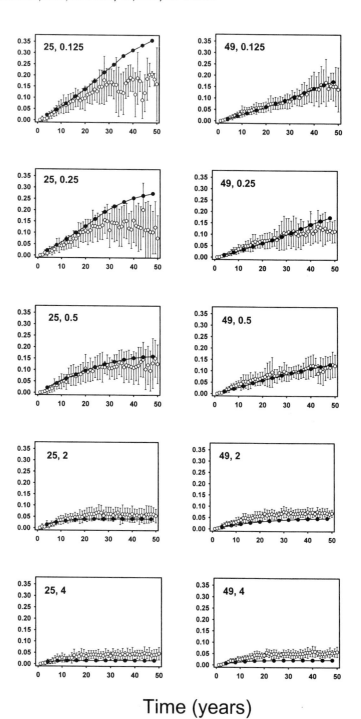

Time (years)

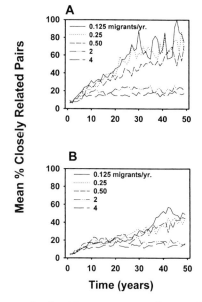

Fig. 8.4. Mean percentage of successfully breeding pairs of red-cockaded woodpeckers that were closely related in populations of (A) 25 and (B) 49 initial territories with 0.125, 0.25, 0.5, 2 and 4 migrants per year.

biology. However, this last result has an important caveat: when estimating inbreeding based on population-genetics theory, we accounted for simulated declines in population size. These results are discussed below, followed by an exploration of their implications for conservation.

Declining population size

All populations of 49 initial territories or fewer declined in size unless the number of immigrants was two or more per year. Prior use of the simulation model (Letcher *et al.*, 1998; J. R. Walters *et al.*, unpublished data) identified the dispersion of territories within the model landscape as well as population size as primary factors influencing population growth. Letcher *et al.* (1998) concluded that severely reduced dispersal, especially that of helper males, has profound influence on the population dynamics of red-

Fig. 8.3 (*opposite*). Mean kinship coefficient between pairs of red-cockaded woodpeckers calculated from simulated pedigrees (open circles, ± SD) compared to inbreeding coefficients based on Wright's mathematical model for populations of 25 and 49 initial territories with immigration rates of 0.125, 0.25, 0.5, 2 and 4 migrants per year.

cockaded woodpeckers. Such a constraint has strong implications for management and reinforces the need to use a spatially explicit model to investigate population processes. We used a moderate amount of dispersion, or territory clumping, in our simulations. Dispersion of territories in wild populations has not yet been measured but probably varies a great deal, and so we chose an intermediate value. The decline exhibited by several of our simulated populations is, sadly, a characteristic of many natural populations of red-cockaded woodpeckers at the present time (Costa & Escano, 1989; Conner & Rudolph, 1991; James, 1991, 1995).

Accumulation of inbreeding

A small, declining and isolated population of red-cockaded woodpeckers will rapidly accumulate extremely high levels of inbreeding, despite avoidance of parent–offspring matings. Populations at risk of extremely high inbreeding have roughly 50 or fewer territories, mean annual growth rates of roughly 0.97 or less and immigration rates below two migrants per year (one effective migrant per year). Populations containing greater than 50 but less than 100 initial territories were not evaluated here.

For small declining populations (less than 50 initial territories), closely related pairs contributed a significant portion of mean kinship between pairs. A larger, stable population (100 initial territories) consisted mainly of moderately related pairs, as may be expected. Reproduction of closely related pairs of red-cockaded woodpeckers in North Carolina is severely compromised (Daniels & Walters, in press), but whether the fitness of moderately related pairs is reduced in proportion to their kinship coefficient remains in question. Hatching rate, for example, was reduced for closely related pairs but did not appear to be a linear function of the kinship of pairs in our study population (Daniels & Walters, in press). Further research into the genetic basis of inbreeding depression and its impact on moderately related pairs is required before the relationship between fitness and inbreeding is fully understood for this species. In the meantime, we suggest that the percentage of closely related pairs in a population may be a better measure of vulnerability to inbreeding than mean levels of inbreeding or kinship. This suggestion has no use in situations wherein kinship is unknown, but it may be important in future models or for researchers and managers of marked populations.

Comparison of pedigree analysis with population-genetics theory

Differences between kinship estimated through pedigree analysis and inbreeding estimated using Wright's model suggest non-random mating in both very small and fairly large populations of red-cockaded woodpeckers. Kinship of pairs in very small populations may be lowered by inbreeding avoidance. The level of inbreeding avoidance built into the model mimics that observed in nature: females avoid mating with fathers and sons. Such inbreeding avoidance may result in greater mating success for immigrants than for residents in these extremely small populations. Conversely, kinship of pairs in fairly large populations is higher than that expected under random mating, probably because reduced dispersal distance results in a subdivided genetic structure. Blackwell *et al.* (1995) found similar evidence of subdivision by comparing inbreeding calculated through pedigree analysis with that expected from Reed *et al.*'s (1993) estimate of effective population size for a large portion (roughly 150 groups) of the North Carolina population. Finally, fairly small populations (near 50 territories) appear to be randomly mating. However, viability of these populations is still sensitive to changes in territory dispersion (Letcher *et al.*, 1998; J. R. Walters *et al.*, unpublished data), suggesting that disrupted dispersal can affect demography even though mating appears to be random. Together, these results underscore the need to use a spatially explicit model to address population viability and population genetics in this species.

When used to assess the effect of immigration on inbreeding levels, both pedigree analysis and Wright's model lead to the conclusion that inbreeding is fairly low if there are two or more migrants per year (one effective migrant or more per year). This is in close agreement with Wright's island model of equilibrium inbreeding (Wright, 1951; Crow & Kimura, 1970),

$$F_{eq} = 1/(4N_em + 1) \qquad (8.3)$$

where N_em is the number of effective migrants per generation in a stable population. Four effective migrants per generation leads to an equilibrium inbreeding level of 0.06, very close to what we calculated from pedigree data after 50 years for both population sizes.

Wright's island model is generally used to say that differentiation between two populations will be minimal if they exchange one migrant per generation in each direction (Falconer, 1989). More specifically, for ideal populations that exchange one migrant per generation, fixation of alleles is minimised while allele frequencies are allowed to diverge (Wright, 1931; Crow & Kimura, 1970; Allendorf & Phelps, 1981; Mills & Allendorf, 1996). In natural populations, more migrants may be necessary to achieve this level of connectivity (Wright, 1931; Mills & Allendorf, 1996). Recent applications of the one-migrant-per-generation concept (reviewed by Mills &

Allendorf, 1996) include its recommended use to eliminate inbreeding depression (Triggs et al., 1989; Gogan, 1990). Our study shows that, for red-cockaded woodpeckers, such a low rate of immigration will not reduce inbreeding. Rather, an effective immigration rate of four migrants per generation or higher is necessary to stabilise small, declining populations and keep inbreeding rates at reasonable levels.

Implications for conservation

Declining population sizes and rapidly increasing percentages of closely related pairs in many of our simulations suggest that small populations of red-cockaded woodpeckers are highly vulnerable to extinction from interacting genetic, demographic and environmental effects. To show this more clearly, we need to incorporate documented levels of inbreeding depression into population simulations. In a viability analysis of an extremely small, heavily managed population of red-cockaded woodpeckers, Haig et al. (1993b) found that hypothetical inbreeding depression dramatically increased the population's probability of extinction. The model they used was individual-based but not spatially explicit.

Our analyses of inbreeding depression in this species were performed on data collected from one of the largest remaining populations of woodpeckers. It is possible that inbreeding depression may be reduced in small populations through selection against detrimental recessives, although the ability of selection to lower genetic load is under current debate (Fu et al., 1988; Hedrick, 1994; Ballou, 1997; Willis & Wiese, 1997; Lacy & Ballou, 1998). The reduction of genetic load through selection is entirely dependent on the genetic basis of inbreeding depression, which is not well understood and may vary among taxa (Hedrick, 1994; Lacy & Ballou, 1998). Further research is required to determine the extent of inbreeding depression in the many small populations of red-cockaded woodpeckers (Daniels & Walters, in press). Although previous studies of allozyme variation (Stangel et al., 1992) and fluctuating asymmetry (Stangel & Dixon, 1995) in red-cockaded woodpeckers have failed to find strong evidence of highly inbred populations, our simulations indicate that high inbreeding is likely; therefore we feel management recommendations are justified.

Earlier research has yielded powerful management tools, such as periodic burning and intensive cavity management, that can be immensely effective in reversing the decline of red-cockaded woodpecker populations if properly used (Walters, 1991; Conner et al., in press). Prior application of the individual-based, spatially explicit model has underlined the woodpeckers' need for tightly clustered territories (Letcher et al., 1998), and managers can fulfil this need by appropriate placement of new cavities. Prior use of the model also indicated that very small populations can be surprisingly persistent, given a tight clustering of territories, and therefore should not be abandoned as lost causes (Letcher *et al.*, 1998; J. R. Walters *et al.*, unpublished data). Haig *et al.* (1993*b*) came to a similar conclusion concerning the viability of a critically endangered woodpecker population, and Stangel *et al.* (1992) stressed that small populations are important both as genetic reservoirs and as stepping-stones between other populations.

Thus, small populations are important and management tools are available to give small populations a reasonable chance of persisting. Our study indicates, however, that close inbreeding may threaten these populations. Specific steps to reduce inbreeding include: (1) maintenance of a stable or increasing population through prescribed burning, cavity management and clumping of territories where possible; (2) enhancing dispersal by retaining as many small populations in the region as possible and by linking disjunct inhabited areas with new cavity trees, which if in suitable habitat will become occupied (Walters *et al.*, 1992); and (3) possibly translocating individuals to a small population with no immigration in a conscientious manner. Conscientious use of translocation will ensure that the most successful methods of moving individuals are used, that the population receiving translocated individuals is closely monitored, that the donor population is within the same geographic region (Stangel *et al.*, 1992), and that impacts to the donor population are carefully considered (Haig *et al.*, 1993*b*).

9

Genetic erosion in isolated small-mammal populations following rainforest fragmentation

SUKAMOL SRIKWAN & DAVID S. WOODRUFF

ABSTRACT

The effects of genetic erosion on the viability of small populations following habitat fragmentation are understood in theory but the critical early stages of the process have gone undocumented as the changes are rapid and difficult to monitor. We found it is possible to monitor genetic erosion in recently fragmented populations by genotyping with panels of 6–7 hypervariable nuclear microsatellite loci as markers of variability. We studied changes in variability in populations of three small mammals isolated on forest fragments in Thailand when the creation of Chiew Larn reservoir flooded the forested Khlong Saeng valley in 1987 and left about 100 rainforest fragments as islands in the lake. Mark–recapture surveys in years 5–8 post-fragmentation on island and matched undisturbed mainland sites showed that habitat fragmentation led to the onset of genetic erosion in surviving populations of a forest rat, *Maxomys surifer*, tree mouse, *Chiropodomys gliroides*, and tree shrew, *Tupaia glis*. Demographic and genetic responses to fragmentation were species-specific, reflecting differences in life history and behaviour. Allelic variation was invariably lost faster than heterozygosity and, in *C. gliroides*, genetic erosion preceded demographic decline. We found that small, recently isolated populations lose variation faster than allowed for in current conservation practice and that genetic erosion may commence before the onset of obvious demographic decline. The project has great generality throughout the increasingly fragmented humid tropics and the methods may be used to monitor genetic erosion in isolated populations of the larger mammals that are typically the focus of conservation efforts. The policy implications of our research are that populations in fragmented forests may require both ecological and genetic management if they are to survive and provide ecological services.

INTRODUCTION

Genetic erosion, the decrease in genetic variation in an isolated population due to random genetic drift and inbreeding, is a phenomenon well understood in terms of population genetic theory but supported by very few empirical studies. Actual observations on the patterns and processes of genetic erosion following habitat fragmentation are almost non-existent. This is partly due to a paucity of situations in nature suitable for experimental analysis and partly due to the difficulty of monitoring genetic changes, until recently. The purpose of this paper is to demonstrate how the critical early stages of genetic erosion in recently fragmented populations can be monitored by multilocus genotyping with highly variable nuclear loci called microsatellites.

Habitat fragmentation has significant negative effects on populations as it reduces their genetic effective size and increases their isolation (Frankel & Soulé, 1981; Gilpin & Soulé, 1986; Lande & Barrowclough, 1987; Young *et al.*, 1996). The reduction of a large outbreeding population to several small isolates and the maintenance of small population size for a number of generations will lead to a decrease in heterozygosity, loss of alleles due to random genetic drift and reduction in gene flow. Drift is normally insignificant in large populations and in metapopulations, but becomes important when populations crash following range destruction and fragmentation.

The maintenance of genetic variation and the avoidance of inbreeding depression are two major goals of conservation geneticists (Frankel & Soulé, 1981; Schonewald-Cox *et al.*, 1983; Woodruff, 1989; Loeschcke *et al.*, 1994; Frankham, 1995b; Avise & Hamrick, 1996; Smith & Wayne, 1996). Although the specific mechanisms underlying the interaction between genetic variation and fitness remain ambiguous (Hedrick, 1996; Mitton, 1998), it is clear that genetic variation is vital to the evolutionary potential of a population, especially in small isolated populations that are increasingly common throughout the rapidly changing world (Laurance & Bierregaard, 1997).

The decrease of heterozygosity, due to drift, is often discussed in relation to the concept of genetic effective population size (N_e). The theoretical relationship between the loss of heterozygosity and effective population size is well known, and has been modelled by various population geneticists since it was first derived by Sewall Wright (Wright, 1931; Crow & Kimura, 1970; Lande & Barrowclough, 1987; Nei, 1987). Wright (1978) showed that in closed populations average heterozygosity decreases at a constant rate of $1/(2N_e)$ per generation. An effective population of 10 is predicted to lose heterozygotes five times faster than a population of effective size 100; 50% of its heterozygosity will be lost in approximately 20 generations. Therefore, in theory, small isolated populations have a higher rate of loss of heterozygosity, and are expected to have lower levels of genetic variation, than large continuously distributed populations. As variability is inherently related to evolvability, genetic erosion in small, recently fragmented populations may thus contribute to their endangerment.

As pre-fragmentation estimates of innate genetic variation are not available for most fragmented populations it should not surprise us that most previous studies have focused on populations that are already threatened or endangered and for which there are no historical data. Only a few investigations of genetic erosion have been designed to include these important baseline data on variation from the outset (Ouborg & van Treuren, 1994; Newman & Pilson, 1997), or have been able to use museum specimens to document the loss of genetic variation during the last 100 years (Mundy *et al.*, 1997; Westemeier *et al.*, 1998). While such genetic studies of small populations are important to the management plans of particular species there are, so far, too few of them to improve our understanding of the process of genetic erosion and its relation to population viability more generally. The research described here takes advantage of the special circumstances involving a 10-year-old hydroelectric reservoir in southern Thailand that provided a remarkable opportunity for a genetic comparison of pre- and post-fragmentation populations. In contrast to several previous studies of the effects of fragmentation, this is among the first to demonstrate genetic stochasticity during the first decade following fragmentation.

Our demonstration of this new method for monitoring genetic erosion was made possible by recent technical advances in molecular genetics: the development of the polymerase chain reaction (PCR), the development of less invasive and non-invasive methods of genotyping, and the discovery of a class of hypervariable markers called microsatellites (Bruford & Wayne, 1993; Woodruff, 1993; Morin & Woodruff, 1996). We have used these techniques to study pedigree relationships, population variability, population structure and gene flow in wild populations of mammals and birds (Woodruff, 1990, 1992; Morin *et al.*, 1993; Gagneux *et al.*, 1997; Mundy *et al.*, 1997) and, as demonstrated here, genetic erosion.

METHODS

Study site and experimental approach

In 1987, the completion of the Rajaprabha hydroelectric dam on the Saeng River (Khlong Saeng), southern Thailand, flooded 165 km² of lowland rainforest, and created approximately 100 permanent islands (formerly hilltops) in the Chiew Larn reservoir (Fig. 9.1) (Electricity Generating Authority of Thailand, 1980; Nakasathien, 1988, 1989). The reservoir is located in the forested hills of peninsular Thailand in the valley between Khlong Saeng Wildlife Sanctuary and Khao Sok National Park. The area is contiguous with other wildlife sanctuaries and national parks and constitutes the largest protected region (3500 km²) in southern Thailand (Brockelman & Baimai, 1993). The vegetation is semi-evergreen tropical forest (Whitmore, 1984, 1990; Richards, 1996) and the monsoon winds bring rainfall from April to November, with the remaining four months comprising a short hot dry season.

We studied faunal collapse in communities isolated on islands in the reservoir and genetic erosion in selected fragmented populations. Small non-volant mammals were chosen as study organisms because they have poor over-water dispersal ability, short generation times and high population turnover, and are important in ecosystem functioning as dispersers of tropical plants, significant seed predators and key dispersers of mycorrhizal fungi. Seventeen species of small mammals were found on the Chiew Larn islands: three arboreal squirrels and 12 ground-frequenting species comprising eight murid rodents, two insectivores, one sciurid rodent and one tree shrew (Table 9.1) (Lynam, 1995, 1997). Three of these 12 species were relatively abundant in the forest fragments and are the focus of this demonstration project: the yellow-bellied rat (*Maxomys surifer*), the pencil-tail tree mouse (*Chiropodomys gliroides*) and the tree shrew (*Tupaia glis*). Ecological observations reported below are from Lynam (1995, 1997) and other sources (Harrison, 1956, 1958; Rudd, 1979; Langham, 1982, 1983; Medway, 1983; Kemper & Bell, 1985; Payne et al., 1985; Lekagul & McNeely, 1988; Corbet & Hill, 1992; Walker & Rabinowitz, 1992).

The genetic study of these small mammal communities during 1992–5 was paralleled by an ecological study (Lynam, 1995, 1997). Small mammals were surveyed by live-trapping at 24 forest sites: 12 islands that ranged in size from 0.7 to 110 ha, and 12 equivalent-sized areas of continuous forest on the adjacent mainland. It was assumed that these mainland areas contained undisturbed populations of small mammals that were representative of pre-fragmentation populations now on the islands. Surveys were

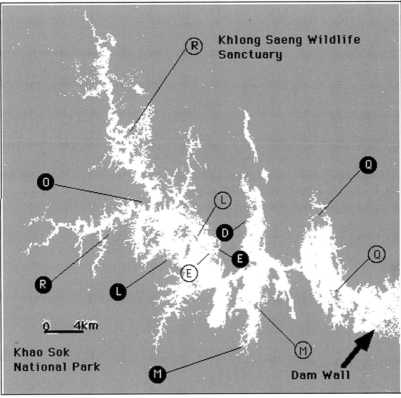

Fig. 9.1. Location of genetic study sites at the Chiew Larn reservoir, Surat Thani Province, Thailand. Hollow circles indicate island sites in the reservoir, filled circles indicate continuous forest mainland sites. Matched pairs of island and mainland sites of equivalent area and similar topography have the same identification letters.

conducted each year during the dry season (January–March) from year 5 to year 8 post-fragmentation and during the wet season of year 6 (1993). A subset of these islands yielded samples that were large enough to warrant genetic characterisation (Fig. 9.1). Areas of the five islands included in this genetic study are: large island R (109 ha), medium island Q (73.11 ha), and small islands M (10 ha), L (6 ha) and E (1 ha).

Study species

The yellow-bellied rat (*Maxomys surifer*) is one of the more abundant species in Thai small-mammal communities. It is restricted to undisturbed interior forest and was among the most sensitive to forest fragmentation.

Table 9.1. Small-mammal species live-trapped at Khlong Saeng, Thailand

Species	Abundance rank[a]	Forest habitat[b]	Density change on islands[c]	Extinction proneness[d]
Rodentia: Muridae				
Chiropodomys gliroides	8	Primary/secondary	increase	10
Leopoldamys sabanus	2	Primary forest specialist	extinction	1
Sundamys muelleri	10	Edges close to water	increase	10
Maxomys surifer	1	Primary forest specialist	decrease	4
Maxomys whiteheadi	3	Primary/secondary	decrease	5
Niviventer bukit	9	Primary/secondary	decrease	3
Niviventer cremoriventer	6	Primary/secondary	no change	7
Rattus rattus	4	Edges close to water	increase	8
Rodentia: Sciuridae				
Menetes berdmorei	7	Primary/secondary	no change	9
Scandentia: Tupaiidae				
Tupaia glis	5	Primary/secondary	decrease	6
Insectivora: Erinaceidae				
Echinosorex gymnurus	12	Edges close to water	no change	12
Hylomys suillus	11	Primary forest	extinction	1

[a] Rank order of species abundances recorded at 12 mainland forest sites in 1992 (dry season, year 5) (from Lynam 1995: Table 1.3). Ties among the four rarest species in 1992 were ranked by abundances at the same sites in 1993 (dry season, year 6).
[b] Lynam (1997).
[c] Data for 12 island study sites surveyed 1992–5 by Lynam (1995: Table 2.4); changes were significant at $P < 0.10$, Wilcoxon's signed-ranks test.
[d] Data for 12 island study sites surveyed 1992–5 by Lynam (1995: Table 2.4).

Smaller than the house rat (*Rattus rattus*), it has a mean weight of 155 g and a mean longevity of about 7 months, and reproduces at two generations per year. *M. surifer* was common on the mainland and rare on all but the larger islands by year 5; this species declined on the islands through time during the first eight years post-fragmentation (Table 9.1). As a habitat specialist, the species suffered the consequences of habitat-area reduction and disturbance, and exhibited changes in dispersion from hyperdispersed on the undisturbed mainland to clumped on the islands.

The small (mean adult weight 26 g), pencil-tail tree mouse (*Chiropodomys gliroides*) was chosen for this study because it differed from the other common small mammals in its demographic response to fragmentation. *C. gliroides* was favoured by fragmentation. This species is rare on the mainland but became more abundant on the islands by year 5 (Table 9.1). Longevity is about 8 months and breeding occurs throughout the year; there are probably three generations per year. This species is primarily arboreal and restricted to bamboo clumps; hence one of its common names is 'bamboo mouse'. It thrives in disturbed forest as bamboos quickly invade disturbed primary forest and dominate the edges. It is rare and has a patchy distribution in mainland forest but is randomly dispersed on islands.

Tree shrews (*Tupaia glis*) are similar in their body size to the forest rats (mean adult weight 140 g). They have a wide ecological range, being found (elsewhere) in dense lowland and montane rainforest, in bamboo clumps and in trees near villages. Diurnal and (despite their common name) not exclusively arboreal, they are typically solitary omnivores with densities of two to five adults per hectare. Data on reproductive behaviour, longevity and ecology suggest that the generation time is about two years (Langham, 1982). In the Khlong Saeng valley, the species was randomly distributed both on the mainland and on islands (Table 9.1). After an initial increase in abundance on the islands, it became relatively rare between year 5 and year 7, and had reduced densities on islands compared to those in continuous forest. Although the populations declined on the islands, this habitat generalist replaced the originally more common forest interior species in disturbed habitats.

Sampling methods

The trapping protocol was the same as that used for the parallel ecological study (Lynam, 1995, 1997). Traplines with cage and Sherman traps were placed at each study site so as to cover each sample 0.25 ha and all habitat types. Each island was trapped simultaneously or within one week of its comparative mainland site. Traps were checked every morning and mammals were quickly identified, sexed, weighed, individually marked by toe-clipping and released. Tissue biopsies (tail clip, hair or toe clip) were obtained during handling and stored in 70% ethanol. Trapping was done for a minimum of five days, after which trapping was continued until the cumulative species number did not change for three consecutive days. In some cases, trapping was continued at pairs of sites beyond the asymptote point of either individual site to increase the sample size for genetic purposes. Samples described here are the result of approximately 40 000 trap–nights of effort.

DNA extraction, amplification and interpretation

Total genomic DNA was extracted from tissue samples (approximately 1 mm^3 in size) using a hair lysis buffer and Chelex incubation (Srikwan et al., 1996). For Maxomys surifer and Chiropodomys gliroides we used heterologous primers developed originally for the laboratory rat, Rattus norvegicus, and laboratory mouse, Mus musculus (Research Genetics, Inc.). Primer pairs were identified which amplified seven different unlinked polymorphic microsatellite loci in each these two species (Srikwan et al., 1996). In the case of the tree shrew species-specific microsatellite primers were developed de novo. Total DNA was extracted from liver and heart tissue of a voucher specimen preserved in 70% ethanol, using a standard phenol–chloroform extraction method. A 300–600 bp partial genomic library was constructed using ligation of size-selected Mbo-I digested total DNA into a pTZ18U/BamHI vector, then transfected into JS5 E. coli competent cells, following standard cloning procedures (Sambrook et al., 1989). Recombinant colonies were probed with dinucleotide polymers $(CA)_{15}$ and $(GA)_{15}$. High-stringency positives were then PCR-amplified and sequenced using pTZ18U primers. Seven primers were designed using the program PRIMER version 5.0, provided by Whitehead Institute for Biomedical Research, from unique sequences flanking a range of microsatellite repeat types found (di-motifs: perfect and imperfect arrays). Six out of seven primer pairs amplified unlinked polymorphic products of the predicted size, and were used to characterise the Khlong Saeng tree shrews. Primer sequences and amplification specifics will be published elsewhere (Srikwan, 1998).

PCR conditions are those described by Srikwan et al. (1996). Forward primers were end-labelled with radioactive ^{32}P and, following amplification in a 15-ml PCR process with Taq polymerase (Promega), the products were electrophoresed on 8% superdenaturing formamide/urea acrylamide sequencing gels. Following electrophoresis at 30 mA for 4–6 h, the gels were dried and exposed to autoradiography film for 2–48 h. An M13 control sequencing reaction was run next to the samples to provide a size-standard marker for the microsatellite alleles. Genotypes were scored by eye. Scoring errors due to artefactual bands (Morin & Woodruff, 1992) were overcome by modifying PCR conditions, and allelic dropout (Gagneux *et al.*, 1997) was not encountered with the DNA concentrations available.

Statistical methods for the analysis of population genetic variation

Genetic variability in each population was measured as the mean number of alleles per locus (A), and the unbiased expected heterozygosity (H_e) calculated from Hardy–Weinberg assumptions (Nei, 1987). Expected heterozygosities (H_e), θ (Weir & Cockerham, 1984) and R_{ST} (Slatkin, 1995) were computed using a program written by Dr Trevor Price (University of California – San Diego). Data on population structure, as revealed by θ and R_{ST} will be described elsewhere. The Wilcoxon's signed-ranks test was used to test for significance in differences in overall expected heterozygosities between pairs of populations (i.e. mainland vs. island, and year vs. year). The same non-parametric test was used for the significance tests of differences in overall allelic diversity between pairs of populations. The genetic effective population sizes (N_e) of the island populations were estimated from the standardised variance in allele frequencies (F) between samples from the same population taken at different times (pre-fragmentation and post-fragmentation). Using GENEPOP (Raymond & Rousset, 1995), we computed F_k for each locus (Weir & Cockerham, 1984). For each population, we calculated a mean estimate of F across loci. These F values were used in the following equation to estimate N_e (Luikart *et al.*, 1998):

$$N_e = \frac{t}{2\left(F - \dfrac{1}{S_o} - \dfrac{1}{S_t}\right)}$$

where t was the duration of the fragmentation in generations, and S_o and S_t were the sample sizes of individuals at year 0 (pre-fragmentation) and year t after fragmentation. Data from mainland populations were assumed to represent the pre-fragmentation state.

Estimates of the rates of genetic erosion expected depend on knowledge of the generation times of the species studied at these study sites. Generation time is defined as the ratio of the natural log of the net reproductive rate to the intrinsic rate of increase, and unfortunately estimates of these parameters are unavailable for these three species. We consequently used published data on reproductive rates and longevity to establish rough estimates of generation time: *M. surifer*, 6 months; *C. gliroides*, 4 months; *T. glis*, 24 months. These estimates are conservative in that, for example, we have assumed that the dry season reduces the number of generations per year by one in both the rat and the mouse, relative to that observed at perhumid sites in Malaysia. These estimates are probably within 50% of their true values but the predictions that flow from them can clearly be modified if site-specific data become available.

RESULTS

Multilocus microsatellite genotypes were obtained for samples of all three species from undisturbed mainland and nearby islands for years 5–8 post-fragmentation (Srikwan, 1998). Allelic diversity and heterozygosity for each sample are reported in Tables 9.2 and 9.3 and Fig. 9.2 and more detailed analyses are in preparation. Genetic erosion was observed in all three species and as expected (Nei *et al.*, 1975; Allendorf, 1986) heterozygosity was found to be less sensitive than allelic diversity to marking its course.

Maxomys surifer

Multilocus microsatellite genotypes of 331 rats were obtained for samples from two island populations (large island R and medium-sized island Q, 110 and 73 ha, respectively) and two size-matched mainland populations (mainland R and mainland Q). Four other samples were too small to provide statistically useful comparisons. For each mainland and nearby island site, the comparable genotypes were obtained from three to four samples representing the populations during years 5–8 post-fragmentation. All seven microsatellite markers were highly polymorphic in all populations, and the mean number of alleles was 9.8 in the mainland and 7.9 in the island populations (Table 9.2).

There was no significant difference in the number of alleles per locus across mainland populations and across different years (Fig. 9.2, Table 9.2). The average number of alleles per locus across loci ranges from 7.6 to 13 in the mainland and from 6 to 10 in the island populations. The allele numbers per locus in the medium-size island were significantly lower than those in the mainland populations in year 7 ($T=1$, $P < 0.02$) and year 8 ($T=0$, $P < 0.01$). Similar significant declines were seen from the comparison of the large island (R) and the adjacent mainland in years 7 and 8

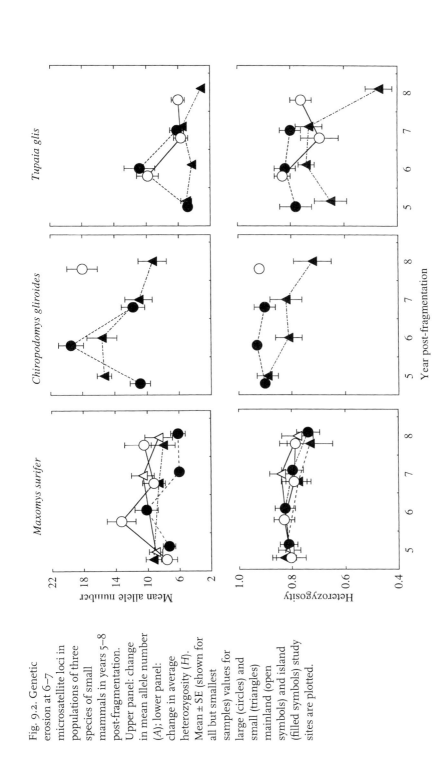

Fig. 9.2. Genetic erosion at 6–7 microsatellite loci in populations of three species of small mammals in years 5–8 post-fragmentation. Upper panel: change in mean allele number (A); lower panel: change in average heterozygosity (H). Mean ± SE (shown for all but smallest samples) values for large (circles) and small (triangles) mainland (open symbols) and island (filled symbols) study sites are plotted.

Maxomys surifer   *Chiropodomys gliroides*   *Tupaia glis*

Mean allele number

22   18   14   10   6   2

Heterozygosity

1.0   0.8   0.6   0.4

5   6   7   8

Year post-fragmentation

Table 9.2. Allelic diversity (A) at seven microsatellite loci in mainland and island populations

Locus	Mainland R (110 ha)				Mainland Q (73 ha)			Island R (110 ha)				Island Q (73 ha)			Island L (6 ha)				Island M (10 ha)			
	Year 5[a]	Year 6	Year 7	Year 8	Year 5	Year 7	Year 8	Year 5	Year 6	Year 7	Year 8	Year 5	Year 7	Year 8	Year 5	Year 6	Year 7	Year 8	Year 5	Year 6	Year 7	Year 8
	n=13	n=46	n=24	n=36	n=23	n=20	n=32	n=12	n=30	n=10	n=12	n=24	n=24	n=25								
Maxomys surifer																						
IO23	8.000	15.000	13.000	11.000	12.000	11.000	9.000	9.000	12.000	8.000	7.000	11.000	10.000	10.000								
ACE	9.000	12.000	10.000	10.000	8.000	10.000	8.000	4.000	8.000	6.000	7.000	10.000	8.000	9.000								
R24	9.000	16.000	9.000	15.000	9.000	10.000	10.000	8.000	11.000	7.000	8.000	13.000	11.000	10.000								
R1150	2.000	7.000	4.000	3.000	5.000	4.000	2.000	6.000	6.000	3.000	2.000	5.000	6.000	2.000								
D18Mit7	7.000	12.000	9.000	8.000	9.000	10.000	9.000	8.000	11.000	6.000	6.000	8.000	8.000	9.000								
R514	13.000	22.000	15.000	22.000	11.000	17.000	17.000	10.000	17.000	8.000	9.000	10.000	11.000	13.000								
R53	5.000	11.000	6.000	5.000	9.000	12.000	5.000	6.000	6.000	4.000	4.000	8.000	6.000	3.000								
Mean A	7.571	13.286	9.143	10.429	9.000	10.571	8.571	7.286	10.143	6.000	6.143	9.286	8.571	8.000								
p[b]									0.931	0.017	0.018		0.914	0.019	0.012							
p[c]									0.039	0.101	0.244		0.163	0.23								
Chiropodomys gliroides																						
				n=30				n=11	n=30	n=12					n=25	n=28	n=21	n=17				
D1Mit4				23.000				7.000	18.000	9.000					17.000	14.000	11.000	11.000				
D1Mit24				18.000				11.000	17.000	15.000					13.000	14.000	9.000	7.000				
D1Mit4				19.000				16.000	23.000	10.000					15.000	18.000	12.000	10.000				
D1Mit36				19.000				11.000	17.000	13.000					15.000	14.000	10.000	8.000				
D12Nds2				24.000				14.000	26.000	18.000					19.000	25.000	19.000	16.000				
D3Mit22				13.000				7.000	14.000	7.000					12.000	9.000	4.000	1.000				
D17Mit20				10.000				10.000	21.000	10.000					16.000	15.000	12.000	12.000				
Mean A				18.000				10.857	19.428	11.714					15.290	15.570	11.000	9.286				
p[b]								0.028	0.611	0.026					0.202	0.234	0.048	0.018				
p[c]								0.017	0.496						1.000	0.046	0.018					

Tupaia glis

	n = 19	n = 8	n = 12	n = 5	n = 33	n = 9	n = 8	n = 6	n = 8	n = 6
TG 1	7.000	5.000	7.000	4.000	7.000	4.000	6.000	4.000	5.000	4.000
TG 4	6.000	3.000	3.000	5.000	6.000	5.000	3.000	4.000	5.000	3.000
TG 16	13.000	4.000	4.000	5.000	12.000	4.000	3.000	6.000	6.000	1.000
TG 19	11.000	7.000	7.000	3.000	18.000	8.000	5.000	6.000	7.000	3.000
TG 21	8.000	6.000	7.000	6.000	8.000	7.000	6.000	3.000	4.000	3.000
TG 22	14.000	9.000	8.000	6.000	14.000	9.000	7.000	3.000	6.000	5.000
Mean A	9.833	5.667	6.500	4.833	10.833	6.167	5.000	4.333	5.500	3.167
p^b					0.655	0.257		0.027	0.705	0.038
p^c					0.028	0.197		0.527	0.457	0.034

[a] Year post-fragmentation.

[b] Significance of the rank sum differences between allele numbers for paired mainland–island sites.

[c] Significance of the rank sum differences between allele numbers in year 5 and following years at this island site.

Table 9.3. Expected heterozygosity (H) at seven microsatellite loci in mainland and island populations

Locus	Mainland R (110 ha)				Mainland Q (73 ha)			Island R (110 ha)				Island Q (73 ha)			Island L (6 ha)				Island M (110 ha)			
	Year 5[a]	Year 6	Year 7	Year 8	Year 5	Year 7	Year 8	Year 5	Year 6	Year 7	Year 8	Year 5	Year 7	Year 8	Year 5	Year 6	Year 7	Year 8	Year 5	Year 6	Year 7	Year 8
Maxomys surifer																						
	n=13	n=46	n=24	n=36	n=23	n=20	n=32	n=12	n=30	n=10	n=12	n=24	n=24	n=25								
1023	0.890	0.829	0.873	0.853	0.853	0.899	0.842	0.909	0.902	0.879	0.725	0.899	0.820	0.836								
ACE	0.886	0.854	0.792	0.854	0.834	0.856	0.799	0.721	0.759	0.821	0.808	0.894	0.811	0.868								
R24	0.892	0.901	0.860	0.885	0.885	0.845	0.887	0.859	0.889	0.863	0.844	0.896	0.882	0.824								
R1150	0.508	0.626	0.531	0.509	0.574	0.594	0.507	0.710	0.689	0.563	0.518	0.567	0.511	0.458								
D18Mit7	0.800	0.801	0.810	0.827	0.796	0.811	0.863	0.757	0.879	0.821	0.743	0.872	0.854	0.874								
R514	0.926	0.949	0.920	0.951	0.907	0.938	0.933	0.926	0.938	0.856	0.844	0.867	0.819	0.902								
R53	0.729	0.843	0.751	0.627	0.825	0.993	0.617	0.801	0.719	0.778	0.681	0.810	0.726	0.352								
Mean H	0.804	0.829	0.791	0.786	0.810	0.838	0.778	0.805	0.825	0.797	0.737	0.829	0.775	0.731								
p^b								0.917	0.735	0.237	0.128	0.310	0.063	0.310								
p^c									0.499	0.612	0.091		0.018	0.091								
Chiropodomys gliroides																						
							n=30	n=11	n=30	n=12					n=25	n=28	n=21	n=17				
D1Mit4							0.954	0.792	0.926	0.874					0.947	0.825	0.895	0.859				
D1Mit24							0.935	0.909	0.912	0.960					0.913	0.931	0.878	0.847				
D1Mit4							0.901	0.974	0.953	0.931					0.919	0.905	0.871	0.888				
D1Mit36							0.944	0.939	0.917	0.935					0.909	0.779	0.830	0.642				
D12Nds2							0.958	0.958	0.966	0.975					0.929	0.952	0.950	0.923				
D3Mit22							0.859	0.861	0.867	0.688					0.677	0.355	0.456	0.000				
D17Mit20							0.864	0.870	0.937	0.913					0.925	0.927	0.890	0.897				
Mean H							0.915	0.899	0.921	0.887					0.883	0.798	0.830	0.720				
p^b								0.753	0.735	1.000					0.310	0.128	0.043	0.043				
p^c									0.398	0.672						0.316	0.028	0.018				

Tupaia glis

	$n=19$	$n=8$	$n=12$	$n=5$	$n=33$	$n=9$	$n=8$	$n=6$	$n=8$	$n=6$
TG 1	0.804	0.758	0.827	0.800	0.723	0.725	0.791	0.712	0.533	0.788
TG 4	0.758	0.592	0.649	0.800	0.789	0.784	0.242	0.803	0.775	0.621
TG 16	0.885	0.442	0.743	0.756	0.701	0.653	0.433	0.818	0.717	0.000
TG 19	0.761	0.625	0.645	0.511	0.932	0.869	0.767	0.818	0.867	0.318
TG 21	0.825	0.783	0.837	0.911	0.871	0.892	0.842	0.682	0.725	0.318
TG 22	0.925	0.925	0.844	0.889	0.886	0.902	0.850	0.621	0.780	0.803
Mean H	0.826	0.688	0.757	0.778	0.832	0.821	0.654	0.742	0.773	0.475
p^b					0.753	0.116		0.116	0.463	0.028
p^c					0.375	0.463		0.753	0.600	0.116

[a] Year post-fragmentation.

[b] Significance of the rank sum differences between expected heterozygosity for paired mainland–island sites.

[c] Significance of the rank sum differences between expected heterozygosity in year 5 and following years at this island site.

($T = 0$, $P < 0.01$). There was no significant decrease in allelic diversity through time in either island population.

No significant difference in heterozygosity was found in any pairwise comparison of mainland populations or in any annual comparison of rats from the same island site (Table 9.3). No significant difference in heterozygosities was found in the pairwise mainland–island comparisons in years 5, 6 or 8, but significantly lower heterozygosity characterised island Q compared with the mainland in year 7.

The loss of alleles originally present in the island populations is apparent when the distribution of allele frequencies at all seven loci is plotted for each population (data not shown; see Srikwan, 1998). The frequency of rare alleles remained constant over time on the mainland, but fell drastically on the islands. Values of θ and R_{ST} are low, both for mainland and island populations, but the degree of differentiation on the islands increases slightly with time, while no trend was evident on the mainland.

Lynam's (1995) demographic study of the Chiew Larn populations suggests the generation time of *M. surifer* is two generations per year. Consequently, the time between pre-fragmentation and year 8 post-fragmentation involves 16 generations. Accordingly, for the large island, the value of N_e was 516, and for the medium-sized island $N_e = 226$. Given these effective population sizes, the expected loss of genetic variation [$1/(2N_e)$ per generation] over 16 generations would be 1.6% for the large island and 3.5% for the medium-sized island. The overall loss of genetic variation by year 8 was 6% for the large island, and 8% for the medium-sized island.

Chiropodomys gliroides

Multilocus microsatellite genotypes were obtained from 174 individuals from island R (110 ha, years 5–7) and island L (6 ha, years 5–8), and one mainland population (Q, 73 ha, year 8 only).

Allelic diversity in the three populations was compared and significant differences were found between the mainland and the small island in years 7 and 8 ($T = 2$, $P < 0.05$; $T = 0$, $P < 0.01$). The mean allele number observed on the large island in year 7 was also significantly lower than that observed on the mainland ($T = 0$, $P < 0.05$) (Table 9.2). The smaller island lost allelic diversity through time, significantly in years 7 and 8 ($T = 2$, $P < 0.05$ for year 7; $T = 0$, $P < 0.01$ for year 8). The mean allele number observed in the large-island population increased significantly and unexpectedly from year 5 to year 6 ($T = 0$, $P < 0.05$) but returned in year 7 to a value close to that measured in year 5.

Unbiased expected heterozygosities were computed for these three populations; all were highly heterozygous, $H_e = 0.7-0.9$ (Table 9.3). In the comparisons between the mainland and the large island, no significant differences were found. The comparisons between the mainland and small-island population were significant in years 7 and 8 (with $T = 2$, $P < 0.05$). Three loci on the small island exhibited lower heterozygosity than on the mainland in year 6. Four loci (of which the previous three were a subset) had lower heterozygosity in year 7 and year 8 in the same comparison. Comparing the large island to the small island, heterozygosity on the small island was significantly lower than that on the large island in year 7 ($T = 3$, $P < 0.02$) (no comparison was possible for year 8). By year 8, the small-island population showed a significant loss of heterozygosity relative to the mainland ($T = 2$, $P < 0.05$).

Mainland population Q was assumed to represent the pre-fragmentation state, and the value of heterozygosity for the mainland was used to calculate the expected loss of heterozygosity in the island populations over 24 generations post-fragmentation. The estimated effective population size is $N_e = 744$ for the large island and $N_e = 195$ for the small island. Accordingly, the expected rate of loss of heterozygosity over 24 generation was 1.6% for the large island and 6.1% for the small island. These predictions may be compared with the observed values, which were 2.8% for the large island and 19.5% for the small island.

Tupaia glis

Genotypes were determined for one mainland (mainland R, 110 ha, years 6–8) and two island (island R, 110 ha, years 5–7, and island M, 10 ha, year 5–8) populations. As there were no other mainland sites with sufficiently large samples, mainland R was assumed to represent the regional population before fragmentation.

Allelic diversity in the three populations was compared and a significant difference was found between the mainland and the small island in year 8 ($T = 0$, $n = 6$, $P < 0.05$) (Table 9.2). The average number of alleles per locus in the small-island population was less than those observed in the large-island and mainland populations (3.2–5.5 for the small island, 4.8–10.8 for the large island, 5.7–9.8 for the mainland population).

The expected heterozygosities were high, $H_e = 0.6-0.8$, in all but one sample from these three populations (Table 9.3), and no statistically significant decline in heterozygosities was found in the mainland or large-island populations over time. The exceptional, less variable sample ($H_e = 0.48$) was from the small-island population in year 8. The differences in heterozygosities between the large- and the small-island populations were not significant (Srikwan, 1998). By year 8, one of the six loci was fixed on the small island and two others were significantly less variable than on the mainland. The mean heterozygosity on the small island was significantly lower than that on the mainland in year 8 ($T = 0$, $P < 0.05$).

Although no statistically significant evidence of a reduction of allele numbers was found in either large or small islands, θ and R_{ST} increased dramatically over time in all pairwise comparisons between the mainland population and the small island (Srikwan, 1998).

Tree shrews are relatively long-lived in comparison to the other small mammals. Average longevity is four years and the generation time is about two years (Langham, 1982). From these data we estimated $N_e = 28$ for the large-island, and $N_e = 26$ for the small-island populations respectively. The rate of loss of heterozygosity in the large-island population (at the constant rate of $1/(2N_e)$ per generation) is therefore predicted to be 7.1%, and 7.9% for the small-island population, after four generations. We observed no loss of genetic variation in the large-island population and a 28% decline in heterozygosity of the small-island population.

As only four generations have elapsed since the initial isolation, and as the population has not declined in size, it is not surprising that the year-8 island populations have retained their innate level of heterozygosity. Although the heterozygosity of the fragmented populations remains similar to that in the unfragmented mainland population, a significant decrease in allelic diversity was observed in the small-island population. Rare alleles were apparently lost quickly following population fragmentation, and this could explain the significantly lower allelic diversity of the small-island population in the last year of observation (year 8). However, small sample sizes (5–12 individuals) preclude detection of subtle losses of heterozygosity so soon after fragmentation.

DISCUSSION

Detecting genetic erosion

Our results show that genetic erosion occurs following habitat fragmentation but that its onset may be missed if one relies on the traditional metric, overall population heterozygosity. Although changes in heterozygosity were insignificant we found that alleles were lost from fragmented populations during each year of observation. The rapid loss of rare alleles and the preservation of intermediate- and high-frequency alleles results in the slower decline in heterozygosity than allele number, and in the genetic similarity of fragmented subpopulations derived from the same founder population. Accordingly, allelic diversity appears to be the more sensitive way of monitoring the loss of genetic variation.

Area and distance effects

The area of a habitat patch has a direct effect on genetic variability, as small islands were found to lose genetic variability significantly faster than larger islands and mainland patches. Whether distance has a similar effect could not be established here as the study site was unsuited to addressing this question. However, with time, those small-mammal populations that are effectively isolated will undoubtedly begin to diverge whereas those that continue to exchange individuals by over-water dispersal will not.

The absence of significant declines of variation in large-island populations, when compared to mainland populations, may be explained by the hypothesis that the mainland populations were also losing variation, albeit at a slower rate. As the mainland samples were obtained mostly from sites adjacent to the reservoir, such populations may show the impact of fragmentation due to edge effects. This hypothesis is supported by the results of a bottleneck analysis (S. Srikwan, in preparation) in which the methods of Cornuet & Luikart (1996) revealed recent reductions of the N_e in all mainland and island rodent populations, except the small-island population of *C. gliroides*. (The same test cannot be applied to *T. glis* as the samples are too small.) Using the infinite-allele model and Wilcoxon's signed-ranks test, significant heterozygosity excess was detected in all samples with the exception of the small-island population of *C. gliroides*. These results suggest that our mainland populations were also affected by habitat disturbance.

Rates of genetic erosion

The Chiew Larn populations appear to be losing variability faster than predicted by simple theory. The observed loss of variability in eight years in all three species was generally at least twice that predicted by Wright's model. The difference between observed and expected losses may be attributable to the more rapid loss of rare alleles at these highly variable microsatellite loci. However, the unexpectedly high losses may also be due to an overestimate of the initial effective population sizes (N_e) based on hypervariable loci. The reasons for the apparently accelerated genetic erosion require further investigation.

A compression effect may further complicate the simplistic interpretation of our data. Faunal compression may contribute to the absence of significant loss of genetic diversity until seven years after fragmentation. Compression may occur when an original population is drawn from a larger area than the present fragment. Animals emigrating from adjacent populations, as the water flooded the valley, converged on a fragment (hilltop) at an early stage of fragmentation. These colonists, with their diverse genetic backgrounds, made up compressed founder populations with enhanced variability. The compression hypothesis cannot be tested directly as we were unable to sample during years 1–4, but is supported by demographic and genetic findings in the small-island population of *C. gliroides*. This population increased in density and showed no sign of demographic decline, suggesting that it originated as a compressed founder population. Furthermore, application of tests by Cornuet & Luikart (1996) reveals heterozygosity deficiency in both year 5 and year 6 samples, indicating that this population expanded in the first few years following fragmentation even though it is not growing today and now shows a heterozygote excess.

Species-specific responses

In terms of maintaining their genetic variation, the tree shrews have coped better with the changes wrought by habitat destruction than the other two species studied. Their populations declined but the level of genetic variation was stable over the course of the observations on the mainland and large island and only declined on the small island in year 8. The persistence of high genetic variation found in some other studies of fragmented populations (e.g. Wauters *et al.*, 1994; Sarre, 1995) has been attributed to high immigration rates and the ability of the species to form a metapopulation (Hastings & Harrison, 1994; Hanski & Gilpin, 1997). We do not believe gene flow accounts for our observation, however, as the tree shrews are poor over-water dispersers and their longevity appears to account for the retention of variation. With their longer generation time and higher tolerance of habitat alteration, it is possible that declines in population size may lead island populations of tree shrews to extirpation before one can detect the erosion of genetic variability with the sample sizes available. Nevertheless, the loss of allelic diversity observed in the small-island population in year 8 suggests that genetic erosion is now detectable in even this species.

In contrast to the population declines seen in the forest rat and tree shrew, three other species of rodents, including *C. gliroides*, increased in abundance on the Chiew Larn islands (Table 9.1). This response to fragmentation is well known in other insular rodents and has been termed the island syndrome. Species showing this response often achieve greater densities than mainland populations, have more stable populations, show reduced reproductive output, differences in behaviour, and higher survival rates (Adler & Levins, 1994). The island syndrome is thought to be due to a release from historical predation and competition. Population density and other manifestations of the island syndrome are predicted to increase with island isolation and to decrease with island area (higher on the islands than in the mainland forest), and dispersion should become more random. On the islands studied the mesopredators had disappeared by year 5. Densities of *C. gliroides* were higher on the islands than in the mainland forest, and the species became randomly distributed on the fragments while its distribution was clumped on the mainland (Lynam, 1995). In this species, genetic erosion commenced before demographic changes indicated there were any problems.

Demographic and genetic responses

In some species, demographic effects of habitat fragmentation may be observed before the effects of genetic erosion, but in others the reverse may be found. The longevity and small litter size of *T. glis* resulted in a population decline before the effects of genetic erosion became significant. On the other hand, *C. gliroides* populations showed no demographic signs of declining viability even though genetic erosion was already under way. This is one of our more important findings, given that genetic studies are usually only performed after demographic studies indicate that there is a problem. This asynchrony between demographic collapse and genetic erosion is illustrated by all three cases examined here. Clearly, managers must monitor both demography and genetics, and their interaction, if they are to assess accurately a fragmented population's viability.

Although we believe the overall population trends reported by Lynam (1995) are basically correct, future studies of this type may need to consider a different sampling design. Our estimates of abundance were based on fixed transects in a landscape that changed over the five-year period. The traplines and traps were set out initially to sample all available habitat types and equal areas (0.25 ha). However, these sampling areas may not be comparable across study sites of different size. It is possible, therefore, that some cases of apparent population decline on large islands and fragments were actually the consequences of changes in species distribution following changes of resource distribution.

The genetic erosion observed at Chiew Larn is due mainly to genetic drift and an increase in population subdivision (Srikwan, 1998). As the process continues the increase in relatedness in small populations will also increase the potential for inbreeding as often observed in insular populations (Frankham, 1998). One recent study (Saccheri *et al.*, 1998) found a strong association between low heterozygosity and extinction in a butterfly metapopulation and attributed the observed decline in viability to inbreeding. In the case of the species studied here, genetic drift played a larger role than inbreeding in the onset of genetic erosion. The effects of increased inbreeding in recently fragmented populations cannot be ignored, however, and will be considered in the context of a more detailed analysis of changes in population genetic structure (Srikwan 1998, in preparation). Although significant inbreeding was not detected in this five-year study it is clear the interaction between genetics and demography is complex and deserves more attention (Lande, 1988; Hastings & Harrison, 1994; Frankham, 1998).

CONCLUSIONS

Habitat fragmentation and genetic erosion go hand-in-hand yet there have been very few studies of the critical early stages of the interaction (Robinson *et al.*, 1992; Collins & Barrett, 1997) and, with the exception of Saccheri *et al.* (1998) and Bouzat *et al.* (1998*b*), none has focused on genetics. Our study of the process of genetic erosion in the first 20 generations post-fragmentation is an attempt to address this deficiency. We were able to show that:

- genetic erosion can be monitored in free-ranging natural populations without undue disturbance
- microsatellites are powerful markers of genetic variation in natural populations and that they are well suited to studies of genetic erosion
- as predicted by theory, genetic erosion affects allelic diversity faster than it affects overall heterozygosity
- each species is likely to respond to habitat fragmentation in a different manner, both demographically and genetically
- although the process of genetic erosion can accompany demographic collapse, the process can begin before any demographic decline is detected.

When this study was conceived in 1990, the study site, the mammal populations and the microsatellite markers were all almost completely unknown. With hindsight, we should have anticipated the impact of dramatic environmental changes on our experimental design: of changes in the habitat fragments due to edge effects, of increased incidence of fire, and of the invasion of disturbance-tolerant species. Furthermore, we did not fully appreciate the fact that microsatellites' great variability requires that large sample sizes be available for statistical analyses. This latter problem can only be partly overcome by increasing the number of loci monitored.

Future assessments of the effects of habitat fragmentation should also address two larger questions left unanswered by this demonstration project. Firstly, at what point (in terms of census size and genetic effective population size) does genetic erosion threaten a population's viability? Secondly, what level of gene flow can protect a population from the negative effects of genetic erosion? Ultimately, wildlife managers may have to counter genetic erosion with the judicious translocation of selected individuals. Although we do not advocate or anticipate such intervention on behalf of the forest rodents of Chiew Larn our methods are applicable to larger mammals that are typically the focus of wildlife-conservation efforts.

Finally, we must point out that many of our fragmented study populations (11 species on 12 pairs of matched island–mainland sites) were extirpated (Lynam, 1997) before genetic erosion became a detectable phenomenon. This result may have considerable generality. So although we have successfully demonstrated that microsatellites are ideal for monitoring the loss of allelic variation their utility in the conservation biologist's toolkit remains to be proven. Only when managers are prepared to intervene to replenish an eroded population's variability will we be able to justify large-scale monitoring of populations of conservation concern. One does not have to await the flooding of another rainforest valley to address this challenge. Habitat fragmentation is such a ubiquitous phenomenon that there are very many situations, throughout the world, suitable for studies of genetic erosion and its mitigation. The demonstration that microsatellite variability can be used to test the predictions made by Sewall Wright 68 years ago opens up numerous research opportunities that should ultimately provide managers with better guidelines for conserving ecosystem integrity in fragmented forests.

ACKNOWLEDGMENTS

This study was conducted with the permission of the Thai National Research Council and grants from the US National Science Foundation. Numerous people including Warren Brockelman, Louis Lebel, Tony Lynam, Chira Meckvichai, Wina Meckvichai, Ronglarp Sukmasuang, Schwann Tunhikorn and our dedicated team of Thai assistants facilitated our fieldwork. We thank Lori Eggert, Philip Morin and Nick Mundy for advice in the laboratory, and Ron Burton, Margo Haygood, Markus Jakobsson, Josh Kohn, Trevor Price and Barb Taylor and two anonymous referees for their comments on the research programme or manuscript.

10

The Tumut experiment – integrating demographic and genetic studies to unravel fragmentation effects: a case study of the native bush rat

DAVID LINDENMAYER & ROD PEAKALL

ABSTRACT

This paper describes preliminary results of integrated demographic and genetic studies of the bush rat (*Rattus fuscipes*) that are being conducted as part of the Tumut fragmentation experiment in southern New South Wales, south-eastern Australia. The experiment is focused on assessing habitat fragmentation and landscape-scale context effects through examining the response of vertebrates to three broad types of sites: sites in remnant patches of native eucalypt forest of different size and shape and embedded within extensive stands of radiata pine (*Pinus radiata*) plantation, sites in large contiguous areas of native eucalypt forest, and sites in stands of radiata pine.

Extensive field-based demographic studies using trapping and hairtubing field techniques showed that the bush rat was virtually absent from exotic softwood stands. Strong positive relationships also were observed between the area of remnants and the probability of patch occupancy by the species; small patches were significantly more likely to be unoccupied.

We examined genetic variation within and among three pairs of populations in remnant patches (0.5 to 1 km apart), and three populations from continuous forest. Using six microsatellite loci we found considerable genetic variation with a total of 55 different alleles detected in the study of 145 individuals. The number of alleles detected per population was correlated with sample size and varied from 27 to 42. The mean expected heterozygosity across all populations and loci was 0.71. Significant genetic differentiation across populations was detected by analysis of molecular variance ($F_{ST} = 0.044$, $P < 0.0001$). However, this relatively low average obscured considerable variation at the population-by-population level, with pairwise F_{ST} values varying from zero to 0.13. Unexpectedly, despite their geographic proximity (< 0.5 km apart), two of three remnant population pairs showed significant and above-average genetic differentiation, relative to other more distant populations. Significant genetic heterogeneity was also detected amongst bush rat populations in continuous native forests, but not necessarily between pairwise combinations of other populations. Neither random mating nor isolation by distance can explain the genetic patterns, nor can habitat fragmentation alone account for the findings, suggesting that other processes have contributed to the genetic patterns.

Our attempts to integrate the results of demographic and genetic studies have identified some potentially important conservation implications and a more complete picture than would have been derived from an investigation confined to any single discipline. Our findings also highlight the complexity of both natural and modified landscape systems, and the difficulty of unravelling the consequences of habitat fragmentation. Further fine-scale genetic studies combined with population manipulations and precise knowledge of the recent history of the landscape fragmentation will provide important new insights into the patterns and extent of bush rat movements and relationships to habitat fragmentation.

INTRODUCTION

Habitat loss and habitat fragmentation are two of the most important processes threatening the loss of species around the globe and driving the present extinction crisis. Identifying the impacts of fragmentation (and how to mitigate them) is essential because so many landscapes are now heavily modified by humans and the original vegetation cover is extensively fragmented as a result (Saunders *et al.*, 1987; Burgman & Lindenmayer, 1998). The effects of fragmentation are complicated because many impacting factors do not act in isolation, but rather they are interacting or cumulative in their influence on the dynamics of populations (Gilpin & Soulé, 1986; Lindenmayer, 1995; Young *et al.*, 1996). For example, habitat loss and fragmentation may reduce population size and change the spatial distribution of remaining subpopulations by confining them to remnant patches. Reduced population size and the isolation of subpopulations may, in turn, result in increased genetic drift and inbreeding, leading to the loss of heterozygosity and genetic variation, and increasing genetic differentiation among populations. Under increased inbreeding, the interplay between inbreeding depression and demography (e.g. juvenile fitness and mortality rates among offspring) (see Ralls *et al.*, 1988; Lacy, 1993*a*; Lacy &

Lindenmayer, 1995) may reduce population growth rates and overall population size, problems further exposing populations to increased inbreeding and genetic drift (Gilpin & Soulé, 1986). Thus, interdisciplinary empirical studies remain crucial to improving our understanding of these types of potentially important (and complex) interactions between demography and genetics.

A common problem in assessing the genetic consequences of habitat fragmentation is the implicit assumption that the pre-fragmentation conditions are represented by the patterns in neighbouring contiguous habitats (e.g. Leung *et al.*, 1993; Sarre, 1995; Cunningham & Moritz, 1998). However, previous historical processes rather than contemporary processes may be the major contributors to extant genetic patterns in 'control sites'. Cunningham & Moritz (1998) found this to be the case in their study of a rainforest skink, where the pronounced historical genetic effects of glacial retreat and expansion prevented detection of recent human-induced habitat fragmentation. Similarly, an apparent lack of genetic divergence among populations does not necessarily indicate unrestricted extant gene flow, since the genetic patterns may reflect previous connectedness (Larson *et al.*, 1984).

The confounding effects of historical processes are likely to be most severe at larger geographic scales. Yet most studies of habitat fragmentation in animals have focused at the scale of 10s to 100s of kilometres (e.g. Gaines *et al.*, 1997). The effects of historical processes will also be accentuated when genetic patterns are assessed with markers that have conservative mutation rates, such as allozymes and mitochondrial DNA (mtDNA) which have been the markers most used to date. In most cases, these markers are unlikely to be in equilibrium with current processes of habitat fragmentation (Larson *et al.*, 1984). New hypervariable genetic markers, such as microsatellites, with much faster mutation rates will track current processes more quickly, making them ideal candidates for genetic studies of habitat fragmentation (Cunningham & Moritz, 1998).

We have commenced an interdisciplinary study that combines the study of population distribution, population demography and population genetics to examine the response of native bush rats (*Rattus fuscipes*) to habitat fragmentation in the Tumut fragmentation experiment, a landscape study that has been under way in south-eastern Australia since mid-1995 (Lindenmayer *et al.*, 1996, 1997*a*, *b*, 1998*a*, *b*, 1999*a*, *b*, *c*, *d*; Cunningham *et al.*, 1999). Small mammals were one of the major groups targeted for detailed study in the Tumut fragmentation experiment. This is because other investigations have indicated they are a group impacted by landscape change and habitat fragmentation (Lindenmayer *et al.*, 1994; Dunstan & Fox, 1996). Furthermore, the short generation time of small mammals increases the likelihood of detecting genetic changes due to recent habitat fragmentation.

In this paper, we briefly outline the background to the Tumut fragmentation experiment and describe both the demographic effects of fragmentation and the preliminary findings of our genetic study of the bush rat. By focusing on a local scale of 10 × 15 km, with populations 0.5 to 3 km apart, and by using new high-resolution microsatellite genetic markers we have attempted to minimise the confounding effects of natural long-term landscape processes. Our findings from the integration of demographic studies and genetic analysis have together provided a more complete picture of the species' response to landscape change.

The Tumut fragmentation experiment: study area and background information

The general area targeted for detailed investigation is shown in Fig. 10.1 and encompasses the Buccleuch State Forest, Bondo State Forest, Bungungo State Forest and Kosciuszko National Park in southern New South Wales, south-eastern Australia. The area is characterised by an extensive plantation of 50 000 ha of exotic softwood conifer trees (primarily radiata pine, *Pinus radiata*). There are numerous (> 190) remnant patches of native *Eucalyptus* forest embedded within the boundaries of the softwood plantation and these vary in size (0.2–125 ha), shape, forest type and a range of other characteristics. Extensive continuous areas of native forest and woodland occur at the northern, eastern and southern margins of the softwood plantation.

The three broad types of sites have been used in major large-scale studies of landscape context and habitat fragmentation effects on populations of arboreal marsupials (Lindenmayer *et al.*, 1997b, 1999b), small mammals (Lindenmayer *et al.*, 1998a, 1999c) and birds (Lindenmayer *et al.*, 1996, 1997a, 1999d; Cunningham *et al.*, 1999). These are: (1) 86 sites in eucalypt remnants stratified on the basis of patch size, patch shape, forest type and time since isolation, (2) 40 sites in stands of radiata pine stratified by topographic position, stand age and thinning condition, and (3) 40 sites in large continuous areas of native *Eucalyptus* forest stratified on the basis of forest type and a range of other variables (see Lindenmayer *et al.*, 1999c).

The study area has particular features which make it especially suitable for fragmentation studies:

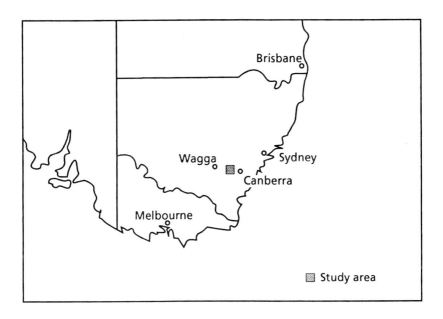

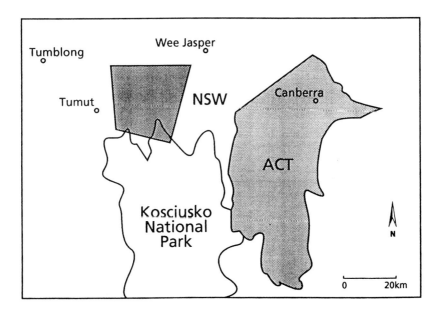

Fig. 10.1. The location of the Tumut study area in south-eastern Australia.

(1)   The fragmentation history of the landscape is well known as the time when compartments of exotic pine replaced native forest has been carefully documented.

(2)   The time since fragmentation varies from 10 to 60 + years so that the longer-term effects of landscape change can be examined; an important advantage as isolation time can influence species loss (Suckling, 1982; Bennett, 1990).

(3)   Climate, soils, geology and vegetation data are available for the entire Tumut region as a result of detailed past surveys and GIS databasing efforts. This information has enabled field sites in the three broad landscape context classes listed above to be matched on the basis of potentially important underpinning environmental conditions that may influence plant and animal distribution.

(4)   There are numerous patches in the study area which provided for substantial site replication, a feature that has strengthened the power of the experimental design.

(5)   The patches in the system at Tumut vary from 0.2 to 125 ha making them a relevant size for the vertebrates targeted for investigation (birds and mammals).

(6)   There has been a number of earlier studies of the vertebrate fauna in the Tumut region (Tyndale-Biscoe & Smith, 1969a, b; Gall, 1982; Rishworth & Tanton, 1995; Rishworth et al., 1995). This provided background information on the species that used to occur in the study area and which may or may not still persist there.

(7)   The landscape matrix at Tumut is dominated by stands of radiata pine and this provides a relatively uniform type of system for sampling and reduces the level of complication for assessing fragmentation effects. This is important because the status and condition of the landscape matrix surrounding habitat patches can be an important factor influencing the distribution and abundance of animals in fragmented landscapes (Diamond et al., 1987; Laurance, 1991a).

METHODS

Field sampling

A major trapping study for the bush rat was completed in January and February 1997 (see Lindenmayer et al., 1998a, 1999c). A total of 59 sites was targeted for detailed sampling: 41 in eucalypt remnants, 10 in large contiguous areas of native forest, and eight in stands of radiata pine. All sites were located in, or close to, a 5000-ha subsection of the radiata pine plantation at Tumut (see Fig. 10.2) except one of the sites within a large contiguous area of native forest that was approximately 10 km to the east of the eastern boundary of the study area.

The patches of eucalypt forest sampled by trapping varied from 0.4 to 40.4 ha in size, and they have been surrounded by radiata pine plantation for 15–40 years. For remnants larger than 3 ha and all continuous eucalypt and radiata pine sites, we set out a 600-m-long flagged transect. Trapping points, each with a single aluminium box or Elliott trap, were established at 50-m intervals along each flagged transect. For remnants covering less than 3 ha, the length of the transect was scaled according to patch size; 400 m for patches of 2–3 ha, 200 m for 1–2 ha and 100 m for 0.4–1 ha. The spacing between traps was the same as for larger remnants, viz. 50 m.

Elliott traps remained open for four consecutive nights. Animals were weighed, sexed and marked prior to release (Lindenmayer et al., 1999c). Data were compiled for the presence and abundance of the bush rat in each of the field-survey sites and used in subsequent statistical analysis of fragmentation effects. In addition, a small piece of ear tissue was collected from each individual bush rat captured during the study. Marking of animals ensured that no more than one ear-tissue sample was gathered from any given individual at a survey site. Ear-tissue samples were stored in long-term storage buffer (100 mM Tris pH 8, 50 nM EDTA pH, 2% SDS) for use in subsequent genetic analyses (see below). In 1998 and 1999, additional targeted trapping of bush rats was conducted in an attempt to increase sample size for the genetic study.

A second major study using hairtubing methods (*sensu* Scotts & Craig, 1988; Lindenmayer et al., 1994) was used to detect the presence of the bush rat in 40 sites dominated by radiata pine forest throughout the softwood plantation at Tumut. Hairtubes are a 'remote' sampling technique that can detect mammals by attracting them to an open cylinder containing a food bait held within a closed chamber. Fur from mammals that enter the hairtube adheres to double-sided tape fixed to the inside of the device (Suckling, 1978; Scotts & Craig, 1988). Hair samples can then be analysed in the laboratory to identify the species (see Brunner & Coman, 1974).

Four types of hairtubes were deployed at the 40 radiata pine sites in our study (see also Lindenmayer et al., 1999c):

(1) Small tubes with an entrance diameter of 32 mm (see Suckling, 1978; Lindenmayer et al., 1994). These were placed on the forest floor and also fixed to trees 2–3 m above the ground.

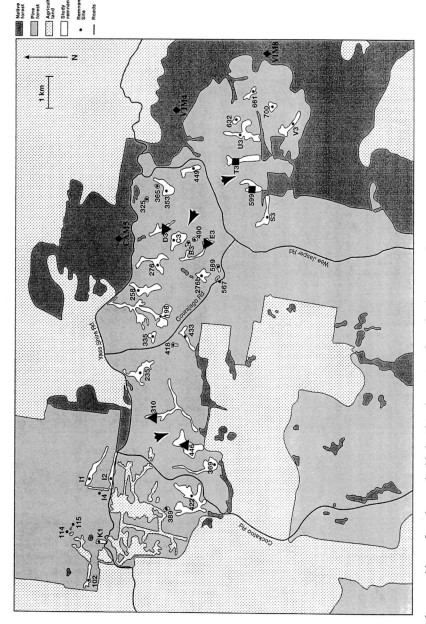

Fig. 10.2. The assemblage of patches embedded within the exotic softwood radiata pine plantation at Tumut targeted for small-mammal trapping. Sites for the genetic sampling are marked with bold symbols (♦, ▼, ▲, ■). Arrows point to the replicate population pairs within the fragmented landscape.

(2)   A modified form of the small hairtube described above in which there were crimped depressions ('dimples') along their length. These were limited to placements on the ground.

(3)   Large tubes (entrance diameter = 105 mm). The design of the large tubes is described by Scotts & Craig (1998). The lack of suitable at-tachment points on the large tubes meant they were confined to positions on the forest floor.

(4)   A new form of hairtube was devised for our investigation (Faunatech Pty Ltd, Eltham, Victoria). The 'hair funnel' was a tapered device 22 cm long, forward opening 13 cm, with a closed bait chamber at its narrowest end (4 cm wide). The contact area for collecting samples of mammal hair in the funnel was a thin wafer insert on its upper wall that was lined with an industrial adhesive.

The various types of hairtubes were set out at 100-m intervals on the 600-m-long flagged transect lines which had been established on the 40 radiata pine sites. This gave seven plots and a total of 42 hairtubes per site.

Populations sampling for genetic analysis

Based on consideration of the DNA samples at hand, we designed a genetic sampling strategy that: (1) contained populations with the largest number of DNA samples available, (2) replicated proximate pairs of populations from the remnant patches and (3) replicated populations from within the surrounding continuous native forest. In total, three pairs of populations within the fragmented landscape were sampled (310 and 466, D3 and E3, 599 and T3), with each pair being 0.5 to 1 km apart at their nearest points. All six populations from the remnant patches were surrounded by pine forest, although several (310, 599 and T3) were less than 0.5 km from con-tinuous native forest. Paired proximate populations within the continuous forest were not available. Instead, three populations distributed 3.5 to 4.5 km apart along the east and north-eastern edge of the fragmented land-scape were used (Fig. 10.2). Two of these populations, VIM4 and CAM5, had relatively small samples sizes of seven and nine individuals respective-ly. Nevertheless, they were retained for genetic analysis, so as to provide preliminary estimates of genetic differentiation among populations in con-tinuous forest for comparison with populations in the remnant patches. Figure 10.2 shows the distribution of the study sites within the Tumut experiment.

Genetic marker selection

Microsatellites, also known as simple sequence repeats (SSRs) and short tandem repeats (STRs), have been widely recognised as powerful and informative genetic markers in both animals and plants. Microsatellites consist of tandemly repeated units of short nucleotide motifs, 1–6 bp long. Di-, tri- and tetranucleotide repeats are the most common [e.g. $(CA)_n$, $(AAT)_n$, $(GATA)_n$, respectively] and are widely distributed throughout the genomes of plants and animals (Jarne & Lagoda, 1996). By virtue of their extreme polymorphism, microsatellite loci are considered to be ideal markers for forensic identification, paternity analysis, gene mapping, conservation biology and population genetics (Weber, 1990; Gupta *et al.*, 1996; Jarne & Lagoda, 1996; Peakall *et al.*, 1998). However, because DNA sequence knowledge is required to design appropriate primers for the PCR assay, the development and application of microsatellites usually requires a considerable investment in DNA library construction, cloning, sequencing and primer design.

Cross-species amplification can facilitate more widespread use of microsatellites. Increasing numbers of animal studies are reporting some successful cross-species amplification of microsatellite loci, with occasional informative transferability of loci to very divergent taxa (see review in Peakall *et al.*, 1998). In this study, we took advantage of the large number of microsatellite loci developed for genetic mapping in the laboratory rat *Rattus rattus* by Serikawa *et al.* (1992) to select a set of six informative microsatellite loci in the bush rat *Rattus fuscipes*, via cross-species amplification. A high proportion of rat microsatellite loci is known to be transferable to mice, and vice versa (Kondo *et al.*, 1993). Therefore, good transferability among rat species was expected.

From the study of Hewitson (1997), two of our six microsatellite loci (PB and PL) were known to be polymorphic in bush rats. Four additional loci were selected from a trial of 10 loci for which primer sequences are provided in Serikawa *et al.* (1992). These 10 loci were initially selected based on the following criteria: (1) the loci were either tetra- or trinucleotide repeats known to be polymorphic in *R. rattus* with 10 or more repeats, and (2) where possible, the loci were located on different chromosomes. Tetra- and trinucleotide repeats were initially targeted in preference to dinucleotide repeats because they tend to produce fewer stutter bands and because they require less stringent electrophoresis conditions. Full results of our tests for cross-transferability will be reported elsewhere. The four loci used here were selected in this initial study because they produced clear and readily interpretable DNA profiles, with minimal optimisation of PCR conditions. Primer sequences, locus identification and the chromosome location in *R. rattus* are shown in Table 10.1 for the six loci. Sequence knowledge is desirable for microsatellite loci generated by cross-species amplification, since there can be considerable sequence change among species, in both the SSR and the flanking regions (Peakall *et al.*, 1998). So far, sequencing at four of the six loci has confirmed that they contain the same class of microsatellite repeat as reported in laboratory rats, with the loci PB and TT being tetranucleotide repeats, CR a trinucleotide repeat and PL the only dinucleotide repeat (Table 10.2). For the two loci that still remain to be sequenced, alleles at CP vary by 4 bp as expected for a tetranucleotide, while at FG allelic variation was more complex than expected for a tetranucleotide, with alleles differing by 2 and 4 bp. Allele size ranges and putative repeat sizes are also shown in Table 10.2.

DNA extraction

DNA extraction followed the procedure of Bruford *et al.* (1992: 228) for mammalian tissue, except that following proteinase K digestion, 4M ammonium acetate rather than 6M NaCl was used as the precipitating salt.

Microsatellite PCR, visualisation and scoring

Microsatellite variation was visualised using high-resolution polyacrylamide electrophoresis of fluorescently labelled PCR products on an ABI Prism 377 automated sequencer. This facility enables the accurate sizing of DNA molecules labelled with any one of four different fluorescent dyes in the same lane. One of the four fluorescent dyes is reserved for an internal marker, enabling precise sizing of alleles and reliable comparisons of data from lane to lane and gel to gel.

The laboratory procedures followed those described in Peakall *et al.* (1998) with some minor modifications. Briefly, all six loci were amplified under the same standard PCR conditions consisting of 10 μl reactions containing 10–20 ng of template DNA, 0.2 μM of primer, 200 μM of each dNTP, 1.5 mM $MgCl_2$, 50 mM KCl, 10 mM Tris-HCl pH 8.3, 0.001% (w/v) gelatin, and 1 U of AmpliTaq (Perkin Elmer). To fluorescently label the DNA fragments, one of three fluorescent dFUTPs – TAMRA, R6G or R110 (Perkin Elmer) – was added to the standard PCR mix at a final concentration of 0.6 μM for TAMRA and 0.2 μM for R6G and R110. Primers for two pairs of loci (PB and PL, CR and FG) were combined in a single PCR, with the remaining two loci, CP and TT, requiring separate PCR runs.

PCR was performed on a Corbett Research FTS-960 thermal sequencer with a touch-down thermal profile consisting of 5 min denaturation at 95 °C, followed by 94 °C for 1 min, 50 °C for 1 min and 74 °C for 45 s, for two cycles; subsequently the annealing temperature was reduced in steps of

Table 10.1. *Rattus rattus* microsatellite primer sequences, locus code, primer number and the chromosome location for the six microsatellite loci used in bush rats, *R. fuscipes*

Code	Locus	Primer number	Chromosome[a]	Left primer	Right primer
CP	CPB	R132	2	GGTGCTAGTAGACAATAAGATAGAT	TTCATGAGTTTCACTGTTTGC
FG	FGA	R123	2	CGTGTGGAAATACTTACAAGCA	CTGCAGACTGATTTGCTCATAA
CR	CRYG	R28	9	CCCAGAAATATGTATTTTTACAAGC	GCCAGAGCTATGTAGAGAGACC
TT	TTR	R138	18	GTGGAAAGCCTTCTGTTCAA	AGAATTCAATAATAACAGTCCCACT
PB	PBPC2	R50	1	TCTGACCCATACTTGTACTTTGC	AATTTCTGCCTCTTTTTCTCAG
PL	PLANH	R137	12	GGGATCTTGCCAAGGTGA	CGGCTTCTGAATGTATTGGA

[a]Based on Serikawa *et al.* (1992).

Table 10.2. Comparison of the microsatellite repeats in laboratory rats, *Rattus rattus*, with those generated by cross-species amplification in bush rats, *R. fuscipes*

Code	Laboratory rats		Bush rats			
	Repeat	Size	Repeat	Size	Size range	Inferred repeat range
CP	$(ATAG)_{13}$	145	—	—	90–118	—
CR	$(TTC)_{20}$	321	$(TTC)_4TAC(TTC)_{19}$	254	242–302	$(TTC)_{17}$ to $(TTC)_{35}$
FG	$(TTTA)_{10}$	116	—	—	88–114	—
PB	$(AGGA)_{12}$	198	$(AGGA)_{11}$	223	199–235	$(AGGA)_5$ to $(AGGA)_{14}$
PL	$(GA)_{34}+(AG)_5$	196	$(GA)_8,(TC)_8$	132	124–132	$(GA)_5,(TC)_7$ to $(GA)_8,(TC)_8$
TT	$(AAAG)_{12}$	147	$(AAAG)_{15}$	137	101–149	$(AAAG)_6$ to $(AAAG)_{18}$

1°C, every two cycles, from 50°C to 45°C, followed by 30 cycles at 45°C, with a final extension at 72°C for 10 min.

With the appropriate combination of different-coloured nucleotides, up to five loci could be combined in a single lane, although in practice electrophoresis was generally performed in sets of three loci. Following PCR, the appropriate samples were combined and then cleaned with sequencing-grade phenol–chloroform mix (Perkin Elmer) following the manufacturer's instructions. The final formamide, DNA standard and dye mix was added directly to the dried DNA pellet, dissolved and then denatured at 95°C for 5 min, then cooled on ice before loading on 5% denaturing polyacrylamide gels, containing 6M urea and run with a 1X TBE buffer on an ABI Prism 377 automated sequencer. Scoring was performed with the aid of ABI Genescan software.

Genetic analysis

A range of new statistical procedures has recently been developed for improved estimation of genetic parameters. Many of these procedures utilise permutational or Markov chain procedures to test significance (Rousset & Raymond, 1997). Exact tests of Hardy–Weinberg equilibrium and population differentiation were performed using GENEPOP software version 3.1b (Raymond & Rousset, 1995). Default parameters for the dememorisation number (1000), number of batches (50) and number of iterations per batch (1000) were used for all analyses in GENEPOP. Genetic differentiation as estimated by F_{ST} was calculated using the software program Arlequin version 1.1 (Schneider *et al.*, 1997), with tests of significance performed by permutation. Arlequin uses the analysis of molecular variance framework (AMOVA) developed by Excoffier *et al.* (1992) for haplotype data and subsequently extended to codominant genotypic data by Peakall *et al.* (1995) and Smouse & Peakall (1999). All permutational tests were performed 1000 times. Within Arlequin, we also calculated R_{ST}, an analogue of F_{ST} developed for microsatellite loci under the assumption of a stepwise mutation model, which is likely at many microsatellite loci (Jarne & Lagoda, 1996). However, tests of this model require larger sample sizes than available here, and the allelic distributions observed in this study suggest that mutation may be more complex than a simple stepwise model (see below). Therefore, comparisons of F_{ST} and R_{ST} will be reported elsewhere following more extensive study at the genotype and sequence levels. The software program GenAlEx (Peakall & Smouse, 1998) was used to calculate standard population-genetic parameters (allele frequency, observed and expected heterozygosity, fixation index and F statistics) and to perform Mantel tests and principal coordinates analysis.

While F_{ST} and related parameters are the most frequently used estimators of genetic differentiation for microsatellite and other loci, Hedrick (1999) has recently shown that due to the large number of alleles, and the extensive heterozygosity typical at microsatellite loci, estimates of F_{ST} (and related parameters) maybe an order of magnitude lower than for less-variable loci such as allozymes. Thus caution needs to be applied when comparing values amongst the different genetic markers. While F_{ST} values may be lower for microsatellites than less-variable loci, the power to detect statistically significant differences is often considerably greater, which can lead to another problem, viz. that statistical significance may not equate with biological significance. Thus a cautionary interpretation may be required when F_{ST} values are very low, but significantly different (Hedrick, 1999). Notwithstanding the limits of microsatellite loci interpretation, it is clear that with appropriate use this class of genetic marker promises to improve considerably our understanding of a wide range of previously intractable evolutionary and other biological questions (Jarne & Lagoda, 1996; Peakall *et al.*, 1998; Hedrick, 1999). Microsatellites particularly hold promise for studies of habitat-fragmentation effects, since their faster mutation rates should allow them to reach equilibrium quickly following recent change (Cunningham & Moritz, 1998).

RESULTS

Field trapping and data analyses

The results of field trapping indicated that the probability of capture of the bush rat was significantly lower ($P < 0.002$) at sites within stands of radiata pine than those located in the eucalypt remnants or large areas of contiguous native forest. Hairtubing surveys produced only six detections from three radiata pine sites ($= 0.3\%$ of hairtubes deployed in the study). Thus, populations of the bush rat were extremely rare in stands of radiata pine (Lindenmayer *et al.*, 1998a, 1999c).

The application of logistic regression analysis (McCullagh & Nelder, 1989) indicated that the probability of capture by Elliott trapping of the bush rat was significantly higher ($P = 0.05$) in larger remnants (see Fig. 10.3). To ensure that this finding was not a sampling artefact recorded because smaller remnants supported fewer traps than larger ones, the analysis was repeated excluding sites where there were four or fewer sampling plots. The final result did not change, confirming both the initial context

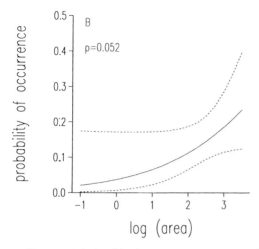

Fig. 10.3. Relationships between remnant area and the probability of patch occupancy by the bush rat, derived from the application of logistic regression modelling to the trapping data (modified from Lindenmayer *et al.*, 1999*c*).

effects (i.e. significantly fewer captures in radiata pine) and patch area effects (i.e. increased chance of capture in larger remnants).

Within-population genetic variation

Table 10.3 summarises the patterns of genetic variation detected within populations. The locus PL showed the least variation with three alleles represented in all populations. Three loci – FG, CP and PB – had seven or eight alleles in total, while 12 alleles were detected at TT and 17 at CR, but at all five loci the number of alleles per population was generally less than the total. Of the total alleles detected, five alleles were unique to a single population. At the TT locus, one unique allele each was observed in populations 446 and VIM4. Two unique alleles were detected at the CR locus at VIM4 and 599, while at the locus PB, one unique allele was found in the E3 population.

Across all loci, a total of 55 different alleles was detected with a maximum of 42 alleles detected in population 310 which had the largest sample size. However, while a positive relationship between sample size and the number of alleles was found ($R^2 = 0.67$, $F = 14.75$, $P = 0.006$), the smallest population, CAM5 with just seven samples, contained 27 alleles, while E3 with only 12 samples contained 39 alleles, exceeding all but the largest population in total number of alleles.

Frequency distributions of allele frequencies are shown in Fig. 10.4. A notable feature is the multimodal nature of the distributions at several loci.

Table 10.3. Number of samples (n), number of alleles per locus (A), observed heterozygosity (H_o), expected heterozygosity (H_e) fixation index (F), and F statistics (F_{IS}, F_{IT} and F_{ST}) across the nine populations for the six microsatellite loci

Loci	TT	FG	CR	CP	PB	PL	Mean	SE
310								
n	25	25	25	25	25	25		
A	9	7	11	5	7	3	7.00	1.15
H_o	0.72	0.64	0.92	0.68	0.76	0.52	0.71	0.05
H_e	0.85	0.72	0.86	0.72	0.67	0.61	0.74	0.04
F	0.16	0.12	−0.06	0.06	−0.13	0.14	0.05	0.05
446								
n	18	18	18	18	18	18		
A	8	4	11	3	5	3	5.67	1.31
H_o	0.67	0.67	0.67	0.39	0.89	0.56	0.64	0.07
H_e	0.73	0.58	0.82	0.55	0.73	0.64	0.68	0.04
F	0.08	−0.15	0.19	0.30	−0.23	0.13	0.06	0.08
D3								
n	10	10	10	10	10	10		
A	6	6	8	4	5	3	5.33	0.71
H_o	0.60	0.90	0.80	0.60	0.70	0.60	0.70	0.05
H_e	0.77	0.70	0.85	0.61	0.69	0.62	0.71	0.04
F	0.22	−0.29	0.05	0.02	−0.01	0.03	0.00	0.07
E3								
n	12	12	12	12	12	12		
A	9	6	9	6	6	3	6.50	0.92
H_o	0.50	0.67	0.75	0.67	0.83	0.50	0.65	0.05
H_e	0.86	0.77	0.86	0.65	0.70	0.62	0.75	0.04
F	0.42	0.14	0.13	−0.03	−0.19	0.20	0.11	0.08
599								
n	21	21	21	21	21	21		
A	8	4	12	8	6	3	6.83	1.33
H_o	0.57	0.57	0.67	0.71	0.71	0.67	0.65	0.03
H_e	0.81	0.62	0.89	0.70	0.64	0.64	0.72	0.04
F	0.30	0.08	0.25	−0.02	−0.11	−0.04	0.07	0.07
T3								
n	20	20	20	20	20	20		
A	8	5	9	6	7	3	6.33	0.88
H_o	0.50	0.75	0.90	0.85	0.70	0.60	0.72	0.06
H_e	0.77	0.66	0.82	0.82	0.63	0.62	0.72	0.04
F	0.35	−0.13	−0.10	−0.03	−0.11	0.03	0.00	0.07
CAM5								
n	7	7	7	7	7	7		
A	6	5	5	4	4	3	4.50	0.43
H_o	0.86	0.71	0.57	0.71	0.86	0.57	0.71	0.05
H_e	0.78	0.72	0.55	0.60	0.70	0.44	0.63	0.05
F	−0.11	0.01	−0.04	−0.19	−0.22	−0.30	−0.14	0.05

Table 10.3. (*cont.*)

Loci	TT	FG	CR	CP	PB	PL	Mean	SE
VIM4								
n	23	23	23	23	23	23		
A	11	7	12	7	5	3	7.50	1.41
H_o	0.78	0.65	0.70	0.83	0.70	0.74	0.73	0.03
H_e	0.84	0.70	0.81	0.76	0.58	0.66	0.73	0.04
F	0.07	0.07	0.14	−0.08	−0.20	−0.12	−0.02	0.05
VIM8								
n	9	9	9	9	9	9		
A	8	5	8	7	4	3	5.83	0.87
H_o	0.78	0.78	0.78	0.78	0.67	1.00	0.80	0.04
H_e	0.83	0.78	0.85	0.83	0.64	0.65	0.76	0.04
F	0.06	0.00	0.08	0.06	−0.04	−0.54	−0.06	0.10
F_{IS}	0.18	−0.01	0.08	0.00	−0.14	−0.05	0.01	0.04
F_{IT}	0.23	0.05	0.16	0.07	−0.08	0.04	0.08	0.04
F_{ST}	0.07	0.06	0.09	0.06	0.05	0.08	0.07	0.01
Means across populations								
n	16.11	16.11	16.11	16.11	16.11	16.11	16.1	0.00
SE	2.23	2.23	2.23	2.23	2.23	2.23		
A	8.11	5.44	9.44	5.56	5.44	3.00	6.17	2.28
SE	0.51	0.38	0.77	0.56	0.38	0.00		
H_o	0.66	0.70	0.75	0.69	0.76	0.64	0.70	0.05
SE	0.04	0.03	0.04	0.05	0.03	0.05		
H_e	0.81	0.70	0.81	0.69	0.66	0.61	0.71	0.08
SE	0.02	0.02	0.03	0.03	0.02	0.02		
F	0.17	−0.02	0.07	0.01	−0.14	−0.05	0.01	0.11
SE	0.06	0.05	0.04	0.04	0.03	0.08		

All calculations are based on the methods described in Hartl & Clark (1997).

For example, the CR locus shows a very strong bimodal pattern. This may reflect an interruption in the long trinucleotide repeat. Only a short allele has been sequenced so far, with an interruption detected close to one end of the repeat sequence (Table 10.2); repeat expansion on both sides of the interruption may well generate a bimodal allelic distribution, but this remains to be confirmed by further sequencing. Sequencing of two of the three alleles at the PL locus revealed that the bimodal distribution at that locus is due to variation at two separate short microsatellites within the DNA fragment (Table 10.2). Pure repeats are apparent at the loci TT and PB (Table 10.2), nevertheless multimodal allele frequency distributions are

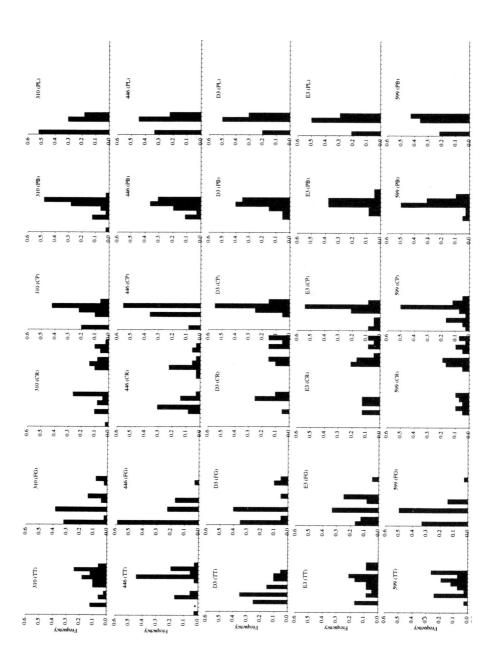

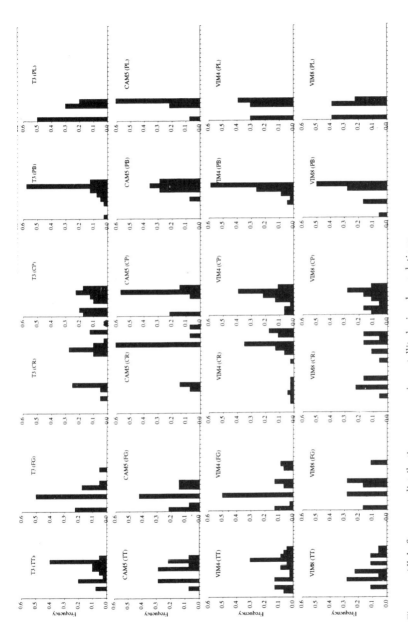

Fig. 10.4. Allele frequency distributions across microsatellite loci and populations.

Table 10.4. Population by population pairwise F_{ST} values and outcomes of tests for significance (below diagonal) versus the number of loci with significantly different allele frequencies ($P < 0.05$) as assessed by exact tests for each pairwise population comparison (above diagonal)

Population	310	446	D3	E3	599	T3	CAM5	VIM4	VIM8
310	—	4	3	2	4	3	3	2	1
446	0.055***	—	4	4	5	5	6	3	4
D3	0.033*	0.058**	—	0	2	1	1	1	1
E3	0.018	0.062**	−0.010	—	2	2	1	1	1
599	0.044***	0.068***	0.007	0.016	—	4	0	4	2
T3	0.091***	0.078***	0.049***	0.036*	0.060***	—	2	1	3
CAM5	0.091***	0.128***	0.054*	0.039	0.034*	0.088***	—	1	3
VIM4	0.023**	0.077***	0.024*	0.010	0.033**	0.022**	0.032*	—	2
VIM8	0.012	0.080***	0.020	0.010	0.032*	0.035*	0.070**	0.019*	—

Key: * $P < 0.05$, ** $P < 0.01$, *** $P < 0.001$.

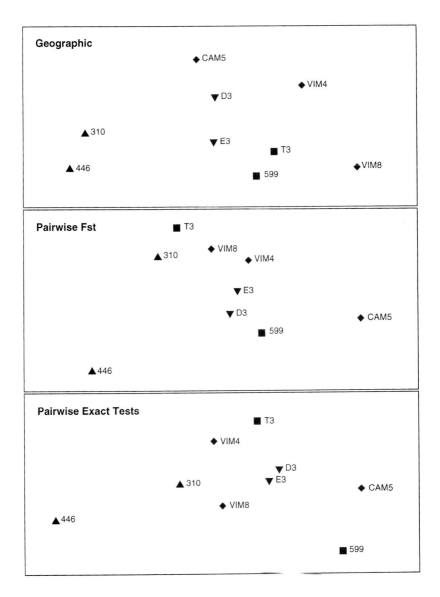

Fig. 10.5. Plots of the first two axes of principal coordinates analysis for pairwise geographic distances, pairwise population F_{ST} values and pairwise population exact tests. The two axes account for 100%, 69% and 61% of the total variation respectively.

also observed in some populations at these loci (Fig. 10.4). Collectively, these patterns may suggest that mutational processes at the microsatellite loci in this study are more complex than a simple stepwise mutational model.

The mean expected heterozygosity across all populations and loci was 0.71 which was close to the mean observed heterozygosity of 0.70, with a mean F of 0.05, suggesting that overall there was little departure from Hardy–Weinberg expectations (Table 10.3). Nevertheless, exact tests across all populations for each locus did detect significant deviation from Hardy–Weinberg expectations at the two most variable loci, TT ($\chi^2 = 70.5$, df $= 18$, $P = 0.00$) and CR ($\chi^2 = $ infinity, df $= 19$, $P = $ highly significant). At both loci, an excess of homozygotes accounts for the deviations from expectation. At the population level, only population 599 showed highly significant deviations from Hardy–Weinberg expectations ($\chi^2 = $ infinity, df $= 12$, $P = $ highly significant). However, this result is largely a consequence of large heterozygous deficiencies at both the TT and CR loci. The populations 446, E3 and T3 showed significant departures from Hardy–Weinberg expectations across loci at the $P < 0.05$ level, but in all cases this was due to heterozygous deficiency at the TT locus. No significant departures from expectation were detected across loci at the remaining populations. Tests of Hardy–Weinberg expectations using pooled alleles (data not shown) reduced the number of significant departures, suggesting the detectable deviations may be artefacts of the large number of alleles.

Population genetic differentiation

Significant genetic differentiation across the nine populations was detected by AMOVA ($F_{ST} = 0.044$, $P < 0.0001$). This procedure yields an estimate of F_{ST} very close to the method of Weir & Cockerham (1984) (mean $F_{ST} = 0.04$, range across loci of 0.040 to 0.054 as calculated in GENEPOP). Marginally higher locus-by-locus values and overall mean F_{ST} are produced by standard estimates of F_{ST} (mean $F_{ST} = 0.07$, range across loci 0.05 to 0.09; Table 10.3). Exact tests across all loci and populations also revealed significant differences in both allele and genotype frequencies ($\chi^2 = $ infinity, df $= 12$, $P < 0.0001$ for both analyses). The AMOVA results show that on average 4.4% of the total genetic variation represents among-population differences. However, this relatively low average difference obscures variation at the population-by-population level, with pairwise F_{ST} values varying from zero to 0.13 (Table 10.4). The results of pairwise population exact tests of population differentiation also varied considerably from population pairs showing no differences at any loci to significant differences at all six loci (Table 10.4). As expected, the results of the pairwise F_{ST} analysis and the pairwise exact tests were congruent (Table 10.4).

Figure 10.5 shows the first two axes of principal coordinates analysis (PCA) for pairwise population F_{ST} values and pairwise exact tests (data shown in Table 10.4) compared with the PCA of pairwise geographic distances. For the genetic analyses, two outlying populations are apparent, 446 and CAM5. The location of the proximate remnant population pairs (shown with the same symbols) is also of particular interest. Neither the pair 310 vs. 446 (▲, $F_{ST}=0.055$, $P<0.001$) nor the pair 599 vs. T3 (■, $F_{ST}=0.060$, $P<0.001$) are located next to each other in the PCA space. Pairwise exact tests showed a similar pattern with four out of six loci being significantly different for both pairs (Table 10.3). Thus, despite their geographic proximity (<0.5 km apart), these remnant population pairs show significant and above average genetic differentiation, relative to other more distant populations. By contrast the proximate remnant population pair D3 vs. E3 showed no genetic differentiation with AMOVA (▼, $F_{ST}=0.000$, $P=0.83$), and no loci were significantly different with exact tests for this pairwise comparison (Table 10.4).

The pairwise F_{ST} values between populations within the continuous native forest (♦, CAM5, VIM4 and VIM8) were significant, varying from 0.019 to 0.07 (Fig. 10.5, Table 10.4). Thus heterogeneity also exists amongst bush rat populations in continuous native forests. By contrast, several of the remnant populations within the fragmented landscape (e.g. 310, D3 and E3) exhibited no significant genetic differentiation in pairwise comparisons with the populations in the continuous native forest (Fig. 10.5, Table 10.4). As might be expected, given these results, a Mantel test of the relationship between pairwise population F_{ST} values vs. pairwise geographic distance was not significant ($R_{xy}=0.28$, $P=0.14$), indicating there was no relationship between genetic differentiation and geographic distance.

DISCUSSION

Ecological consequences of habitat fragmentation?

The bush rat was almost totally absent from stands of radiata pine – a result returned both from intensive trapping and hairtubing surveys. The paucity of animals from exotic softwood forests contrasts with earlier studies of small mammals in radiata pine plantations. Warneke (1971), Barnett et al. (1977), Suckling & Heislers (1978) and Smith (1982) found the species was not uncommon in plantations. Unlike softwood stands examined in earlier studies, the plantation at Tumut has experienced prolonged use of heavy machinery to thin and harvest trees. Recurrent intensive and extensive disturbance by these machines, particularly damage to, and dispersal of, residual windrowed piles of eucalypt logs left following clearing of the original native forest to plant radiata pine trees, has made conditions unsuitable for the bush rat. Notably, the extremely rare detections of the bush rat in our study (e.g. < 0.3% of detections in hairtubes) were confined to older stands of radiata pine where intact windrows of logs remain.

Our trapping and hairtubing results indicate that populations of bush rats are extremely rare in stands of radiata pine and that very few animals live in these areas. However, these methods do not provide information on whether individuals move through stands of radiata pine to colonise patches of remnant eucalypt forest. However, the paucity of animals captured in these areas suggests that the softwood plantation may inhibit the movements of the bush rat, although further empirical work is required to establish whether this is true.

The results of our field studies revealed some negative effects of habitat fragmentation on the bush rat, particularly the fact that smaller patches of remnant native vegetation were significantly more likely to remain unoccupied by the species than larger ones (Fig. 10.3). Relationships between patch size and occupancy are well known in landscape ecology (reviewed by Hanski, 1994a) and they underpin many widely applied models for metapopulation dynamics that attempt to forecast patch occupancy by wildlife in fragmented habitats (e.g. Hanski, 1994c; Possingham & Davies, 1995). On this basis, patch size and occupancy relationships may be due to a number of factors, such as: (1) larger remnants may have a greater chance of supporting suitable habitat (Simberloff, 1988); (2) larger remnants may be capable of supporting higher populations of a given species and these, in turn, may have a reduced chance of local (i.e. patch) extinction due to factors such as demographic and environmental stochasticity (Burgman *et al.*, 1993; Lande, 1993); (3) larger remnants may have a greater chance of being contacted by animals during dispersal resulting in more animals successfully colonising such areas (Stamps *et al.*, 1987; Forman, 1996); and (4) larger remnants have a higher probability of supporting some suitable refuge habitat than smaller ones in the event of a catastrophic event such as a fire (McCarthy & Lindenmayer, 1999). Our present data do not allow us to determine which (if any) of these four factors is important in influencing the patch-occupancy patterns we observed from analyses of field data we gathered for the bush rat.

Genetic consequences of habitat fragmentation?

At the scale of our genetic study (approximately 10 × 15 km), and in a uniform landscape, an appropriate null model is either a panmictic genetic structure (i.e. lacking differentiation among populations) or a pattern of isolation by distance (i.e. increasing genetic differentiation with distance). Habitat fragmentation in this hypothetical uniform landscape is expected to impede gene flow, such that populations begin to develop some genetic differentiation due to genetic drift. If a metapopulation structure exists, founder effects associated with repeated extinction and colonisation may further increase genetic differentiation. At this stage in the fragmentation process, neither random mating nor isolation by distance will characterise the genetic patterns. Instead, it is likely that heterogeneity will characterise the system, possibly with little or no underlying pattern since the degree of genetic differentiation among populations will be strongly influenced by a series of complex and unpredictable processes. In the longer term, or when fragmentation is severe, the cumulative effects of genetic drift, inbreeding and local extinction and recolonisation events will lead to serious losses of genetic diversity (Lacy, 1987; Hedrick & Miller, 1992; Lacy & Lindenmayer, 1995; Young et al., 1996).

Neither random mating nor isolation by distance can explain the genetic patterns observed in this study of bush rats. The most interesting finding was the detection of significant genetic differentiation between two of the proximate population pairs within the fragmented landscape. Also of interest is the findings of pairwise population F_{ST} values of up to 0.13 (for the population pair 446 vs. CAM5 which are less than 6 km apart), as well as significant genetic differentiation among populations within continuous native forest. Few if any similar microsatellite studies of native rat species are available for comparison with our study. Ruscoe et al. (1998) report the results of an allozyme study of the canefield rat, *Rattus sordidus*, a native rat of Queensland, Australia that seasonally colonises sugar-cane crops from refugia in native vegetation. Within two replicated areas of a similar scale to our study (approximately 10 km across) but separated by 500 km they found little or no genetic differentiation, with an overall mean F_{ST} of 0.0135. As expected, significant genetic differentiation between the two areas was found, with an overall mean F_{ST} of 0.155. However, this mean value was substantially inflated by an outlying locus, whose exclusion gives a mean F_{ST} value of 0.047. [Note that Ruscoe et al. (1998) report separate mean values for each year of the two years of the study, but given their similarity we have reported the mean of these yearly averages.] Based on these findings, Ruscoe et al. (1998) concluded that despite potential impediments including a river and habitats generally avoided by the rats, gene flow was extensive. Sherwin & Murray (1990) reported that F_{ST} values based on allozyme studies ranged from 0.10 to 0.28 in four species of rodents and marsupials sampled over a large geographic scale. By contrast with the findings of other investigations, given the local scale of our study and given also that estimates of F_{ST} at microsatellite loci may be as much as an order of magnitude lower than for allozymes (Hedrick, 1999), genetic differentiation appears to be above average in bush rats within the Tumut fragmentation study area.

Initially it may appear that the genetic heterogeneity among proximate remnant populations of bush rats reflects the genetic consequence of habitat fragmentation. However, we also detected heterogeneity among populations within continuous native forest. Also, some of the remnant populations were found to exhibit little or no genetic differentiation relative to other populations in both the fragmented landscape and the continuous forest. Thus, genetic divergence among some pairs of populations cannot be solely due to isolation as a result of habitat fragmentation. This suggests that other processes have contributed to the genetic patterns.

Natural heterogeneity in the landscape combined with peculiarities of bush rat biology may be sufficient to generate the genetic patterns we have observed, even in the absence of habitat fragmentation. Several studies have found that the bush rat favours moist environments with abundant ground cover such as the vegetation typically associated with watercourses (e.g. Warneke, 1971; Bennett, 1990). If the bush rat disperses only within suitable habitat, as is known to occur in many species of animals (e.g. Garrett & Franklin, 1988; Lorenz & Barrett, 1990; Baur & Baur, 1992), drainage lines characterised by dense mesic vegetation may be important for guiding movements through the landscape. It is of interest that both the proximate population pairs that exhibited genetic differentiation (446 vs. 310, T3 vs. 599) are located on parallel creek systems with the confluence of their creeks at a considerably greater distance than the direct geographic distance between the populations. Thus the direct geographic distance may provide a poor indicator of connectivity between such populations (see also Wegner & Merriam, 1979).

Implications for animal conservation

The results of our integrated demographic and genetic study of the bush rat at Tumut have some potential implications for nature conservation, particularly the ways existing softwood plantations are managed and the design of new ones presently being established. Steps to enhance the conservation value of exotic softwood forests in southern New South Wales are important because the plantation estate will be large (it presently exceeds 200 000 ha) and is likely to expand considerably over the next decade (Clark, 1992). The expansion of the softwood resource will take place on semi-cleared grazing lands that presently support patches of remnant native vegetation (D. Hobson, State Forests of New South Wales, personal communication). Indeed, although native vegetation has declined in some states in recent years, it is set to increase again as a result of proposals to establish radiata pine plantations on private land (The Commonwealth of Australia & Department of Natural Resources and Environment, 1997). Therefore appropriate plantation design, including the exemption of existing native eucalypt remnants from clearing during site preparation, may assist the conservation of an array of species that might otherwise be lost (Lindenmayer *et al.*, 1997*b*, 1998*a*).

Large patches of remnant native vegetation should not be cleared as part of plantation establishment and will be the key ones for retention during landscape modification. This conclusion is based on the strong relationships we observed between patch area and patch occupancy (Fig. 10.3). In addition, it appears possible that the bush rat uses watercourses and drainage lines to move and disperse between patches of suitable habitat. Thus, functional connectivity (*sensu* Bennett, 1998) between subpopulations may occur not via the closest geographic routes across stands of radiata pine, but rather along particular topographic features associated with streamlines. Water-quality considerations mean that prescriptions for the careful management of riparian areas are now in place within the softwood plantation estate in New South Wales. Part of these prescriptions should include efforts to re-establish strips of native forest and woodland in streamside areas where it was formerly cleared. These efforts may restore functional connectivity to populations of some species (like the bush rat) and promote their persistence in highly modified landscapes. Thus, larger patches may be more likely to support suitable habitat for the bush rat and, in turn, the offspring of resident animals may use riparian vegetation to disperse between remnant areas. Therefore, the combined outcomes of the demographic and preliminary genetic studies completed at Tumut indicate there is merit in both conserving existing large remnants and connecting them with riparian vegetation.

Further work

Our attempts to integrate the results of demographic and genetic studies have identified some potentially important outcomes and a more complete picture than would have been derived from an investigation confined to any single discipline. Our findings also highlight the complexity of both natural and modified landscape systems, and the difficulty of unravelling the consequences of habitat fragmentation. Few other studies appear to have investigated demography and genetics patterns in fragmented landscapes on such a local geographic scale. For example, most other genetic studies of habitat fragmentation in animals tend to focus on larger geographic areas, often by the comparison of 'island'-like fragments with continuous 'mainland' populations (e.g. Leung et al., 1993; Gaines et al., 1997; Cunningham & Moritz, 1998). In addition to their larger geographic scale, many of these studies have been predominantly concerned with assessing the relationship between population size and genetic diversity (often with mixed results). While the loss of genetic diversity associated with reductions in population size is arguably the most important and potentially serious long-term outcome of habitat fragmentation, the formation of genetic differentiation among fragmented populations is likely to precede detectable loss of genetic diversity, since allele-frequency divergence will usually precede allelic loss (unless there is a very severe bottleneck), and since heterozygosity estimates are insensitive to lower-frequency alleles. Thus the formation of genetic differentiation may be an important early indicator of habitat-fragmentation effects. Yet this effect of has been largely neglected in both animal (Gaines et al., 1997) and plant studies (Young et al., 1996). This suggests that further study of the patterns of population differentiation may be a fruitful area of habitat-fragmentation research, particularly given the increasing availability of high-resolution genetic markers, such as microsatellites.

Notwithstanding the potential insights patterns of local genetic differentiation may reveal about the effects of habitat fragmentation, for bush rats it is clear that even on the small scale of the Tumut fragmentation study (10×15 km), unknown natural long-term processes have generated genetic heterogeneity. This suggests that further studies must focus at an even finer scale (perhaps 0.5×3 km) to minimise these confounding effects. At this smaller scale, it may be feasible to measure migration rates by combining detailed genetic analysis with mark–recapture experiments and other population manipulations. Other studies have already highlighted the potential of microsatellite markers for determining paternity and for estimating migration rates (e.g. Double et al., 1997; Favre et al., 1997), and given the extensive variability already detected with just six microsatellite loci, and the large numbers of microsatellite loci available for the genus *Rattus*, it will be feasible to undertake such studies with minimal laboratory development. Funding permitting, we will shortly commence a study that will compare fine-scale genetic structure along replicate parallel creek systems in both remnant and continuous forest. Such a study will allow us to test whether bush rat movements are restricted to riverine systems and whether habitat fragmentation further impedes gene flow. Combined with precise knowledge of the recent history of the landscape fragmentation, our ongoing studies will provide important new insights into the patterns and extent of bush rat movements and relationships to habitat fragmentation.

ACKNOWLEDGMENTS

The Tumut fragmentation study would not be possible without the contributions of many colleagues, particularly Ross Cunningham who was responsible for the experimental design. Matthew Pope, Ryan Incoll, Chris McGregor and Craig Tribolet have made major contributions through their dedicated field work. Others to have contributed significantly to work at Tumut include Christine Donnelly, Henry Nix, Karen Viggers, Hugh Possingham, Bob Lacy, Mick McCarthy, Bruce Lindenmayer, David Patkeau, Craig Moritz, Barbara Triggs, Hugh Tyndale-Biscoe and Ian Ball. The study at Tumut is supported by Land and Water Resources and Research and Development Corporation, Rural Industries Research and Development Corporation, Forests and Wood Products Research and Development Corporation, New South Wales Department of Land and Water Conservation, State Forests of New South Wales, NSW National Parks and Wildlife Service, Environment Australia, CSR Ltd, Visy Industries, The Australian Research Council, the Canberra Ornithologists Group and a private donation from Jim Atkinson.

We also wish to thank Heidi Hewitson, whose honours project provided the impetus for this study, Jeanette Cashin, Daniel Ebert and Stephen Murphy for technical assistance in the laboratory.

11

Demographic evidence of inbreeding depression in wild golden lion tamarins

JAMES M. DIETZ, ANDREW J. BAKER
& JONATHAN D. BALLOU

ABSTRACT

Golden lion tamarins are small, arboreal primates endemic to lowland Atlantic coastal rainforest of Rio de Janeiro State, Brazil. About 600 individuals are found in 14 forest fragments, the largest of which is Poço das Antas Reserve. This reserve is a forest island surrounded by cattle pasture and contains about 347 tamarins. The authors monitored the behaviour and demography of about 110 individuals in 20 breeding groups for 13 years in this reserve. All individuals in the study population were individually marked and habituated to the presence of human observers. Dates and locations of all births, deaths, emigrations and immigrations and the identities of dam and probable sire for all offspring were recorded. Analysis of these data was used to test for the presence of inbreeding and inbreeding depression in this population. A total of 481 offspring were born during the study, and 47 of these were classified as inbred. Mortality of inbred offspring was significantly greater than that of non-inbred offspring (3.4 lethal equivalents per individual). Although the effects of inbreeding depression were most acute during the first six months of life, survivorship of inbred tamarins remained low relative to that of non-inbred individuals for at least the first two years of life. Inbreeding was thought to result when a daughter failed to disperse and bred with a close relative in her natal group or when an individual dispersed into another group and mated with a relative therein. The frequency of the latter type of inbreeding suggests that tamarins do not recognise relatives outside their natal group or do not reject them as mates. Computer models suggest that inbreeding is not a direct threat to survival of the Poço das Antas population but may be a significant problem for all the other wild populations of golden lion tamarins.

INTRODUCTION

Although there is considerable evidence from breeding experiments and captive populations of plants and animals demonstrating that inbreeding results in the relative decrease in fitness of offspring at a variety of age classes (inbreeding depression), data from wild populations have been slow to accumulate (Jiménez et al., 1994; Frankham, 1995b; Pusey & Wolf, 1996). Without the benefit of molecular genetic techniques, it is often difficult to differentiate between inbred and non-inbred offspring under field conditions. In addition, few field studies have had the scope and duration necessary to compare the lifetime performance of a representative number of inbred and non-inbred progeny (see review by Lacy, 1997). Thus it is not surprising that conservation biologists struggle to determine the extent to which genetic factors contribute to the endangerment or extinction of wild populations (Lande, 1988; Frankham & Ralls, 1998).

In a summary of infant mortality rates for inbred and non-inbred primates in captivity, Ralls & Ballou (1982) found significantly higher mortality for nearly all populations of 16 species. Studies documenting inbreeding depression in wild primate populations, although few and sometimes equivocal, generally support the conclusion that there is potential for inbreeding depression in primates (Moore, 1993). However, in the Callitrichidae (lion tamarins, tamarins and marmosets), in which fraternal twins are the normal litter, blood chimerism is common in uterine siblings (Dixson et al., 1992). Pope (1996) suggested that low levels of allozyme diversity observed in lion tamarins (Forman et al., 1986) may reflect an adaptation resulting from chimerism between twins. She hypothesised that these primates may be adapted to a homozygous genetic background, and thus less likely to suffer inbreeding depression than species with more heteroselected genomes.

Golden lion tamarins (*Leontopithecus rosalia*) are small, arboreal primates endemic to lowland Atlantic coastal rainforest of Rio de Janeiro State, Brazil (Coimbra-Filho & Mittermeier, 1973). Extensive deforestation has reduced the number of tamarins in the wild to about 600 individuals in 14 isolated forest fragments, all in central Rio de Janeiro State (Kierulff, 1993). Following translocation of the smallest populations, seven native (not reintroduced) populations remained, all but one containing fewer than 100 animals (Kierulff & Oliveira, 1996; Ballou et al., 1998). The species is classified by the World Conservation Union (IUCN) as 'critically endangered' (Baillie & Groombridge, 1996).

The 5200-ha Poço das Antas Biological Reserve, one of two legally protected areas containing *L. rosalia*, is located 70 km north-east of Rio de Janeiro, and holds the largest wild population of golden lion tamarins, about 347 individuals (Ballou *et al.*, 1998). The habitat of the reserve includes a mosaic of forests in various stages of secondary succession, the result of clear-cutting and selective logging during the past 100 years. There has been no significant deforestation in the reserve since its creation in 1974. However, the periphery of the reserve is almost entirely surrounded by cattle pasture and farmland. About 2000 ha of private forest are connected with the reserve by narrow forest corridors.

We continuously monitored the demography and behaviour of golden lion tamarins in Poço das Antas Reserve for 13 years (see 'Methods'). In this paper we use these long-term data to test for the presence of inbreeding and inbreeding depression in this population. We then conduct computer simulations to estimate the effect of inbreeding on the viability of the Poço das Antas tamarin population.

The social organisation and mating system of golden lion tamarins
Most golden lion tamarins live in territorial groups. Group size ranges from two to 14, with a mean of 5.6 individuals (Dietz & Baker, 1993; Dietz *et al.*, 1997). Most groups contain one reproductive female (Dietz & Baker, 1993) and one or two non-natal (potentially reproductive) males (Baker *et al.*, 1993). Although both males in two-male groups may exhibit reproductive behaviour, several lines of evidence suggest that only the dominant male is likely to father offspring (Baker *et al.*, 1993). Genetic analysis is necessary to verify this assumption.

Approximately 10% of reproductive groups are polygynous (Dietz & Baker, 1993). Polygynously reproducing females are generally mother–daughter duos resulting from the philopatry of a natal female into adulthood. Although polygyny is more likely to arise when a male unrelated to either mother or daughter is present in the group, polygyny and polygynous reproduction are occasionally observed when the dominant male is closely related (father, brother or uncle) to the reproductive daughter. In a previous study offspring of apparently incestuous matings had a lower probability of surviving to weaning than did non-inbred offspring (Dietz & Baker, 1993).

Reproduction in this population is seasonal, with most births occurring from September to November (Dietz *et al.*, 1994). Females produce one or occasionally two litters per year, with a modal litter size of two (Dietz *et al.*, 1994). Golden lion tamarins, like all members of the Callitrichidae, are co-operative breeders. All group members care for all infants in the group, with potentially reproductive males being the most actively involved in this behaviour (Baker, 1991). Infants are fully independent before the onset of the June–August dry season.

Most tamarins disperse from their natal groups, usually by 3 years of age (Baker, 1991, Baker & Dietz, in preparation). However, in contrast to cotton-top tamarins (Savage *et al.*, 1996) immigration into lion tamarin groups is rare and usually takes place in the context of same-sex replacement of a breeding individual. Intact groups appear to be closed to immigration by females but not by males (Baker & Dietz, 1996). About 31% of male and 76% of female emigrants disappear and are assumed to have died before finding a breeding opportunity (Dietz & Baker, 1993, unpublished data). Thus, variance in female reproductive success is greater than that in males. A similar pattern has been documented for other callitrichids (Goldizen *et al.*, 1996).

In light of the social organisation and group territoriality characteristic of this closed population, inbreeding might take place by two general mechanisms. Firstly, close relatives, e.g. father and daughter, might mate either within a group or through extra-group copulations. Secondly, dispersing individuals might enter a group containing their kin and breed with them. In both cases tamarins might mate with relatives either because they do not recognise them as kin or because they do not reject them as mates.

METHODS

Monitoring of the study population and systematic habituation of tamarin groups to observation began in 1983 (see Baker *et al.*, 1993; Dietz & Baker, 1993; Baker & Dietz, 1996). The number of tamarins under observation grew to about 110 individuals in 20 breeding groups and has remained relatively stable for the past decade. From 1983 to 1996 we amassed partial or complete life-history data on 614 individuals (1360 animal–years).

Twenty groups were observed at least once per week to note changes in group composition (emigrations, immigrations, births or deaths). For about half these groups we systematically collected data on individual behaviour (e.g. Martin & Bateson, 1993), including information on dominance relations and sexual behaviour. At intervals of about six months we non-injuriously captured the majority of these tamarins to affix radiocollars to one or more adults in each group. At that time we also tattooed new individuals in the population, and collected hair and blood samples for genetic studies. The mothers of all infants were assigned based on prepartum external palpation of the mother's abdomen or by observation of lactation or elongate nipples post-partum (Neyman, 1977; Baker *et al.*, 1993; Dietz *et al.*, 1994).

Dominant males in multi-male groups were assumed to be sires of offspring born in that group. Dominance was assessed based on the direction of aggressive interactions (Baker *et al.*, 1993). For the 18 offspring born in multi-male groups where data on dominance were inconclusive we calculated inbreeding coefficients using all possible combinations of potential sires. The choice of sire affected the level of inbreeding in the population in only one of these cases.

As mentioned above, about 10% of groups contained two reproductive females, usually mother and daughter. Over 95% of documented sexual behaviour by 'mothers' in these polygynous groups took place with an unrelated male of the same group (A. J. Baker & J. M. Dietz, unpublished data). Likewise, sexual behaviours of 'daughters' in polygynous groups involved unrelated same-group males, when available. However, when these groups did not contain an unrelated male, sexual behaviour by 'daughters' often involved extra-group males (A. J. Baker & J. M. Dietz, unpublished data). In the latter cases, inbreeding coefficients were calculated in two ways, one assuming a same-group (related) sire and the other assuming an extra-group (unrelated) sire.

Dates and locations of all births, deaths, emigrations and immigrations and the identities of dam and probable sire for all offspring were recorded in SPARKS (Single Population Animal Record Keeping System, International Species Information System) version 1.42, a commercial software program used to manage genealogical data for populations in captivity (see Ballou *et al.*, 1998). Tamarin breeding groups were classified as 'institutions' in SPARKS. When an animal transferred from group to group (i.e. 'transactions') the 'date of removal' was the last date that the individual was observed in the original group. The 'accession date' was defined as the date when the individual was first observed in its new group. Intervening time was not accounted for unless the animal's whereabouts were known (e.g. the tamarin was known to be in a floating group). Animals that disappeared were placed in a group named 'disappeared' and later listed as 'lost-to-followup'. The ages of animals with unknown birthdates were estimated based on body weight and measurements, and tooth wear.

Lowland forest in Rio de Janeiro State was reduced to relatively small patches in the past century. However, in 1971 the vegetative cover around Poço das Antas was over 70% forest. Additional deforestation in the region took place following the creation of the Reserve and before its effective protection (Magnanini, 1977). Given the recent and incomplete isolation of the Poço das Antas population, and lack of molecular genetic evidence necessary to construct pedigrees, we assumed that all adults in the population were unrelated at the beginning of our study.

We modelled population viability using VORTEX version 7.41 computer simulations (Lacy, 1993b; Lacy et al., 1995). VORTEX simultaneously models the effects of demographic stochasticity, environmental variation in the annual birth and death rates, catastrophes and inbreeding depression in small populations. We modelled the viability of the Poço das Antas population for 100 years under the various scenarios of inbreeding presented herein. Other model input parameters were those used in the 1997 population viability assessment for this species (Ballou et al., 1998).

RESULTS

A total of 481 offspring were born during the study. Of these, 434 were classified as non-inbred, and 47 were classified as inbred at some level (see the assumptions regarding parentage, below). Documented inbreeding coefficients ranged from 0.016 to 0.250. The frequencies of inbreeding coefficients for offspring born during the study are presented in Fig. 11.1.

Mortality of inbred offspring was significantly greater than that of non-inbred offspring (inbred offspring: 12 of 47 died = 25.5%, non-inbred offspring: 23 of 434 died = 5.3%, $\chi^2 = 25.7$, $P < 0.0001$). We used logistic regression to test for the effects of inbreeding in this population by regressing survival of offspring to 7 days of age against their assigned inbreeding coefficient (Fig. 11.2). The slope of the regression line was significantly different from zero (maximum likelihood test: $P < 0.0001$) suggesting that inbreeding across all levels results in reduced survival of offspring. The slope of the regression line indicates that the population contains 3.4 lethal equivalents per individual. That is, assuming that all deaths resulting from deleterious or lethal combinations are due to independently acting lethal alleles, individuals in this population carry, on the average, 3.4 alleles in the heterozygous condition that if made homozygous would kill the bearer.

Daughters reproducing in natal groups in which the dominant male was father or brother produced 17 infants. As discussed above, classification of these infants as inbred or non-inbred is in doubt due to documented extra-group sexual behaviour by dams. In the preceding calculations these infants were classified as inbred. If the 17 suspect infants are excluded from the analysis, the mortality of inbred offspring remains significantly greater than that of non-inbred offspring (inbred offspring: 5 of 30 died = 16.7%, non-inbred offspring: 23 of 434 died = 5.3%, $\chi^2 = 6.4$, $P = 0.011$). If the sus-

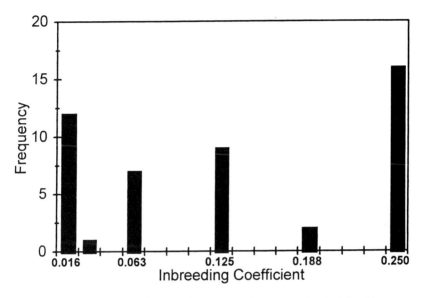

Fig. 11.1. Inbreeding coefficients of the 47 tamarins assumed to be inbred born during the study.

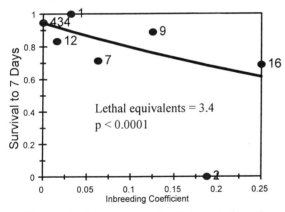

Fig. 11.2. Logistic regression in which survival to 7 days of age is regressed against inbreeding coefficient for 481 offspring born during the study.

pect infants are included with non-inbred infants in the analysis, the mortality of inbred infants still remains greater than that of non-inbred offspring (inbred offspring: 5 of 30 died = 16.7%, non-inbred offspring: 30 of 451 died = 6.7%, $\chi^2 = 4.2$, $P = 0.04$).

To visualise longer-term effects of inbreeding in this population we graphed survivorship (L_x) of inbred and non-inbred infants to 24 months of

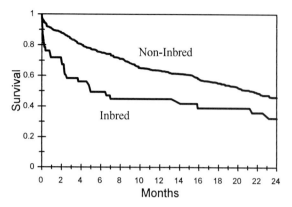

Fig. 11.3. Survivorship (L_x) to 24 months of age for non-inbred and inbred offspring.

age (Fig. 11.3). Results suggest that the effects of inbreeding depression are most acute during the first six months of life.

DISCUSSION

Our results indicate significant inbreeding depression in this population. Survival of newborn inbred infants was lower than that of non-inbred infants. In addition, infant survival decreased with the severity of inbreeding. Finally, the survivorship of inbred tamarins was low relative to that of non-inbred individuals for at least the first two years of life.

We found that about 10% of all tamarin offspring born in Poço das Antas Reserve were inbred. Because our assumption that all adults were unrelated at the beginning of the study is likely to be incorrect, this number may underestimate the level of inbreeding in the population. Inbreeding was suspected to have occurred through two routes. First, adult females that had not dispersed from their natal group may have bred with a close relative – usually her presumptive father or brother. However, without molecular genetic analyses we can not rule out the possibility of extra-group paternity in these cases. Indeed, it would seem adaptive for daughters breeding in their natal group to seek extra-group copulations rather than mate incestuously.

Second, dispersing individuals entered and bred in groups that contained their relatives or formed new groups comprised of related dispersing individuals. In these cases, 43% of inbred offspring had inbreeding coefficients equal to or less than that of first cousins ($F = 0.063$). Given that Poço das Antas Reserve can only support about 62 tamarin territories (Ballou *et al.*, 1998), it is likely that dispersing individuals will have the opportunity to mate with relatives by chance. The observed frequency of this type of inbreeding suggests that tamarins do not recognise relatives outside their natal group or do not reject them as mates. The facts that golden lion tamarins in Poço das Antas Reserve tend to be genetically monogamous and co-operative breeders (not all adults breed) may exacerbate the increase of inbreeding in the population (see Chesser, 1983).

What, if any, are the effects of inbreeding on viability of this population? Using the VORTEX population-modelling program (Lacy, 1993*b*), we ran simulations including and excluding the inbreeding depression documented herein (3.4 lethal equivalents per individual) and demographic data used in previous population viability analyses (Ballou *et al.*, 1998). We found no extinctions over a 100-year period for a population of 347 individuals in either case. However, when we ran simulations for populations containing 50 individuals, the probability of survival was 89% without inbreeding and 60% when inbreeding was included in the model. While these probabilities can not be interpreted as absolute risks of extinction, the relative increase in the extinction risk when inbreeding is included does suggest that inbreeding depression can have a significant impact in smaller populations. Although inbreeding does not appear to be a direct threat to survival of the Poço das Antas population, in the absence of genetic management it may be a significant problem for all the other wild populations of golden lion tamarins.

ACKNOWLEDGMENTS

We thank the Associação Mico Leão Dourado, National Science Foundation (Grants BNS-8616480, BNS-8941939, BNS-9008161, BNS-9318900), International Environmental Studies Program of the Smithsonian Institution, WWF-US, National Geographic Society, Friends of the National Zoo, Centro de Primatologia do Rio de Janeiro, TransBrasil Airlines and Instituto Brasileiro do Meio Ambiente e dos Recursos Naturais Renováveis for financial and logistic support of this research.

12

Inferring demography from genetics: a case study of the endangered golden sun moth, *Synemon plana*

GEOFFREY M. CLARKE

ABSTRACT

The development and application of quantitative methods for assessing the viability and risk of extinction of populations requires considerable background demographic and life-history data of the modelled species and populations. Typically such data are difficult to obtain for most invertebrate species due to time or resource constraints. Quantitative sampling methodologies are not well developed for the bulk of invertebrate species and field-based estimates of migration, survival, fecundity, etc. are problematic. However the use of genetic-marker technologies such as allozymes and mitochondrial DNA (mtDNA) sequence data has the potential for inferences to be made about underlying demographic processes within and among populations useful for quantitative model development. In this paper I will show how such marker technologies have been applied to an endangered species of grassland-inhabiting moth, *Synemon plana*, to infer some fundamental life-history and demographic parameters.

INTRODUCTION

Effective conservation management of threatened species requires considerable detailed information on the life history, demographics and population structure of the taxa of interest. In addition, the development of quantitative models of population persistence [e.g. population viability analyses (PVA)] almost universally requires parameters such as generation time, fecundity, fertility, adult and juvenile mortality and migration as model inputs (e.g. Burgman *et al.*, 1993; Lacy, 1993*b*). For many species, the acquisition of such data is not overly problematic (although may involve many years of detailed field work), and these are the same types of data used in the original determination of the species' threatened status. Indeed for many threatened species, such data are available at the individual in addition to the population level, and often at fine spatial scales, leading to the developmental of individual-based spatially explicit models (DeAngelis & Gross, 1992; Dunning *et al.*, 1995). This is particularly true for many mammals and birds for which many such models have been developed (e.g. Price & Kelly, 1994; Letcher *et al.*, 1998), and is becoming more common for threatened plants (e.g. Burgman & Lamont, 1993; McCarthy, 1996).

For the vast majority of threatened species demographic and life-history data have been accumulated through direct or indirect observations of field or captive individuals and populations. Identification and tracking of individuals within and between areas is made possible through a variety of now commonly used techniques such as banding, radio-telemetry and remote video. Although genetic data are seldom required or incorporated into most quantitative models, such data are rapidly being used either to directly estimate or to infer many demographic and life-history parameters otherwise unattainable (Goldstein *et al.*, 1999). In particular the development of highly variable microsatellite markers has made it possible to 'genetically' mark individuals (Bruford *et al.*, 1996). In addition, recent advances in DNA extraction techniques have led to the development of a series of non-invasive sampling techniques (e.g. faecal or hair sampling), such that actual sighting and 'capture' of individuals is no longer required (Morin & Woodruff, 1996). In addition to providing routine estimates of levels of genetic diversity and variability within and among populations, genetic data have also been used to determine otherwise difficult demographic parameters, e.g. breeding structure, parentage, home ranges and patterns of migration (see Smith & Wayne, 1996). One of the major advantages of genetic data is that they make it possible to infer historic, as well as recent, population processes, e.g. bottlenecks, fragmentation, gene flow and inbreeding (e.g. Moritz *et al.*, 1996)

The invertebrates are a group of organisms for which the application of genetic data for inferring basic life-history and demographic parameters may prove valuable. For the vast majority of invertebrate species direct field observation is difficult as they are small and/or cryptic, seasonal, and may undergo highly variable population density cycles. This makes even simple parameters such as numbers of individuals within a population very difficult to determine. As a result, criteria such as population size are seldom used in determining the threatened status of invertebrate taxa (number of populations and patch size are the most commonly used criteria). This difficulty in assessing basic life-history and demographic parameters is undoubtedly one of the reasons invertebrates are under-represented on lists of threatened species worldwide. Despite making up almost 80% of the world's named species, invertebrates account for fewer than 10% of threatened species listed in the 1996 *Red List of Threatened Species*. That is, less than 0.2% of the known invertebrate fauna is regarded as being threatened, compared with over 20% for mammals and an average of over 5% for most other groups. Unfortunately, there is a general lack of empirical data to test whether or not this under-representation represents biological reality or not.

There have been very few long-term or detailed studies of the impacts of fragmentation on invertebrate species (Margules *et al.*, 1994; Van Dongen *et al.*, 1998). Again the most probable explanation is the difficulty in accruing basic biological and ecological data. The most notable exceptions are those involving highly conspicuous butterflies (Brookes *et al.*, 1997; Lewis *et al.*, 1997; Thomas & Hanski, 1997). This lack of data is unfortunate as studies involving invertebrates have considerable potential for investigating and testing hypotheses of the demographic and genetic impacts of fragmentation. The relatively small spatial scale and rapid generation time of most invertebrate species make them particularly useful in this regard. Frankham and colleagues have shown how useful invertebrates are under laboratory conditions for understanding and testing of many of the basic tenets of modern conservation biology (Loebel *et al.*, 1992; Spielman & Frankham, 1992; Briton *et al.*, 1994; Woodworth *et al.*, 1994).

In the remainder of this paper I will focus on the genetic analysis of the endangered golden sun moth, *Synemon plana* (Lepidoptera: Castniidae). In particular, I will show how genetic data have been used to infer some basic life-history, demographic and population parameters for this species.

SYNEMON PLANA

The golden sun moth, *Synemon plana,* is a medium-sized (wing span approximately 32 mm) diurnal moth, the larvae of which are thought to feed exclusively on native grasses within the genus *Austrodanthonia.* The species is one of approximately 44 occurring in this Australian endemic genus, many of which are also restricted to, and thus fully dependent on, native grasslands. *Synemon plana* was once widespread throughout south-eastern Australia, matching the distribution of native grasslands and grassy woodlands. Temperate grasslands are the most endangered of all vegetation types in Australia with less than 1% of the approximately 2 million ha existing prior to European settlement still remaining (Kirkpatrick *et al.*, 1995), and have recently been listed as an endangered ecological community under the Commonwealth Endangered Species Act. The remaining grasslands are threatened by urban and agricultural expansion, weed invasion and grazing by stock and rabbits. Consequently the remaining populations of *S. plana* are highly fragmented. *S. plana* is currently known from approximately 48 sites the majority of which are less than 3 ha in area (Clarke, 1999a). The species is listed in all three States in which it occurs under their respective threatened species legislation and has been nominated for listing under the Commonwealth Act.

The life history of the species is very poorly known. The adults are short-lived (one to four days) and do not feed, having no functional mouth parts. Males are capable of active and prolonged flight, but they will not fly long distances (> 100 m) away from suitable habitat. Females have reduced hind wings and are reluctant to fly even when disturbed and are thus rarely encountered in the field. The sex ratio appears heavily male-biased; however, the cryptic nature of the females does not allow for an accurate estimate. Adult emergence is continuous throughout the annual flying season which typically lasts six to eight weeks in early summer (November–December). Males fly only in bright sunshine during the warmest part of the day. The cues initiating adult emergence are unknown. The larval and pupal stages and their duration (and hence generation time) are unknown. Females are known to carry up to 200 eggs (following dissection) (Edwards, 1994). However, it is not known how many are actually laid, or if eggs are laid singly or in clusters. Eggs are thought to be laid at the base of the grass clumps. It is assumed that larvae feed underground on the roots of the grass plants. The number of plants required for the development of a single larvae is unknown. Rates of adult and juvenile mortality are also unknown. Since formal listing of this species it has not been possible to undertake any research which involves habitat disturbance or manipulation, and all attempts to breed the moth *ex situ* have failed.

Detailed census work, using mark–release–recapture techniques, has been conducted at a single 0.4-ha site over a three-year period (Harwood *et al.*, 1995). The population at this site appears relatively stable at approximately 1500 individuals (assuming a 1:1 sex ratio). Assuming females lay their entire complement of eggs this suggests that up to 99% of total potential fecundity is unrealised.

METHODS

Samples

A total of 1200 adult males were collected from 36 of the 48 extant sites during the 1997 and 1998 flying seasons using a hand net (Fig. 12.1). Five of these populations were sampled in both seasons. Although up to 30 males were taken from each site, the timing of collection was designed to enhance the collection of post-reproductive males. In addition, at all sites sampled, *Synemon plana* is locally abundant. Captured individuals were returned alive to the laboratory and placed at $-20\,°C$ until dead. Individual abdomens were then removed and stored at $-80\,°C$ until required.

Electrophoresis

A total of 16 enzyme systems representing 20 loci were analysed by cellulose acetate electrophoresis. Details of protocols and loci can be found in Clarke & O'Dwyer (2000).

RESULTS AND DISCUSSION

Full details of the genetic data for both years' samples can be found elsewhere (Clarke, 1999*b*, *c*; Clarke & O'Dwyer, 2000) and I will only present summary information here in the context of the current discussion.

Generation time

Although not tested, generation time in *Synemon plana* was thought to be in the range of one to three years based on known generation time for other species within the genus (Edwards, 1994). A generation time of two years was considered most likely despite the fact that adult moths can be found each year within *S. plana* populations. It is conceivable that over evolutionary time moths could become present annually, even with a two-year generation time, if larval development displayed some level of plasticity and climatic determinants varied during the period. However, one would also expect that as conditions stabilised then these odd- and even-year cohorts would become genetically differentiated in more recent times.

An examination of both allele and genotype frequencies from five populations sampled in successive years revealed no significant differences between years (Clarke, 1999*c*). Thus the genetic data strongly suggest that generation time is approximately one year. It is difficult to imagine larval development time to be variable enough to maintain such genetic homogeneity across years, particularly as the timing of the flight season varies only marginally each season.

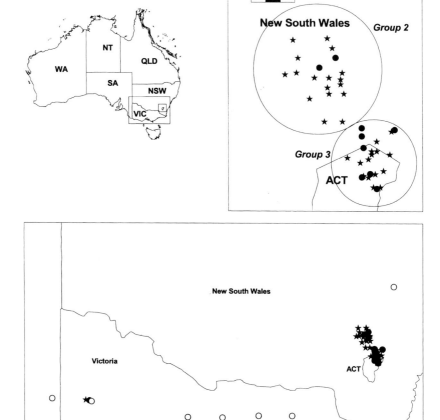

Fig. 12.1. Historic and present distribution of *Synemon plana* populations.
○ represents historic (presumably extinct) populations; ● represents current
extant populations; ★ represents sampled populations.

Inbreeding and population subdivision

An interesting result from the first year of sampling (in which 20 populations were analysed) was the significant level of inbreeding present in almost all populations (mean $F = 0.190$, $P < 0.005$; range 0.045–0.326) (Clarke & O'Dwyer, 2000). Observed heterozygosities were consistently lower than that expected under random mating, primarily due to the fact that the majority of rare alleles were found in homozygous condition, rather than as heterozygotes as expected. This appeared to be independent of population size (although accurate estimates of size were not possible crude relative estimates are available). It was hypothesised that this high level of apparent non-random mating may be due to temporal subdivision of populations. The average life span of adult moths (1–4 days) is considerably shorter than the duration of the flying season (6–8 weeks). Thus moths emerging early in the season have no chance of mating with individuals emerging late in the season. If larval development was reasonably synchronous, such that eggs laid early in one season developed into adults which emerged early in the following season, then it could be imagined how a single 'population' may become temporally subdivided. Each cohort would be effectively isolated from all others and thus would become genetically differentiated over time. As the sampling regime collected adults on a single day these would represent a single cohort (or perhaps two) and thus represent a considerably smaller 'population' which could promote inbreeding.

This hypothesis was tested in the second year's sampling by collecting moths at three times (early, mid and late) throughout the season from four sites. Pairwise comparisons of both allele and genotype frequencies between each sampling date within each population were all non-significant, indicating that temporal subdivision of populations was unlikely (Clarke, 1999c). Again the inbreeding values of each sample were in the same range as that observed previously. It would thus appear as though larval development time is variable enough to maintain genetic homogeneity within the population.

In the context of the current discussion the importance of this result is not the lack of evidence for subdivision, but more the fact that genetic data were used to construct an hypothesis concerning demographic and life-history parameters which could be tested by genetic means. This has led to the development of an alternative testable hypothesis, viz. that the high F values are due to partial parthenogenesis in this species. It is worth noting that *S. plana*'s closest relative (*S. selene*) has both sexual and parthenogenetic forms.

Effective population size

The estimation of effective population sizes is one of the most common uses of genetic data in conservation (see Neigel, 1996). The effective size of a population can be defined simply as the number of individuals making a contribution to the next generation. Factors such as fluctuating population size, unequal sex ratio and differences in adult survival and fitness mean the effective size is typically less than the census size, often by an order of magnitude (Frankham, 1995c). Effective population sizes (N_e) can be estimated from genetic data by looking at changes in either inbreeding coefficients, gene frequencies or levels of heterozygosity between generations. These estimates are based on the fact that finite populations will lose genetic variation each generation due to genetic drift at a rate that is inversely proportional to the effective population size (Wright, 1978). Thus repeated genetic sampling of populations over time presents the possibility of estimating the population effective size (see Lande & Barrowclough, 1987; Neigel, 1996). It must be stressed, however, that these estimates are based on a number of assumptions such as random mating, no mutation, migration and selection, almost all of which are violated in natural populations.

Of the five populations sampled in both 1997 and 1998 only one showed a decrease in heterozygosity over time (Clarke, 1999c). The four remaining populations showed slight increases. In terms of inbreeding coefficient, again only one population showed an increase in F with the remaining populations showing appreciable decreases. These data suggest that effective sizes are close to census size for these populations; however estimates based on differences between two successive generations are subject to considerable error, particularly if the population undergoes natural temporal size fluctuations, as is typical for many insect species. The accuracy of estimates is enhanced if the time between sampling is increased.

Adult and larval mortality

These data on effective population sizes, albeit crude, can be used to make inferences about patterns of adult and juvenile mortality within the sampled populations. If mortality occurred predominantly at the adult stage then one might expect effective population size to be considerably less than census size as only a small proportion of the total females present in the population would mate and oviposit. Alternately, if mortality predominantly occurred at the larval stage, and assuming that such mortality was randomly distributed amongst larval cohorts, then effective sizes might be expected to more closely match census size. Obviously, interpretation of effective population size in this manner is problematic, given that N_e can be

Table 12.1. Estimates of genetic diversity, heterozygosity and gene flow within and among population groups of *Synemon plana*

Group	Genetic diversity (A_p)	Heterozygosity		Migrants per generation (N_m)
		Expected (H_e)	Observed (H_o)	
Group 1	2.60	0.114	0.185	0.99
Group 2	2.40	0.121	0.148	3.52
Group 3	2.20	0.083	0.096	9.55

determined by many population and evolutionary processes other than mortality. However, in the absence of other data, these assumptions may provide a starting point for model parameterisation, particularly in the case in which effective size is close to census size, as this would indicate that adult mortality is not appreciable.

Population structure, fragmentation and colonisation
The analysis of patterns of genetic structuring among sampled populations has been used not only to investigate relationships among populations in the conventional sense, but also for the determination of priority populations, or groups of populations, for conservation management, e.g. evolutionarily significant units (ESUs) or management units (MUs) (see Moritz, 1999b). The genetic relationships among all sampled populations of *S. plana* are shown in Fig. 12.2. The names of individual populations are unimportant in the current context and suffice to say that they fall into three main clusters which correspond closely with geographic location (see Fig. 12.1). This pattern of structuring fits an isolation-by-distance model in which the further apart populations are geographically the more genetically different they are ($R^2 = 0.740$, $P < 0.0001$; Fig. 12.3). This pattern is to be expected due to the limited dispersal ability of the species.

In conjunction with estimates of genetic diversity and gene flow within populations and population clusters, it is possible to use this pattern of structuring to infer patterns of colonisation and fragmentation for this species. The four Victorian populations in Group 1 display the greatest average levels of allelic diversity and genetic variation compared with the other two groups of populations (Table 12.1). The 15 populations within Group 3 display the lowest levels of diversity, with the populations in Group 2 being intermediate. Given that we know that current extant populations of *S. plana* are effectively isolated due to the limited dispersal ability of the adults, gene flow estimates can be used to infer historic patterns of connectedness between populations. Estimates of gene flow among the popula-

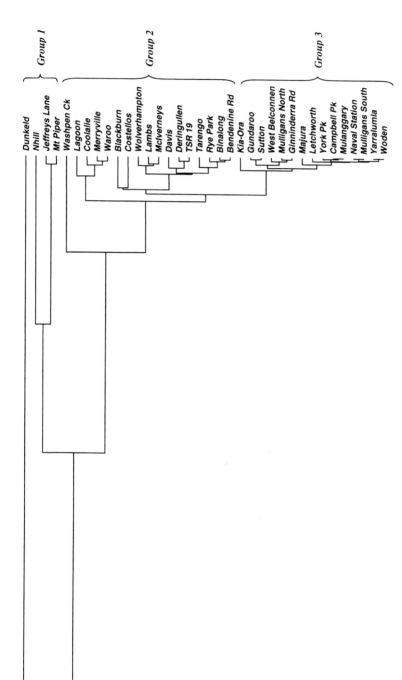

Fig. 12.2. UPGMA phenogram based on Nei's (1978) genetic distance showing relationships among all sampled *Synemon plana* populations. Branch lengths represent genetic distance.

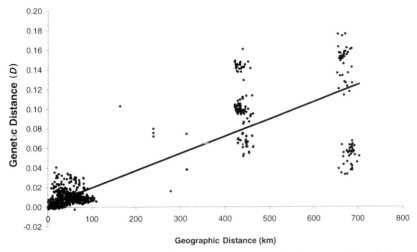

Fig. 12.3. Scatterplot of genetic distance versus geographic distance of sampled populations of *Synemon plana*.

tions in Group 3 are almost 10 migrant individuals per generation. This very high level indicates that these populations were very recently connected and the time since isolation has been insufficient to erase the signature of connectedness. The estimates for populations within Groups 1 and 3 are approximately 1 and 3.5 respectively, indicating longer periods of isolation.

These inferences are supported by the genetic diversity data and what we know of the fragmentation history of the areas where these populations occur. The area containing the Group 3 populations has been heavily fragmented within the last 70 years due to the increase in urban and agricultural expansion associated with the establishment and settlement of the Australian Capital Territory region. In fact much of the fragmentation has occurred in the last 30–40 years. The low levels of genetic variability within these populations are consistent with reductions in population size following fragmentation of near-contiguous habitat into a number of very small patches. The history of fragmentation of the areas containing the populations in Groups 1 and 2 is much older, dating back 150–200 years, with agricultural expansion soon after European settlement. It thus might be expected that these populations would contain higher levels of variation.

Overlying this history of fragmentation is the pattern of colonisation of this species over evolutionary time. The current hypothesis is that the genus *Synemon* originated in central Australia and has since radiated both eastward and westward (E.D. Edwards, personal communication). The closest relatives of *S. plana* occur in South Australia and Victoria. This would suggest that the Victorian populations of *S. plana* are the oldest with colonisation moving eastward resulting in the current distribution. The timing of this is unknown and it is possible the current distribution is quite ancient as no further easterly movement is possible due to an hypothesised altitudinal barrier. Thus again, it might be expected that Victorian populations would contain greater levels of variation and diversity compared with more recent colonists. These conjectures are supported by preliminary DNA sequence data (based on two mitochondrial genes and one nuclear gene) which show Victorian sequences to be ancestral to those from populations in the other two groups. In fact the pattern of population structuring revealed by the allozyme data is matched by the sequence data.

THE FUTURE

Although genetic data have been useful in unravelling some of the life history and demographics of *Synemon plana* populations as well as assisting in the development of testable hypotheses, there remain many unresolved issues. Many of these may be addressed by genetic techniques, particularly through the development and application of highly variable microsatellite markers. Many of the questions still surrounding generation time, effective population sizes and patterns of mortality may be more readily answered by the increased resolution associated with tracking of multilocus microsatellite genotypes over time. Unfortunately, microsatellite markers have proved difficult to isolate within the Lepidoptera in general. There are also some questions which can only be resolved by detailed long-term ecological studies, particularly those relating to habitat requirements and usage. Given the current threatened status of this species, the establishment of an experimental *ex-situ* population may be required for these studies.

CONCLUSIONS

With over 1200 individuals from 36 populations, representing over 24 000 individual genotypes, this study is one of the largest and most comprehensive genetic analyses of any threatened invertebrate. Not only has it provided much-needed information on some of the fundamental life-history and demographic attributes of the species but it has also given insight into the evolutionary, historic and recent population processes that have contributed to its current distribution and population structure. This study has also provided much-needed empirical data on the impacts of fragmentation on invertebrate species. At least in this case it would appear as though insects are not immune to the genetic and demographic consequences of habitat loss commonly observed in other taxonomic groups, viz. reductions in population size, loss of genetic diversity and increased inbreeding.

Given this greater knowledge of *Synemon plana* population structure and dynamics we are now in a much stronger position to implement effective conservation management strategies. In addition, we are better placed to begin developing quantitative models of population viability. Thus I believe this study has shown that it is possible to generate valuable, practical data for use in conservation management in circumstances where conventional ecological and demographic studies are problematic.

Although this paper has been focused on the potential uses of genetic data for inferring demographic and life-history parameters and for understanding past and recent population processes it must be stressed that genetic data alone should not be viewed as being either comprehensive or exclusive. It is only through the integration of studies from genetics, ecology, resource management, economics, politics and sociology that we can ever hope to achieve comprehensive effective long-term management of threatened species.

ACKNOWLEDGMENTS

I would like to thank Ted Edwards for providing much of the background information on *Synemon plana* life history and taxonomy. Cheryl O'Dwyer, Wendy Lee and Suellen Grosse are thanked for assistance in the field. This work has been supported by the New South Wales National Parks and Wildlife Service and Environment ACT.

13

Genetic population structure in desert bighorn sheep: implications for conservation in Arizona

GUSTAVO A. GUTIÉRREZ-ESPELETA, STEVEN T. KALINOWSKI
& PHILIP W. HEDRICK

ABSTRACT

Bighorn sheep populations have been reduced in both distribution and abundance during the last 200 years, mainly due to the introduction of new infectious disease carried by domestic livestock. Translocation efforts to historical habitat have been quite successful, but the expense of such projects, and the importance of selecting appropriate source stock, make an understanding of genetic variation within and among populations very important. Two subspecies of desert bighorn sheep are currently recognized in Arizona: *Ovis canadensis nelsoni* in northern Arizona and *O. c. mexicana* in southern Arizona. From our study of ten microsatellite loci it was found that: (1) all populations have high amounts of genetic variation, (2) populations within northern Arizona and within southern Arizona are genetically similar, (3) northern Arizona populations are genetically different from southern Arizona populations and (4) genetic distance appears to be a function of geographic distance over short distances (< 300 km) in the south-western region of the United States.

INTRODUCTION

Over the past 200 years, bighorn sheep (*Ovis canadensis* spp.) populations have been greatly reduced in both distribution and abundance. Buechner (1960) reviewed the status of this species throughout its range and reported a reduction of nearly 98% (approximately 25 000 animals left). Disease transmission from livestock is considered the most important factor contributing to the population decrease although overhunting, habitat loss and other factors have also been implicated. To rebuild populations, there has been reintroduction into the historic range and augmentation of existing populations. These translocation efforts have been quite successful, but the expense of such projects, and the importance of selecting appropriate source stock, make an understanding of the genetic variation within and between populations very important.

Two subspecies of desert bighorn sheep are currently recognized in Arizona: *Ovis canadensis nelsoni* in the north and *O. c. mexicana* in the south. These designations are based on a limited number of comparisons of skull measurements made by Cowan (1940). He posited the existence of six subspecies of bighorn sheep: Rocky Mountain (*O. c. canadensis*), California (*O. c. californiana*) and four desert subspecies: *O. c. nelsoni, O. c. mexicana, O. c. cremnobates* and *O. c. weemsi* in south-western USA. There have been re-evaluations of Cowan's work and some authors have challenged the validity of these subspecies designations. Wehausen & Ramey (1993) and Ramey (1995) examined skull morphology variation and mitochondrial DNA (mtDNA) variation and found very low genetic variation within and no significant differentiation between the four desert subspecies. They suggested that the four subspecies should be recognised as a single polytypic subspecies (*O. c. nelsoni*). Boyce *et al.* (1997) examined variation at three microsatellite loci and five major histocompatibility complex loci in populations of bighorn sheep in California and New Mexico and found a complex set of relationships between populations of *O. c. nelsoni* and other putative subspecies of bighorn sheep.

Two goals of conservation genetics are to identify evolutionary units to target for conservation and to maintain genetic diversity within such units. Microsatellite markers can contribute to both goals, perhaps better than other techniques that have been applied to conservation, such as allozyme and mtDNA analyses. Due to their high variability, high mutation rate, large number, distribution throughout the genome and codominant inheritance, microsatellite loci are now generally considered the nuclear markers of choice for molecular population genetic studies. They have been useful for measuring variation within populations (Valdes *et al.*, 1993; Bowcock *et al.*, 1994; Estoup *et al.*, 1995) and for studying evolutionary relationships in closely related taxa (Ashley & Dow, 1994). They can potentially be applied to an almost limitless number of ecological and evolutionary genetics issues (Bruford & Wayne, 1993) and some authors considered them as the most powerful Mendelian markers available (Jarne & Lagoda, 1996; Goldstein & Pollock, 1997).

In this study, ten microsatellite loci were used to characterise the genetic variation within and among populations throughout the range of desert bighorn sheep, including populations from Arizona, California and New

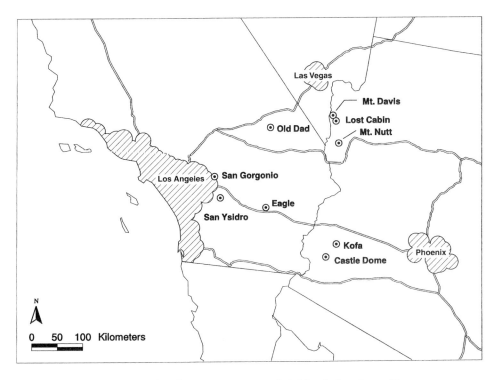

Fig. 13.1. Location of study sites in Arizona and California (locations of Stewart Mountain (AZ), Wheeler Peak (NM), Red Rock (NM) and Alberta, Canada are not shown).

Mexico. For comparative purposes, we also included two populations of Rocky Mountain bighorn sheep from Alberta, Canada. These data were used for determining interpopulation differentiation and relationships between closely related taxa. In particular, we were interested in determining whether the northern Arizona populations were genetically different from the southern Arizona populations.

METHODS

Study populations

We studied 279 bighorn sheep from 13 populations (Fig. 13.1). The location, subspecies and samples sizes for the 13 populations are shown in Table 13.1 (for further details see Gutiérrez-Espeleta *et al.*, in press). Ninety-eight blood samples and four liver or spleen tissue samples from Arizona bighorn sheep were collected by Arizona Game and Fish Department person-

Table 13.1. Study populations of bighorn sheep, current subspecies designation and sample sizes

Population	Subspecies	Sample size
Kofa Mountains, AZ	*O. c. mexicana*	9
Stewart Mountain, AZ	*O. c. mexicana*	14
Castle Dome Mountains, AZ	*O. c. mexicana*	20
Red Rock Refuge, NM	*O. c. mexicana*	25
Mount Davis, AZ	*O. c. nelsoni*	15
Lost Cabin, AZ	*O. c. nelsoni*	16
Mount Nutt, AZ	*O. c. nelsoni*	28
Old Dad Mountains, CA	*O. c. nelsoni*	23
Eagle Mountains, CA	*O. c. nelsoni*	23
San Gorgonio, CA	*O. c. nelsoni*	22
San Ysidro, CA	*O. c. cremnobates*	22
Wheeler Peak, NM	*O. c. canadensis*	7
Alberta, Canada	*O. c. canadensis*	55

nel and hunters. DNA from 122 bighorn sheep from California and New Mexico, including a population from Wheeler Peak, NM, that was derived from Rocky Mountain bighorn sheep transplanted from Banff, Alberta, Canada, was provided by W. Boyce (see Boyce *et al.*, 1997). S. Forbes (Forbes *et al.*, 1995) provided DNA from 55 Rocky Mountain sheep from Alberta, Canada.

DNA isolation of Arizona samples

DNA was isolated from blood samples, using two different methods of DNA extraction: (1) standard proteinase K digestion, followed by phenol–chloroform extraction and ethanol precipitation (Sambrook *et al.*, 1989) and (2) MasterPure Genomic DNA Purification Kit (Epicentre Technologies). Alternatively, whole blood was centrifuged at 8000 rpm for 10 min and separated in three phases and the buffy coat (white cells) was used for DNA extraction using the QIAmp Tissue Kit (Qiagen). This same kit was used to purify genomic DNA from the tissue samples.

Analysis of microsatellite polymorphism

Individuals have been genotyped with nine dinucleotide microsatellite loci (*OarFCB11*, *OarFCB128*, *OarFCB266*, *OarFCB304*, *MAF33*, *MAF36*, *MAF48*, *MAF65* and *MAF209*) characterised in domestic sheep (*Ovis aries*) (see Buchanan *et al.*, 1993; Crawford *et al.*, 1994) and one dinucleotide microsatellite locus (DS52) characterised in cattle (*Bos taurus*) (Steffen *et al.*, 1993). These microsatellite loci were selected because of their high polymorphism and high number of alleles previously detected in sheep and cattle.

Primer pairs were initially tested for amplification using a Perkin Elmer 9600 Thermocycler. PCR reactions (10 µl) contained 50 ng of purified genomic DNA, 1 X Taq buffer (50 mM KCl, 10 mM Tris-HCl), 1.2 or 3 mM MgCl$_2$, 0.2 mM each dNTP, 10 pM of each unlabelled primer, and 1 U Taq DNA polymerase. All amplifications included an initial denaturation step of 3 min at 94 °C, followed by 30 cycles of 30 s at 94 °C, 30 s at the appropriate annealing temperature (see Buchanan *et al.*, 1993; Steffen *et al.*, 1993) and 22 s at 72 °C. Final extension was for 5 min at 72 °C. PCR products were electrophoresed in 2% agarose gels and visualised after staining with ethidium bromide (1.5 µg/ml) against a standard marker (100 bp). To genotype individuals, 1 µCi of ^{32}P dATP was directly incorporated in a new 5-µl reaction volume, under identical conditions. Amplification products were mixed with 4 µl of sequencing loading buffer (95% formamide, 20mM EDTA, 0.05% bromophenol blue and 0.05% xylene cyanol), heated to 85 °C for 3 min and then put in ice for 2 min. Three microlitres of this mix were then run on 6% denaturing polyacrylamide gels. As a size marker, pBSMB-sequencing control (Perkin Elmer) was also loaded. The amplification products were electrophoresed for approximately 2.5 h and the gels were fixed by soaking in 10% methanol / 5% acetic acid for 8 min. The gels were dried under vacuum at 94 °C and exposed to X-ray film overnight at room temperature.

Data analysis

We first tested whether the data at each study site were consistent with Hardy–Weinberg proportions using GENEPOP version 3.0 (Raymond & Rousset, 1995). We tested each locus, each study site and each locus at each study site. The Bonferroni adjustment for multiple comparisons was used as the criterion for statistical significance. Next, we calculated an unbiased estimate of the gene diversity (mean expected heterozygosity) at each study site (Nei, 1987). This statistic is a measure of the amount of genetic variation present at each location and is not affected by differences in sample sizes. Confidence intervals for estimates of gene diversity were obtained using the *t*-distribution. Then we calculated F_{ST} values to measure the extent of divergence over populations (Nei, 1987). Finally, we calculated the genetic distance *D* (Nei, 1978) between each pair of study sites. Randomisation was used to test for the statistical significance of each genetic distance.

Genetic relationships between study sites were summarised with two methods. Firstly, we used PHYLIP (Felsenstein, 1993) to construct a UPGMA dendrogram of the 13 sampling sites. Bootstrapping over loci using the DISPAN software package tested the significance of the nodes in the dendrogram (Ota, 1993). Secondly, we compared the genetic distance between each pair of study sites with the geographic distance. Geographic distances were obtained from the GIS program ARCVIEW version 3.0 (Environmental Systems Research Institute, 1998). For the three study sites of transplanted sheep (Stewart Mountain, Wheeler Peak and Red Rock), we used the original location of their sheep to calculate geographic distances. We used a Mantel test (Sokal & Rohlf, 1995) to test for correlation between genetic and geographic distances.

Here we present values calculated from standard measures following the recommendation of Forbes *et al.* (1995). However, we also calculated size-based measures (Goldstein *et al.*, 1995; Slatkin, 1995) and obtained generally similar results.

RESULTS

Genetic variation within populations

Of the ten loci in the 13 populations, 98% (127 out of 130) of the locus–population combinations were polymorphic (see complete data in Gutiérrez-Espeleta *et al.*, in press), with the exception of *MAF33* in Red Rock and *OarFCB128* in Mount Nutt and Mount Davis, which were fixed for a single allele. *D5S2* had the highest observed gene diversity ($\hat{H}_o = 0.732$) and had the highest average number of alleles per population (4.38), while *OarFCB128* had the lowest gene diversity ($\hat{H}_o = 0.184$) and the lowest average number of alleles per population (2.00). *MAF65* had 10 alleles; every dinucleotide between 115 and 133 was represented in at least one population, while *OarFCB11* had just three alleles. There was no evidence of null alleles for any of the loci (consistent with Forbes *et al.*, 1995).

All populations had high amounts of genetic variation as shown by the average number of alleles and gene diversity. The average number of alleles per locus ranged from 2.4 in Red Rock to 4.4 in Alberta, with a mean of 3.4 alleles per locus. The average gene diversity was 0.51 for the 11 desert study sites, 0.57 for the two Rocky Mountain sites and 0.52 overall. It ranged from 0.36 in Red Rock to 0.63 in Eagle. Some of the alleles were unique to a single population (*MAF65-121* in Castle Dome, *OarFCB128-112* in San Ysidro, *MAF65-133* and *MAF209-111* in Old Dad and *OarFCB128-118*,

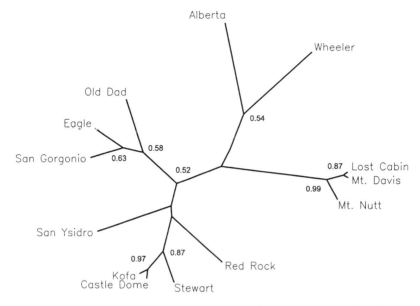

Fig. 13.2. UPGMA dendrogram based on Nei (1978) genetic distance where the numbers indicate the percentage of bootstrap replicates sharing the labelled node.

MAF48-134 and *MAF65*-119 in Alberta). None of the loci or study sites differed significantly from Hardy–Weinberg proportions.

Differences between populations and regions

We calculated F_{ST} values for the 10 loci for different regional grouping of populations. Genetic differentiation among the three northern Arizona populations and among the three southern Arizona populations was low (F_{ST} = 0.069 and 0.064 respectively); among the six Arizona populations it was greater (F_{ST} = 0.204). The F_{ST} for all desert populations was 0.267 and when all 13 populations (desert and Rocky Mountain) were combined, the F_{ST} was 0.264.

  D values between bighorn sheep populations ranged from 0.02 (Mount Davis – Lost Cabin) to 0.87 (San Ysidro – Alberta). All of genetic distances were statistically significant ($P < 0.01$) except for the comparisons Lost Cabin – Mount Davis and Kofa – Castle Dome. We used all the pairwise D values to build a dendrogram (Fig. 13.2), which summarises genetic relationships between study sites. The numbers shown at the nodes of the dendrogram estimate the probability of obtaining the indicated clusters of study sites if the study was repeated with 10 randomly chosen loci. As can

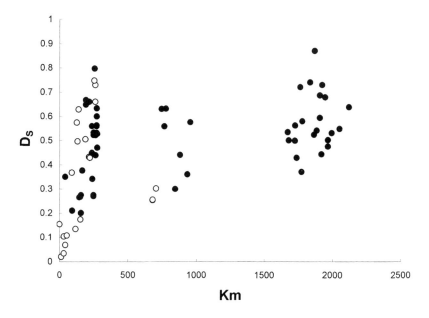

Fig. 13.3. Nei genetic distance (D_S) plotted against geographic distance (km) where comparisons between and within currently accepted subspecies of bighorn sheep are indicated by filled and open symbols, respectively.

be seen from this figure, only two clusters of study sites received reasonable support from the data: the three northern Arizona populations (Lost Cabin, Mount Davis and Mount Nutt) and the three southern Arizona populations (Kofa, Castle Dome and Stewart). These two clusters are composed of neighbouring locations (see Fig. 13.1). The dendrogram clustered together the two Rocky Mountain bighorn sheep populations (Alberta and Wheeler) and it is also consistent with a metapopulation structure for populations in the Mojave Desert (San Gorgonio, Eagle and Old Dad) (Boyce *et al.*, 1997).

We also made comparisons among regional grouping of populations using D values. When we compared distances between northern Arizona samples, D was 0.094 while between southern Arizona samples, D was 0.162. When northern and southern Arizona samples were compared, the average pairwise D value was 0.644. The two highest D values were obtained when southern Arizona was compared to Alberta ($D = 0.668$) and northern Arizona was compared to San Ysidro ($D = 0.786$).

In general, the magnitude of genetic distance between populations increased with geographic distance (Fig. 13.3). The relationship is roughly linear for distances up to about 300 km, then it appears to 'plateau' with D

values between 0.25 and 0.75 for populations > 300 km apart. Focusing on the pairs of populations between 50 and 300 km apart, this figure shows no relationship between genetic distance and currently recognised subspecies.

DISCUSSION

Genetic variation within populations

In general, all populations studied had high amounts of microsatellite genetic variation in terms of their gene diversity and average number of alleles per locus. The Red Rock population had the lowest average number of alleles and gene diversity, while the populations at Eagle and Alberta had the highest values. The Red Rock population is a large captive herd that was derived primarily from animals in the San Andres Mountains (NM), while the Eagle population is of the Mojave Desert where the populations are heterogeneous and they may belong to several different metapopulations (Boyce *et al.*, 1997).

These results are in apparent contrast with Ramey (1995) and Jesup & Ramey (1995), who found low mtDNA nucleotide diversity and low heterozygosities for allozymes, respectively, in bighorn sheep. Microsatellite loci, because they have a relatively high mutation rate, appear to provide more resolution than mtDNA. In addition, because mtDNA is maternally inherited and is haploid, the effective population size determining genetic drift is half the female effective population size.

Differences between populations

F_{ST} values were quite different for comparisons within and across regions. The low F_{ST} values within northern Arizona and within southern Arizona indicate that they are genetically similar. Furthermore, allele frequencies were very similar among adjacent populations separated by short distances (e.g. Castle Dome and Kofa), indicating that there are not extrinsic barriers to gene flow between them. The F_{ST} for all Arizona populations was greater, indicating that northern Arizona populations are genetically different from southern Arizona populations and that there was substantial subdivision of genetic variability among these populations. These high F_{ST} values between northern and southern Arizona populations are largely due to alleles present in one or a few populations and absent in others. The most common allele in northern Arizona was not always the most common allele in the southern Arizona. This indicates that neutral forces such as genetic drift have caused substantial differentiation between northern and southern Arizona populations.

We found a positive correlation between genetic and geographic distance (Fig. 13.3). Genetic distances were relatively low for nearest-neighbour comparisons (e.g. Mount Davis – Lost Cabin), and values tended to increase with increasing geographic distance up to about 300 km. Beyond this geographic distance, D values remained in the range of 0.25 to 0.75. Perhaps constraints on allele size will cause genetic distance measures to plateau, with the level of the plateau being determined by the degree of constraint, the mutation rate and population size (Nauta & Weissing, 1996; Feldman *et al.*, 1997).

Conservation genetics of desert bighorn sheep from Arizona

Our results are consistent with a metapopulation structure both for the three northern and for the three southern populations in Arizona. This indicates that the three populations in northern Arizona and the three populations in southern Arizona form discrete groups with relatively high gene flow within them. On the other hand, both F_{ST} and D values between the three northern Arizona populations and the three southern Arizona populations were much larger than the values within each region. This implies there has been low gene flow between the populations in northern and southern Arizona. From a conservation genetics perspective (Hedrick & Miller, 1992), populations should be managed so that enough genetic variability is retained to provide for future adaptation and successful expansion of native and reintroduced free-ranging populations. Because considerable local differentiation has been detected in northern and southern Arizona populations, we suggest managers follow the recommendation of Wehausen (1991), that reintroduction stock should come from nearby populations to preserve potential local variation and/or adaptations.

ACKNOWLEDGMENTS

We thank the Arizona Game and Fish Department, especially R. Lee, W. Boyce, and S. Forbes for providing samples and K. M. Parker for technical advice. This study was supported by a Heritage grant from the Arizona Game and Fish Department.

PART III

Plant case studies

There has been generally less research on the response of plants to habitat loss and fragmentation, at either the community or population level, than there has been on animals. This is surprising as the sessile growth habit of plants disposes them to direct and immediate changes in population size and structure that might be expected to influence both demographic and genetic population processes. Furthermore plants themselves often represent habitat, so effects on them provide avenues of secondary influence of fragmentation on other organisms.

Early work by Levenson (1981) and colleagues on temperate forest fragments, and subsequently by Kapos (1989) in the tropics, emphasised the influence of fragment size on species diversity and the importance of microclimate edges and their impact on population demography (see Murcia, 1995 for a review). This began to provide useful management guidelines in terms of minimum patch sizes for maintenance of plant community structure. This work was extended by Menges (1991a, b) and others (e.g. Lamont et al., 1993; Aizen & Feinsinger, 1994a, b) to direct examination of effects of population size on fecundity.

More recently, the research focus has shifted to assessment of the genetic implications of fragmentation. This has principally involved assessing the importance of genetic erosion and inbreeding in reducing fitness and population viability (e.g. van Treuren et al., 1991; Young et al., 1993; Prober & Brown, 1994; Oostermeijer et al., 1994a). Results are now available from a diverse range of species (see Young et al., 1996 for a review), and it has become clear that commonalities of response are hard to pick, other than a generally positive relationship between population size and genetic variation. For example some species, such as *Gentiana pneumonanthe* (Raijmann et al., 1994), show fairly dramatic reductions in cross-fertilisation rates in small populations. Others, such as the tree *Metrosideros excelsa*, show no detectable increases in inbreeding despite elimination of two of its major bird pollinators from forest fragments (Schmidt-Adam *et al.*, in press).

There are several possible reasons that effects of habitat fragmentation are so varied for plants. One is the variety and complexity of plant reproductive strategies encompassing as they do both sexual reproduction, either by outcrossing or selfing (or commonly a flexible combination of the two – mixed mating), and asexual reproduction by a number of mechanisms, e.g. production of rhizomes or stolons. Combined with large differences in population sizes and geographic distribution among species this means that pre-fragmentation genetic structures are often very different, even among quite closely related congeners (e.g. Karron, 1991). Some breeding systems also directly constrain both genetic and demographic responses to reduced population size. For example genetically controlled self-incompatibility systems, which are common within the angiosperms, provide a link between genetic diversity and fecundity as populations with low *S* allele diversity can experience severe mate limitation (DeMauro, 1993).

The reliance of many species on animals for pollination and seed dispersal further complicates matters, as post-fragmentation patterns of mating and gene flow, as well as levels of demographic connectedness among populations, will depend largely on the responses of the animals involved. Variations among insect pollinators in behavioural responses to fragmentation, especially regarding 'gap-crossing' ability, are now well documented (Powell & Powell, 1987; Aizen & Feinsinger, 1994*a*, *b*).

The fact that gene flow is mediated by movement of both haploid pollen and diploid seed is also important, as the relative role of these two may differ owing to variation in the genetic contribution they represent, their relative numbers and differences in their dispersal curves. Significant variation in the dynamics of seed- and pollen-mediated gene flow among populations of *Silene alba* has been detected (McCauley, 1997*b*). Seeds also represent a method of temporal genetic transfer within populations which is generally unavailable to animals, with long-lived seed banks providing potential genetic reservoirs that may be realised after disturbance (Levin, 1990). However, seed banks can also promote overlap of generations that may reduce effective population sizes.

Finally, polyploidy is very common within plants. This has significant implications regarding rates of genetic erosion due to founder effects and genetic drift. The effect of these processes is largely determined by the genetic sample size and so the higher number of gene copies per individual in autopolyploids is expected to reduce genetic losses in small populations (Bever & Felber, 1992). Loss of heterozygosity is also expected to be less in polyploids than for equivalent-sized diploid populations, and indeed allopolyploids may well benefit from fixed heterozygosity even when losses of alleles at constituent loci are severe. Even when heterozygosity is reduced, the occurrence of partial heterozygotes means that inbreeding depression may well be lower under a partial-dominance model of inbreeding depression (Husband & Schemske, 1997). Conversely, this very protection from inbreeding depression also reduces the potential for purging of genetic load in polyploid species.

The following section of this book contains six case studies covering a diversity of plant species with varied distributions, ecologies and histories of habitat loss and fragmentation. The studies employ a variety of tools to quantify the impacts of fragmentation on genetic and demographic processes and how these affect population viability. The first two case studies, by Kelly, Ladley, Robertson & Norton (Chapter 14) on the mistletoe *Peraxilla tetrapetala* and by Whelan, Ayre, England, Llorens & Beynon (Chapter 15) on a range of *Grevillea* species, examine effects of fragmentation on reproductive biology. The second two studies, by Richards (Chapter 16) on *Silene alba* and White & Boshier (Chapter 17) on *Swietenia humilis*, use genetic markers to examine the dynamics of interpopulation gene flow and its influence on population genetic structure. In the last two case studies, on *Gentiana pneumonanthe* by Oostermeijer (Chapter 18) and *Rutidosis leptorrhynchoides* by Young, Brown, Murray, Thrall & Miller (Chapter 19), demographic simulation models are used to explore the importance of inbreeding in the first case and mate limitation in the second for determining long-term population viability.

14

Limited forest fragmentation improves reproduction in the declining New Zealand mistletoe *Peraxilla tetrapetala* (Loranthaceae)

DAVE KELLY, JENNY J. LADLEY, ALASTAIR W. ROBERTSON
& DAVID A. NORTON

ABSTRACT

Fragmentation may disrupt mutualisms such as pollination or dispersal, adding indirect negative effects on native plant species to the direct effects of habitat loss. However the effect of fragmentation on mutualisms has been studied only rarely. Here we show that a limited degree of fragmentation improves reproduction in the endemic mistletoe *Peraxilla tetrapetala* (Loranthaceae) in New Zealand.

P. tetrapetala has declined since European settlement 150 years ago; the decline has been attributed partly to weakened pollination and dispersal mutualisms. The decline of native honeyeaters (Aves: Meliphagidae) has caused strong pollen-limitation for *P. tetrapetala* at some sites. A native lepidopteran, *Zelleria maculata*, also limits reproduction by destroying more than half the flower buds in some populations.

Here we report that flower predation by *Z. maculata* decreased and bird pollination increased with fragmentation over four sites at Lake Ohau, South Island. Flower predation decreased from 48% in continuous forest to 8% on isolated trees. Pollination was lowest in forest (14% seed set) and highest on isolated trees (45%). Fruit set therefore increased 4.4-fold with fragmentation. Plant density was also 2-3 times higher on fragment edges. Dispersal was good at all sites.

Therefore, *P. tetrapetala* seems to benefit from the forest edges created by fragmentation, provided that enough forest habitat survives to maintain bird densities. High levels of fragmentation beyond those measured here could possibly result in abrupt failures in the mutualisms. The benefits of moderate levels of fragmentation may partially offset declines in mistletoe numbers from habitat loss and introduced herbivores, which means that small fragments may still be of high value for mistletoe conservation.

INTRODUCTION

Habitat fragmentation has a number of direct and indirect effects on native plants and animals. Direct effects include the removal of vegetation (which alters the size of habitat patches), and physical changes associated with edges such as altered light, humidity and wind (which alter the nature of the patches). Indirect effects include alterations in the interactions between organisms, such as altered risks of predation, or disruptions of mutualisms. In this paper we study the effects of fragmentation on pollination and dispersal mutualisms.

There has been little work on the effect of fragmentation on mutualisms and how this affects the persistence of plant populations. The widely cited work of Aizen & Feinsinger (1994a, b) showed that in an Argentinian dry forest, increasing habitat fragmentation led to decreasing pollination rates (median decrease = 20%) in a range of plant species, due to a number of factors affecting both pollen quantity and quality. Fragmentation appeared to favour the introduced honeybee (*Apis mellifera*) and decrease visits by native pollinators.

Native mistletoes of the genus *Peraxilla* (Loranthaceae) in New Zealand provide a unique opportunity to test the effects of fragmentation on pollination and dispersal. Since European settlement of New Zealand around 1840, all six species of endemic loranthaceous mistletoes have declined in numbers (de Lange & Norton, 1997). The declines have been attributed to herbivory by introduced Australian brushtail possums (*Trichosurus vulpecula*) and destruction of forest habitat for farming. Native forest cover has been reduced from 78% in pre-human times to 23% today (Atkinson & Cameron, 1993). Habitat clearance has lead to an overall decline in the distribution of the mistletoes, and extant populations have been reduced in size (de Lange & Norton, 1997). Perhaps the most graphic example of the effect of habitat loss is the extinction of *Trilepidea adamsii* (Loranthaceae), which may have disappeared primarily because of forest clearance (Norton, 1991), although pollination may also have been involved (Ladley & Kelly, 1995). The introduced brushtail possum has also negatively affected the extant species of mistletoe, by browsing adult plants (Wilson, 1984; Ogle & Wilson, 1985), although at some sites little damage to mistletoes is evident (Owen & Norton, 1995).

However, there has been no work on how habitat fragmentation affects the reproductive processes and population sizes of New Zealand mistletoes. The Loranthaceae is a large family (*c.* 75 genera and 900 species) with a predominantly southern hemisphere distribution (Barlow *et al.*, 1989).

Many mistletoes prefer high-light environments such as edges for germination and growth, both worldwide (Kuijt, 1964) and in Australia (Norton et al., 1995), and anecdotal and distributional information suggests the same may be true in New Zealand (Norton & Reid, 1997). Kuijt (1964) also suggests that forest fragmentation may increase edges and thereby benefit birds, which could benefit mistletoes. However, loranthaceous mistletoes are less common along corridors in Western Australia, which Norton et al. (1995) attributed to reduced numbers of mutualist birds in these corridors. Therefore the net effect of fragmentation on mistletoe abundance is very dependent upon changes in reproductive mutualisms.

Also, *Peraxilla* species have very exacting requirements for reproduction. They rely on native honeyeater birds to open and pollinate their flowers (Ladley & Kelly, 1995; Kelly et al., 1996; Ladley et al., 1997). Without birds, flower buds do not open and very little seed is produced. Several species of small native solitary bees (*Hylaeus agilis* and *Leioproctus* sp.) can open the flower buds and thus act as pollen vectors, but these are less important than birds (Kelly et al., 1996). At several sites on the mainland of New Zealand, bird visits are too infrequent for sufficient pollen transfer, and seed production is much lower (5%–30%) than is achieved through hand pollination (50%–80% seed set: Robertson et al., 1999). *Peraxilla* species also rely on the same native birds for seed dispersal, and without bird dispersal the seeds cannot germinate (Ladley & Kelly, 1996). Thus, changes in bird density or behaviour because of fragmentation could have a large effect on reproduction in *Peraxilla*.

Peraxilla fruit production is also affected by the predation of flower buds. Native caterpillars of *Zelleria maculata* (Yponomeutidae) eat out the interior of *Peraxilla* spp. flower buds (Patrick & Dugdale, 1997), and buds attacked by *Z. maculata* almost never produce seeds. At seven sites throughout the South Island in the 1997/8 season, *P. tetrapetala* lost from 8% to 44% of buds to *Z. maculata* (Crowfoot, 1998). Therefore the two main determinants of seed production at a site are the level of pollination and the extent of bud predation.

The aim of this study was to measure the effect of forest fragmentation on the reproductive biology of the mistletoe *Peraxilla tetrapetala* at Lake Ohau. This site was chosen for four reasons: (1) *P. tetrapetala* is consistently pollen-limited at the site (Robertson et al., 1999); (2) it grows on a single host species (*Nothofagus solandri* var. *cliffortioides*) which is the sole canopy tree there; (3) the previously continuous forest cover has been fragmented into various sized remnants; and (4) *P. tetrapetala* occurs in many of these forest remnants. Robertson et al. (1999) considered that because of declines in native pollinating birds, these mistletoe populations could well be seed-limited. We set out to measure how the density of mistletoe plants per unit area, the rates of pollination and bud predation, the overall fruit-set rate per flower, and the rate of dispersal of ripe fruits varied among four remnants differing in degree of fragmentation.

METHODS

Four mountain beech (*Nothofagus solandri* var. *cliffortioides*) forest patches at Lake Ohau, central South Island, were chosen for comparison (Fig. 14.1). The most intact of the four forest sites, Temple Stream North Branch (NBT), was located within a 690-ha block of continuous forest, about 600 m from the forest edge. The Round Bush (RND) site was a 5.3-ha intact forest fragment, while Parsons Creek (PAR) was on the edge of a 100-m wide 6.5-ha riparian strip of forest containing many gaps and fallen trees. The final study site (ISO) consisted of three isolated groups of three to seven free-standing trees in pasture along the edge of Lake Ohau. The four patches were ranked by degree of fragmentation according to patch size, and how exposed the study mistletoes were to edge effects. At Temple Stream, all of the mistletoes were on host trees well within the interior of the forest. At Round Bush, two-thirds of the mistletoes were in the interior of the forest, and the other one-third were within 5 m of the edge of the patch. At Parsons Creek, all the mistletoes were on hosts on the edge of the patch of forest; the Isolated mistletoes were on single host trees surrounded on all sides by pasture.

The main pollinating birds in this area are bellbirds (*Anthornis melanura*, Meliphagidae), which along with silvereyes (*Zosterops lateralis*, Zosteropidae) are also the principal dispersers of *Peraxilla tetrapetala* (Ladley & Kelly, 1996; Ladley et al., 1997).

In each forest patch, five variables were measured: mistletoe density, rate of predation of flower buds by *Z. maculata*, pollination rates, fruit-set rates, and fruit-dispersal rates. The density of *P. tetrapetala* plants at the different sites was measured by setting up a number of 20 × 20 m plots (Table 14.1). All mistletoes within these plots were mapped and their volume (m³) measured; the total number and trunk diameter of host trees was also recorded. At each site a smaller number of *P. tetrapetala* plants (Table 14.1) were tagged and used to measure the pollination rate, fruit set and rate of insect attack on flower buds. Flowering commenced in early December 1997. On each tagged plant a branch that had approximately 100 flowers on it was marked, and the exact number of flowers was recorded, along with

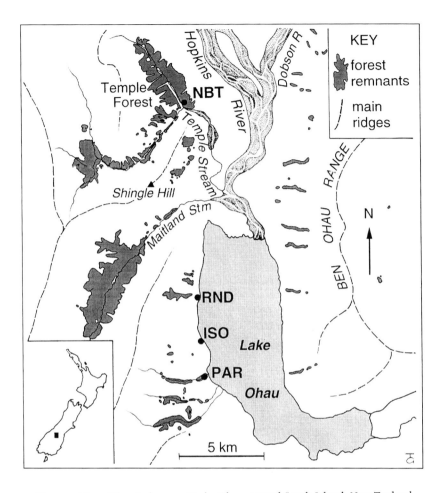

Fig. 14.1. Map of the study area at Lake Ohau, central South Island, New Zealand. Remaining forest areas are shaded, and the four fragments used in this study are labelled. NBT, Temple Stream North Branch (TCP was adjacent); RND, Round Bush; ISO, isolated plants; PAR, Parsons Creek. The inset shows the location of the main map within the South Island.

the number of flowers that had been attacked by *Z. maculata*. Approximately three months after flowering in March 1998 the number of ripening fruits was counted on the tagged branches. Like most of the Loranthaceae, *P. tetrapetala* has single-seeded fruits which are easily scored in the field for success or failure. From these numbers we calculated: the rate of *Z. maculata* attack (number of insect-attacked flowers, divided by the total

Table 14.1. Locations of the study sites, and number of *Peraxilla tetrapetala* mistletoe plants and plots used for the various analyses, at Lake Ohau, 1997/98 flowering season and 1998/99 fruiting season

Site	Longitude E	Latitude S	Altitude (m)	Number of plants, pollination[a]	Number of plants, predation[a]	Number of plants, dispersal[a]	Number of 20 × 20 m plots
Temple (NBT/TCP)	169° 48.9'	44° 06.4'	700	10	10	10	4
Round Bush (RND)	169° 49.0'	44° 12.6'	540	43	46	10	6
Parsons Creek (PAR)	169° 49.2'	44° 15.0'	540	10	8	9	4
Isolated plants (ISO)	169° 49.4'	44° 14.4'	540	7	8	7	3

[a]Not all mistletoes in each plot were used for pollination, predation and dispersal measurements.

number of flowers); the pollination rate (the number of fruits set, divided by the number of flower buds not attacked by Z. *maculata*); and the overall fruit-set rate (number of ripening fruits divided by the total number of flowers).

Dispersal measurements were carried out in February–July 1999. The same four patches were used, except that the Temple Stream site (TCP), while still within the large Temple Forest, was located closer to the edge of the forest due to logistic problems in monitoring the NBT site. At each site branches were tagged on seven to 10 plants and fruits on the branches were recorded on seven occasions at three-week intervals. At each date fruits were classified as unripe, ripe, or overripe (withered: see Ladley & Kelly, 1996). To measure dispersal efficiency, for each plant we summed the number of ripe and overripe fruits seen over all dates through the season, and calculated the overripe fruits as a percentage of all ripe and overripe fruits. With more efficient dispersal, ripe fruits would be removed before they turn overripe (which takes four to six weeks), and the overripe percentage would be lower.

RESULTS

The trend was for density of P. *tetrapetala* plants to be higher at more fragmented sites (Fig. 14.2a). The greatest densities of P. *tetrapetala* mistletoes were recorded in the edge habitats (Isolated and Parsons Creek) with 2- to 3-fold more mistletoes than were located in larger forest patches (Temple Stream and Round Bush). The same result was found for mistletoe volume divided by basal area of host trees per plot, so the result was due largely to changes in mistletoes per host, not hosts per plot. The confidence intervals were wider for the Isolated site, because there were fewest plots at this site (Table 14.1), and perhaps also because of stochastic factors in mistletoe colonisation of isolated trees.

Flower bud predation by Z. *maculata* was markedly lower at the more fragmented sites (Fig. 14.2b). The greatest rate of flower bud destruction by Z. *maculata* larvae was recorded at the continuous forest site (Temple Stream), and there were very low levels of predation at the two edge habitats (Parsons Creek and Isolated).

Pollination rates in unattacked flowers were higher in the more fragmented patches (Fig. 14.2c). The variance was higher for isolated plants, and this cannot wholly be attributed to a smaller sample size, so the exact location of the isolated plants in relation to movement patterns by pollinating birds may be important for adequate pollination.

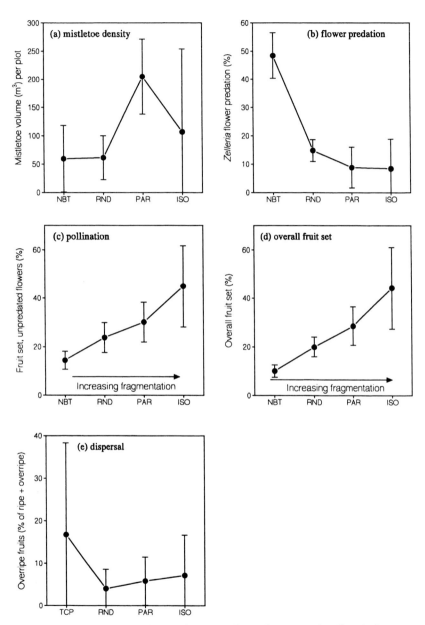

Fig. 14.2. Reproduction of *Peraxilla tetrapetala* mistletoes at Lake Ohau in four forest sites varying in their degree of fragmentation from undisturbed forest (NBT/TCP) to single isolated trees (ISO). (a) Density of adult mistletoes in 20 × 20 m plots (m³ per plot). (b) Percentage of flower buds eaten by the native moth *Zelleria maculata*. (c) Percentage of non-*Z. maculata*-attacked flowers which set seed (a measure of pollinator visitation rates). (d) Overall chance of a flower producing a seed. (e) Rates of dispersal (percentage of ripe and overripe fruits through the season which were overripe). Error bars are 95% confidence intervals; sample sizes are given in Table 14.1.

Because smaller fragments had both a lower rate of flower predation and a higher rate of pollination, the overall fruit-set rate (Fig. 14.2d) was very much higher at fragmented sites (Isolated, mean 44.9%, and Parsons Creek, 28.6%) than in continuous forest (Temple Stream, 10.0%). Combined with the higher density of adult mistletoes on edges (Fig. 14.2a), there will be very much greater numbers of seed produced per unit area of forest on the edges of forest and in small fragments than in undisturbed continuous forest, even though *P. tetrapetala* grows successfully throughout the upper canopy of the tall intact forest at Temple Stream North Branch.

Dispersal service was good at all sites (Fig. 14.2e). The overripe-to-total-ripe ratio was less than 17% at all sites and showed no clear trend with fragmentation. This suggests that fruit removal by birds normally occurs before fruits have withered.

DISCUSSION

Zelleria maculata, the native moth that consumes the flower buds of *Peraxilla tetrapetala*, was less common at more fragmented sites. A parallel study in the same area showed that this was also true at the individual plant level, with mistletoe plants located on edges having lower rates of *Z. maculata* attack than plants located in the interior of patches (Crowfoot, 1998). There may be various reasons why *Z. maculata* attacks more flowers inside forest patches, such as more favourable pupation sites (e.g. Montllor & Bernays, 1993), lower predation by parasitoid wasps (e.g. Rausher, 1979), or preferential oviposition by the female moths, in forest interiors. In any case, the effect is that mistletoe seed set is higher on edges, apparently because the native moth is itself affected by the creation of edges during fragmentation.

Pollination is also more effective on edges, which almost certainly means that bird visits per flower are more frequent there. Previous work suggests that fruit set is essentially determined by the number of pollen grains delivered by animal visitors (Robertson *et al.*, 1999). Therefore, it seems that bird visits per flower were highest on the isolated plants, and lowest in intact forest. The more frequent visits to edge mistletoes could be due to a range of aspects of bird behaviour including: greater accessibility to birds of flowers on forest edges; local increases in the relative attractiveness of mistletoes in predominantly pasture areas with few other nectar-producing plants; or, if mistletoes flower more profusely on edges, preferential foraging by birds in areas with higher densities of flowers. The mistletoes produce showy bright red flowers which are visible from great distances (> 1 km) across open country. Aizen & Feinsinger (1994*b*) also found higher visitation rates in isolated fragments for two hummingbird-pollinated plants in Argentina. At Lake Ohau, the main limit to seed set is pollinator availability (Robertson *et al.*, 1999), and thus enhanced attention by birds to edge mistletoes increases the seed set there.

With lower predation and higher pollination, the chance of a *P. tetrapetala* flower ripening a seed is much higher on edges. Also, adult mistletoes are much more abundant on edges. It is possible that more successful flowering on edges in the past has resulted in a locally elevated seed source for establishing more mistletoes on the edges. However, other factors apart from elevated flowering have probably independently increased the density of mistletoes on edges. Firstly, seeds are more likely to be deposited on an edge host tree if the dispersing birds (bellbirds and silvereyes at Ohau) favour edges (see above). Dispersal did not vary greatly between fragmented and less-fragmented sites, but this was because dispersal was efficient at all sites. We were unable to measure where dispersed seeds were deposited after passing through the bird, and this could still show a bias towards edges if birds roost or perch there. Second, higher light conditions may favour the establishment of seeds, or subsequent growth of *P. tetrapetala*. Overseas studies have often shown better germination or growth of mistletoes in higher light (Kuijt, 1964; Lamont, 1983; Norton *et al.*, 1995). The New Zealand species appear to be relatively unfussy in their general germination requirements, apart from a requirement for removal of the exocarp (Ladley & Kelly, 1996), but have not been tested for light effects. Establishment success of the closely related mistletoe *Alepis flavida* is known to vary with branch size (Norton & Ladley, 1998), but an unpublished experiment shows little effect of light levels on establishment in *A. flavida* or *P. tetrapetala* (D. Norton & J. Ladley, personal communication). There are no published data on growth rates of these species in relation to light intensity but it is possible that edge mistletoes have higher growth rates and therefore will be larger after a given time interval.

Our results are limited to the scale of fragmentation that we studied. Even our most isolated host trees were no more than 2–3 km from other patches of forest containing mistletoes, and birds readily move over these distances. However, if fragmentation proceeded, then fragments might be too far apart to sustain bird populations (or the birds might be unable to move out to the mistletoes from more intact areas where populations are sustained). In those cases, then we would expect a collapse of the mutualism at very high levels of fragmentation, especially since exotic bird species which are more likely to inhabit highly fragmented landscapes (Williams & Karl, 1996) rarely serve as effective pollinators for *Peraxilla* (Ladley *et al.*, 1997). While exotic birds may serve as useful dispersers of mistletoes (Keast, 1958), it is pollination which is limiting reproduction in the New Zealand species (Robertson *et al.*, 1999).

Also, our results are limited to a short time-scale. We have shown that mistletoes on the edges of forest patches produce more seeds per flower, but these mistletoes cannot persist long term without a supply of young regenerating mountain beech trees. Edges and clearings generally favour *Nothofagus solandri* regeneration (Wardle, 1984), but at most of our sites beech regeneration is limited by fires, and grazing by sheep and cattle, and mistletoes seem to be largely establishing on large, old, even moribund trees. In the longer term, without effective beech regeneration, the mistletoe edge population cannot persist even if there is abundant seed produced on edges.

CONCLUSION

A moderate level of forest fragmentation appears to benefit this declining native mistletoe, principally by creating more edge habitat. While fragmentation may sometimes serve to disrupt mutualisms, worsening the performance of the surviving native species (e.g. Aizen & Feinsinger, 1994*b*), clearly this will not always be the case. In our study, fragmentation improved mistletoe reproduction by reducing flower predation by a native moth, and concentrating visits by native pollinating birds, on edges and exposed trees. However, this may partly be an artefact of the extensive changes in the New Zealand biota, with greatly reduced densities of the pollinating birds due to introduced mammalian predators like rats, cats and stoats (Atkinson & Cameron, 1993). Had these predators not reduced densities of bellbirds in the Ohau area, there might have been enough birds to adequately pollinate the *P. tetrapetala* plants at all sites, including in the intact forest. Concentrating the mistletoes on to easily accessible edges may be beneficial now mainly because it gives the highest possible fruit set with the current low densities of surviving birds.

Also, around sites like Parsons Creek and Isolated, the past loss of *P. tetrapetala* forest habitat (> 95%; Fig. 14.1) exceeds the increase in density and reproduction on edges (7- to 9-fold). Consequently, there would still be more mistletoe seeds produced per square kilometre in intact forest than in this mosaic of small fragments and pasture. But given that the forest was cleared in the past, enhanced reproduction of mistletoes in small fragments means that even small remnants can be valuable habitat for the conservation of these endemic plants. The higher seed production per flower on edges partially compensates for the loss of forest habitat and herbivore damage by possums, thus enhancing the survival prospects of the mistletoes. However, extensive regeneration of *Nothofagus solandri* will be required for the long-term persistence of these endemic mistletoe populations at Lake Ohau.

ACKNOWLEDGEMENTS

We thank Lisa Crowfoot, Deirdre Hart and Laura Sessions for help with field work, Laura Sessions and C. F. Williams for helpful comments on a draft, Tim Galloway for drawing the map, and the Public Good Science Fund for support.

15

Ecology and genetics of *Grevillea* (Proteaceae): implications for conservation of fragmented populations

ROBERT J. WHELAN, DAVID J. AYRE, PHILIP R. ENGLAND,
TANYA LLORENS & FIONA BEYNON

ABSTRACT

Predicting the genetic and evolutionary consequences of disturbances such as population fragmentation requires an understanding of pollinator activities within and among population fragments and of the complex interactions of pollen transfer with plant breeding systems. We have been studying these issues using a set of *Grevillea* species which are visited by a wide range of potential pollinators including the introduced honeybee *Apis mellifera*. This series of case studies reveals the following points. There are striking differences between the levels of self-compatibility (estimated from hand-pollination experiments) and realised mating systems (based on genetic analyses), even within species. For a self-compatible species (*Grevillea macleayana*), the realised mating system varies among populations from random mating to almost complete selfing. In outcrossing species (e.g. *G. mucronulata*, *G. sphacelata*), allozyme analyses suggest that outcrossing must have been almost exclusively restricted to pollen exchange with immediate neighbours.

Patterns of pollinator visitation can explain some of this variation. *Apis mellifera* acts as a pollen thief in *G. macleayana*, reducing the level of seed set to below that observed following autogamy (all pollinators excluded).

Patterns of genetic subdivision do not always match predictions based on estimates of current mating systems. In *G. macleayana*, genetic variation [determined using randomly amplified polymorphic DNA (RAPD) markers] was partitioned within and among populations of established plants as expected for a highly outcrossed species, despite the fact that realised mating systems appear to be predominantly selfing. In contrast, for *G. caleyi* plants (which are in an environment recently fragmented by urban subdivision), amplified fragment length polymorphism (AFLP) markers revealed substantial genetic subdivision among population fragments.

The combination of mate choice and a diverse seed bank may buffer populations against loss of genetic variation. Although *G. macleayana* is fully self-compatible, this and other *Grevillea* species have very low levels of seed set and a great potential for mate choice. Pre- and post-zygotic selection therefore have the potential to maintain levels of heterozygosity even within populations in which there is a high level of selfing. There is some evidence for mate choice among flowers in *G. macleayana* inflorescences.

Taken together, the studies reviewed here imply that powerful molecular tools will be needed to understand the evolutionary consequences of variation in mating systems and gene flow, to determine the fine-scale movement of genes within populations and to determine the outcome of mate choice under the more realistic circumstances of natural pollination. We have developed a set of microsatellite markers which shows promise of widespread application within the genus *Grevillea*, and which can be used with DNA from pollen, seed or leaf tissue.

INTRODUCTION

Many potential detrimental consequences of small population size and isolation have been described for fragmented populations of plants, including increased probability of local extinction through stochastic demographic processes, reduced pollinator activities, reduced reproductive success and increased inbreeding (see Ellstrand & Elam, 1993; Frankel *et al.*, 1995; Walker & Bawa, 1996). Decreases in both long- and short-term population viability and average plant fitness are generally expected to result from these demographic and/or genetic effects (Soulé, 1976; Lacy, 1987; Lande, 1988; Frankham, 1996). It is becoming increasingly apparent that the consequences of fragmentation are dependent upon complex interactions of demographic and genetic variables.

The potential genetic effects and the consequent fitness effects have been explored to some extent in the literature, but they require more extensive empirical investigation (see Barrett & Harder, 1996). A recent review by Young *et al.* (1996) of studies of genetic consequences of habitat fragmentation revealed that genetic variability tends to decrease with reduced remnant population size, but not all fragmentation necessarily leads to genetic losses.

Demographic variation within and among population fragments can have direct effects on reproduction and recruitment. Skewed sex ratios, for example, may severely restrict reproductive output if, in a given generation, a population has more males than females. Moreover, skewed sex ratios or variation in reproductive output should have clear genetic consequences since they will reduce the genetically effective size of populations and will increase the rate of loss of variation through genetic drift and inbreeding depression (Wright, 1969; Frankham, 1995b, c).

Plant life histories that include canopy- or soil-stored seed banks add considerable complexity because the stored seed bank will not only buffer the population from disruptions to established plants but it will also buffer against reductions in effective population size by spreading genes across generations (Templeton & Levin, 1979; Falconer & Mackay, 1996).

An integrated combination of animal ecology (altered pollinator activities), plant reproductive biology (factors determining seed set and recruitment) and genetics (patterns of genetic diversity within and between populations, and inbreeding) is required for a thorough understanding of the consequences of declining population size and increased isolation. Effective management to ensure conservation rests on many uncertainties. How are pollinator types and activities affected by reduced population size and isolation? How does this affect patterns of pollen movement within and among remnant fragments of plants? Are there genetic consequences for the plants (e.g. reduced outcrossing)? Does the plant's breeding system buffer a population against, or make it susceptible to, genetic changes? What are the fitness consequences of small population size and/or isolation (reduced reproductive success and/or reduced seed and seedling fitness)? These sorts of questions clearly demonstrate that ecological and genetical processes are interlinked and therefore need to be studied together.

In this field, several techniques are relatively well established, allowing some questions to be addressed readily. For example, several demographic and modelling studies, as well as experimental field studies of breeding systems, are described in this volume. Some genetic techniques, for example allozyme studies, have also been used for some time to give effective answers to questions about the implications of population fragmentation (e.g. Young & Brown, 1999; Young et al., 1999). Recent advances in DNA technology provide an opportunity for much more powerful and/or fine-scale assessment of questions at the interface of ecology and genetics. These techniques are being developed and refined as they are being used to answer questions about the genetics of plant populations.

For some time, we have been studying the ecology and genetics of a range of species of Proteaceae in south-eastern Australia (e.g. Whelan & Goldingay, 1986, 1989; Ayre & Whelan, 1989; Carthew, 1993). There is substantial variability among species in many ecological and genetic features of the family Proteaceae, and a number of challenges to genetic studies:

- there is a range of pollinator syndromes including native and introduced insects, birds and mammals (Collins & Rebelo, 1987; Goldingay *et al.*, 1987, 1991)
- species are predominantly long-lived perennial woody shrubs and trees (Johnson & Briggs, 1975)
- there is a range of fire responses (Whelan, 1995; Bond & van Wilgen, 1996), including resprouting from stems, resprouting from lignotubers and obligate seeding
- there is a range of seed storage and dormancy strategies, with some species storing seeds in woody fruits in the canopy (e.g. Bradstock & Myerscough, 1981; Whelan *et al.*, 1998), some with soil-stored seed banks (e.g. Edwards & Whelan, 1995; Vaughton, 1998) and some with no apparent dormancy
- low fruit: flower ratios are common and widespread across the family, and these can be as extreme as 0.1% (e.g. *Banksia menziesii*: Collins & Rebelo, 1987)
- some species within the family have offered real challenges to traditional (allozyme) approaches to genetic work, because of low levels of allozyme polymorphism (e.g. Scott, 1980; Carthew *et al.*, 1988).

The family Proteaceae is an important element in the Australian flora and it is species-rich in many communities. Many species of Proteaceae feature on the current listing of *Rare or threatened Australian plants* (Briggs & Leigh, 1996) and significant numbers of species are listed in the New South Wales Threatened Species Conservation Act as 'endangered' or 'vulnerable' (43 species). Many of these species occur in population fragments that represent a small subset of their former ranges, and fragmentation is frequently identified as a factor contributing to the risk of extinction (e.g. Lamont *et al.*, 1993; Rossetto *et al.*, 1995; Hogbin *et al.*, 1998).

The purpose of this paper is to use a set of case studies, focusing mostly on our own work on species in the genus *Grevillea* (Proteaceae), to illustrate the importance of the integration of plant and animal ecology, plant reproductive biology and increasingly refined genetic studies in attempts to understand the actual and potential consequences of fragmentation and isolation of plant populations. In our own work, this integration has been a progression from field ecological studies in isolation to an integration of ecological experiments with genetic studies using sensitive molecular markers.

THE GENUS *GREVILLEA*

Grevillea is the largest genus in the family Proteaceae, containing over 250 species, mostly Australian endemics (McGillivray, 1993). This is a good case-study genus for examining the importance of the interaction between pollination and mating systems in determining the responses of populations to fragmentation and isolation. The genus is widespread in all states of Australia. There is a wide range of flower sizes and shapes across species, implicating a range of different animals in pollination, from insects to birds and mammals (McGillivray, 1993).

Flowers are typically presented in inflorescences, and it is usual to find only a small proportion of flowers producing fruits. Individual flowers appear to be protandrous, and the pollen is deposited from the anthers on to a 'pollen-presenter' just prior to flower opening. Inflorescence structure and pollen presentation on the style end provide ample opportunity for self-pollination, and McGillivray (1993) suggested that self-compatibility may be widespread in the genus – though this hypothesis is largely untested.

Of the total of just over 250 species of *Grevillea* in Australia, 163 are listed among the *Rare or threatened Australian plants* (Briggs & Leigh, 1996). Those species which occur in urban or agricultural landscapes are increasingly being fragmented into smaller and more isolated populations, well illustrated by the plight of *G. caleyi* in the Sydney region (Auld & Scott, 1996; see below).

In New South Wales, 18 *Grevillea* species are currently listed as endangered or vulnerable under the Threatened Species Conservation Act, and a further 13 are considered to be 'rare or threatened' (Briggs & Leigh, 1996). The Threatened Species Conservation Act requires that Recovery Plans are put in place for all listed species. Many draft Recovery Plans are identifying a need for 'genetic research' (Peakall & Sydes, 1996) in order to be able to design effective recovery actions. We argue that the potential genetic effects of population fragmentation and isolation, and the consequences of these for population viability, will depend upon complex interactions between pollination systems and plant mating systems. The genus *Grevillea* offers an ideal opportunity for detailed studies of these complex interactions.

POLLINATORS AND POLLINATION

For non-autogamous species, reproductive success may be directly linked to pollinator activities. Pollination may be compromised in disturbed areas, either because pollinator numbers and/or behaviour are altered as a result

of small population size or isolation, or because introduced species interfere with natural pollination systems (see examples in Weller, 1994).
Lamont *et al.* (1993) found that reproductive success in *Banksia goodii* was
correlated with the size of the small, remnant populations of this endangered species, and this pattern was interpreted to be due to a lack of bird
pollinators in the small populations. Aizen & Feinsinger (1994*a*) found
that small size of forest fragments in Argentina was associated with fewer
visits by native pollinators to flowers but a greater number of visits by introduced honeybees.

A range of pollinator species has been observed visiting species of
Grevillea in eastern Australia, including native insects and introduced
honeybees as well as a range of honeyeater (Aves: Meliphagidae) species
(Taylor & Whelan, 1988; Vaughton, 1996; Richardson *et al.*, in press; F.
Beynon, unpublished data).

We found that pollinator activity on *G. caleyi* may vary dramatically with
population size (B. Stewart & T. Llorens, unpublished data) and should
favour increased biparental inbreeding in small populations. In observations of large population fragments (>1000 individuals), 55% and 58% of
inter-plant foraging movements by birds and honeybees respectively were
to distant plants (>20 m distant), compared with only 27% and 31% in
small fragments (<40 individuals). However, in contrast with Aizen &
Feinsinger's (1994*a*) study, we found that the frequency of visits by honeybees was six times greater in large rather than small fragments.

Because *Grevillea* species rely on the services of pollinators to achieve
any exchange of genetic material, a key issue in their pollination ecology is
the determination of the effects of activities of various potential pollinators
on the reproductive success (and subsequent fitness) of plants in fragmented habitats.

The relative effectiveness of birds and insects as flower visitors varies
among species, with insects (including the introduced honeybees) often
contacting pollen presenters in the small 'spider-flowered' species (e.g. *G.
sphacelata*: Richardson *et al.*, in press) but typically 'stealing' nectar and/
or pollen in the species with larger flowers, without contacting pollen
presenters or stigmas (e.g. *G. macleayana*,[1] *G. mucronulata* and
G. × gaudichaudii) (Taylor & Whelan, 1988; Vaughton, 1996; Richardson *et
al.*, in press).

---

[1] *Grevillea barklyana* ssp. *macleayana* is the subspecies that is the subject of the studies
referred to in this paper. Recent taxonomic reviews (Olde & Marriott, 1994, 1995; Makinson, 1999) have elevated the subspecies to species status, thus producing the change of
name from *G. barklyana* to *G. macleayana*.

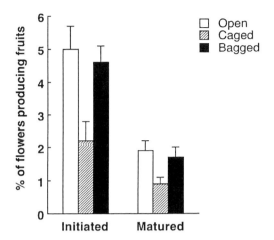

Fig. 15.1. Percentage of flowers of *Grevillea macleayana* initiating and maturing fruits with three levels of pollinator access. 'Open' allows access to all potential pollinators, 'Caged' denies access to vertebrates but allows insects (predominantly honeybees) to visit, 'Bagged' excludes all animals. For both fruit initiation and fruit maturation, fruit set in caged inflorescences (insect access) was significantly lower than in the other two treatments, which did not differ from each other. Data from Vaughton (1996).

In *G. macleayana*, one of the bird-pollinated species, insects (observed to be mostly the introduced honeybee) interfere with the natural pollination system by reducing the amount of pollen available. Vaughton (1996) found that fruit set on inflorescences from which birds were excluded but to which bees had access was reduced to about 50% of that on inflorescences with access for both birds and bees. Surprisingly, this level of fruit set was also significantly below that displayed by inflorescences from which all pollinators were excluded (Fig. 15.1). Thus, honeybees appear to be interfering with the potential for fruit set even through autogamy. F. Beynon (unpublished data) also found that honeybees were frequent visitors to flowers of *G. macleayana*. Their inter-plant movements differed from those of birds (more honeybee movements were short-distance), and they were less effective in depositing pollen on stigmas (Fig. 15.2). The site with the most effective pollen deposition (Honeymoon Bay) was the site with the greatest frequency of honeyeater visits to inflorescences.

Introduced honeybees were the only visitors observed to visit *G. sphacelata,* implying that they were the most important pollinators of this species which has small flowers presumed to be insect-pollinated (Richardson *et al.*, in press). A relatively high proportion of flowers was pollinated (50.3%) and, at one site, fruit set resulting from open pollination (4.1%) did

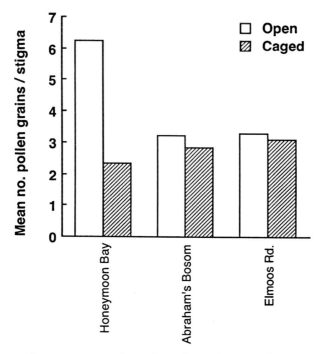

Fig. 15.2. Mean numbers of *Grevillea macleayana* pollen grains deposited per stigma in two pollinator treatments (open pollination, all pollinators with access; caged, only insects with access) at three sites. Pollen-presenters/stigmas were washed clean of pollen in the morning and pollen deposition was measured in the afternoon. The washing procedure resulted in negligible transfer of a flower's own pollen to the stigma (mean of 0.35 grains per stigma).

not differ from fruit set following hand-pollination with outcross pollen (4.7%), implying that the honeybees and honours students were equally effective pollinators, and that honeybees provided sufficient pollination to provide maximal fruit set.

In summary, fragmentation of populations can lead to an alteration of pollinators and their activities. Disruption to the normal pollination systems of some *Grevillea* species can affect pollination success and hence reproductive success. In other species, there may be no apparent effect. Whatever the outcome, it is important to ask whether such alteration to pollinator activities can also affect the genetic composition of seeds and their quality (fitness). The answer to this question would allow a more complete assessment of the potential for fragmentation to alter populations.

The genetic consequences of particular patterns of pollinator abundance and behaviour will depend on the plant's capacity to set seeds from a particular pollination (e.g. selfed, closely related, outcrossed, interpopulation). We define the type of pollination favoured by the plant's breeding system as the 'preferred mating system' because fruit set would be maximal if all pollinations were of this nature. The 'realised mating system' will be the result of the interaction between the plant's preferred mating system and the array of pollinations it actually receives.

PREFERRED MATING SYSTEMS

The degree to which altered pollinator numbers will influence reproductive success or altered pollinator behaviour in fragmented populations will have genetic consequences depends initially on a plant's breeding system (Barrett & Harder, 1996). For example, a non-autogamous species will experience lower fruit set if pollinators disappear; a self-incompatible species would suffer reduced fruit set if pollinator behaviour changed from transferring pollen mostly between plants (outcrossing) to transferring it mostly within plants (selfing). The implications of altered pollination will depend on the preferred mating system.

As indicated above (Fig. 15.1), *Grevillea macleayana* may be able to maintain fruit set even if all pollinators are excluded, indicating that this species can set seeds by autogamy. Harriss & Whelan (1993) found that fruit set in this species did not differ among four pollination treatments: open pollination, self-pollination, cross-pollination and autogamy. Likewise, Vaughton (1995) found that a self-compatibility index (ratio of fruit set on selfed inflorescences to fruit set on outcrossed inflorescences) approximated unity, indicating complete self-compatibility, in two populations: Honeymoon Bay and Abraham's Bosom (see below). Thus, alteration of pollinator abundances and activities may be expected to have little impact on fruit set in this species, unless there is direct interference with pollination as indicated for honeybees (see above).

A truly mixed mating system (Brown, 1990) of this type is not universal among *Grevillea* species; the preferred mating system varies among species and even among populations within a species. Comparing *G. mucronulata* and *G. sphacelata*, Richardson *et al.* (in press) found that neither species matured fruits in bagging treatments testing for autogamy: *G. mucronulata* produced low levels of fruit set after experimental self-pollination (between 3% and 30% of the fruit set resulting from outcrossing) while *G. sphacelata* was completely self-incompatible. Similarly, Hermanutz *et al.* (1998) found variability in self-compatibility both between species and between two populations of *G. oleoides* (Table 15.1). In the self-incompatible species,

Table 15.1. Comparison of percentage of inflorescences setting fruit after experimental hand-pollination with self and outcross pollen for five species of *Grevillea*, including two populations of *G. oleoides*

Species	Selfed	Outcrossed
G. mucronulata	14.3	48.2
G. linearifolia	3.8	69.2
G. sphacelata	9.5	60.0
G. longifolia	61.5	100.0
G. oleoides (A)	3.3	66.7
G. oleoides (B)	46.7	86.4

Data from Hermanutz *et al.* (1998).

reductions in visits by effective pollinators are expected to cause reductions in seed set.

Another dimension of the preferred mating system relates to a more realistic situation in which flowers on a plant simultaneously receive pollen from a variety of sources. This provides the potential for mate choice, which has been demonstrated for *Banksia spinulosa* by Vaughton & Carthew (1993). Harriss & Whelan (1993), Vaughton (1995) and Richardson *et al.* (in press) conducted experiments in which flowers on one half of an inflorescence (divided longitudinally) were hand-pollinated with self pollen and the flowers on the other half were hand-pollinated with outcross pollen. For a highly self-compatible species such as *G. macleayana*, a preference for outcrossing (whatever the mechanism might be) would be revealed by greater levels of fruit set in the outcrossed half of the inflorescence. No such preference is obvious in either *G. macleayana* (Vaughton, 1995) or *G. mucronulata* (Richardson *et al.*, in press). However, Harriss & Whelan (1993) found that when inflorescences were divided laterally, and the flowers on the distal half (i.e. the first flowers to open) were pollinated with outcross pollen, significantly more fruits developed in the distal, outcrossed half. No such bias occurred when the first flowers to open (distal half) were selfed and the later ones were outcrossed (Table 15.2). Where there is a preference for particular pollen, inflorescences receive pollen from many sources, and flower numbers far exceed maximal levels of fruit set, the genetic composition of the seeds produced on an inflorescence will be a non-random subset of the array of pollinations received by the flowers on an inflorescence.

To be more realistic still, future experiments will need to examine the potential for choice between different pollen sources when both arrive on the same flower. Such experiments are yet to be done with *Grevillea*, and

Table 15.2. Fruit set from selfed and outcrossed flowers on inflorescences on which half the flowers were hand-pollinated with self pollen and half with outcross pollen

Plant number	Number of flowers treated[a]		Number of fruits set		Significance[b]
	Self half	Cross half	Self half	Cross half	
(1) Proximal half selfed					
1	73	62	1	2	
2	72	49	2	0	
3	65	46	2	1	
4	65	41	0	0	
5	31	39	2	2	
6	124	112	7	5	
(2) Distal half selfed					
1	33	38	0	1	
2	106	104	2	12	*
3	41	60	0	7	**
4	52	29	1	0	
5	83	58	4	2	
6	73	56	2	0	

[a] In (1), the proximal flowers (these are the first to open and the closest to the branch) were selfed and the distal flowers were outcrossed; in (2) the selfed flowers were on the distal half of the inflorescence, i.e. the last to open (see Harriss & Whelan, 1993).
[b] Fisher's Exact test at * $P < 0.05$ and ** $P < 0.01$. A Wilcoxon paired-sample test indicated a significant trend overall towards the development of cross-fertilised seeds.

they will require genetic confirmation of the source of pollen (e.g. studies on *Banksia* by Carthew *et al.*, 1996).

REALISED MATING SYSTEMS

The interaction of pollinator activities with the preferred mating system of a plant species will produce what we refer to as the realised mating system. A population containing plants that are completely self-compatible and therefore, like *Grevillea macleayana*, produce equal numbers of fruits after autogamy, self or outcross pollination treatments, can nevertheless express a high level of outcrossing if most pollen transport is among, rather than within, plants. A highly self-incompatible plant will display a high outcrossing rate even if many pollinator movements are within plants.

The most common approach to the estimation of outcrossing rates within populations has been to use allozyme markers to detect outcrossing events. Using such an approach with *G. macleayana*, Ayre *et al.* (1994)

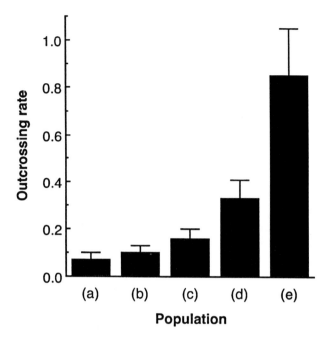

Fig. 15.3. Outcrossing rates estimated for four populations of *Grevillea macleayana*. (a) Abraham's Bosom, 1989; (b) Abraham's Bosom, 1990; (c) Elmoos Road, 1991; (d) Gravel Pit, 1991; (e) Honeymoon Bay, 1991. Data from Ayre *et al.* (1994).

found considerable variation in outcrossing rates among populations (Fig. 15.3). Although two populations (Abraham's Bosom and Elmoos Road) displayed low outcrossing rates, indistinguishable from complete selfing, a third population (Honeymoon Bay) showed a very high outcrossing rate, indistinguishable from random mating. A feature of the data collected from the Abraham's Bosom population, which had outcrossing rates approaching zero, was that individual plants typically produced selfed seeds even where they were closely adjacent to individuals with clearly different electrophoretic phenotypes (Ayre *et al.*, 1994).

In *G. mucronulata*, another bird-pollinated species displaying some self-compatibility, Richardson *et al.* (in press) found striking evidence of outcrossing but with very limited pollen dispersal. Seven homozygous maternal plants (out of a total of 20 plants) each produced progeny arrays (of 10–11 seeds) which were entirely heterozygous, indicating a high level of outcrossed matings but with a selected one (or a few) pollen donors with the alternative homozygous genotype.

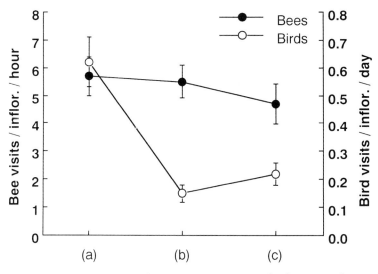

Fig. 15.4. Variation among three sites at Jervis Bay in the frequency of visits of honeybees (*Apis mellifera* – mean visits per inflorescence per hour) and honeyeaters (mean visits per inflorescence per day) to *Grevillea macleayana* inflorescences. (a) Honeymoon Bay; (b) Abraham's Bosom; (c) Elmoos Road. F. Beynon (unpublished data).

CONSEQUENCES OF HABITAT FRAGMENTATION

Pollinators and mating systems

In considering the genetic consequences of population fragmentation, we asked whether a disrupted pollination system might be causing these patterns in the realised mating system of the self-compatible *Grevillea macleayana*. F. Beynon (unpublished data) quantified honeybee and honey-eater visits to *G. macleayana* inflorescences in three of the populations surveyed by Ayre *et al.* (1994) (Fig. 15.4). Visit frequencies were greatest for honeyeaters at Honeymoon Bay, the population displaying the highest out-crossing rate (see Fig. 15.3). In contrast, bird visits were least frequent at Abraham's Bosom, the population displaying the lowest outcrossing rate (see Fig. 15.3). Higher levels of selfing may therefore be associated with lower bird activity and greater honeybee activity. Some pollinator-manipulation experiments are called for, with determination of the genetic consequences of eliminating particular groups of flower visitors.

One limitation of our studies conducted to date is the relative insensitivity of allozymes as a genetic marker in several species of *Grevillea*. Ayre *et al.* (1994) found that only four of the 25 allozyme loci tested in *G. macleayana*

Table 15.3. Numbers of seeds that were detectable outcrosses (determined by AFLP analysis) in seven *Grevillea macleayana* plants in open-pollination and bird-exclusion treatments at Abraham's Bosom, Jervis Bay in 1992

	Open access		Bees only	
Plant number	Detectable outcrosses	n^a	Detectable outcrosses	n^a
1	—	0	0	1
2	2	7	0	3
3	9	24	8	14
4	0	14	0	10
5	1	9	3	7
6	1	24	0	5
7	0	3	—	0

[a] Represents the number of seeds produced as a result of the pollination treatments.

showed any polymorphism. More sensitive markers may be needed in many systems to obtain more detailed information on patterns of mating within populations. We therefore compared AFLP profiles generated from seeds produced on flowers that were (1) open to pollination by both bees and birds, and (2) experimentally manipulated to exclude birds but not bees, in the Abraham's Bosom population (where outcrossing rates were very low and honeybee activity was relatively high).

Twenty polymorphisms were detected from five primer pairs. From a sample of seven plants, 81 seeds from open pollinated flowers and 40 from bird-excluded flowers were tested. With a relatively small and variable number of samples of seeds per plant (Table 15.3), no strong genetic effects of bird exclusion are obvious. Seeds that were apparently selfed (i.e. not detectable as outcrosses) predominated in both treatments on almost all plants and both treatments produced some outcrossed seeds on some plants. This study awaits greater resolution, with larger sample sizes of plants and seeds. In addition, the fact that AFLPs are dominant, diallelic markers reduces the power to detect outcrosses. We are continuing these studies using microsatellite markers (see below). Nevertheless, the AFLP data illustrate that honeybees were responsible for some outcrossing in this system, despite the observations that they rarely contact stigmas.

Genetic differentiation and gene flow

If recent habitat fragmentation has altered patterns of pollinator movements and consequently changed realised mating systems, the spatial distributions of genetic variation revealed in current matings (seeds) would be expected to differ from that expressed in previous generations (adult plants). We have been able to examine this by comparing recent studies of genetic differentiation among populations of adult plants of *G. macleayana* and *G. caleyi* with our realised mating system studies.

Hogbin *et al.* (1998) used randomly amplified polymorphic DNA markers (RAPDs) (see Williams *et al.*, 1990) to generate markers for the quantification of genetic variation within and among three road-verge and three non-verge populations of *G. macleayana*. In this study, three primers (which were all reproducible and readily scored) revealed sufficient variation to assign a unique phenotype to each of the 60 plants sampled. This genetic variation could be partitioned (1) among individuals within populations, (2) among populations within a location (verge vs. non-verge) and (3) between verge and non-verge sites. By far the greatest proportion of the variation (80%) occurred among individuals. Verge vs. non-verge locations explained only 3% of the variation. This is the pattern that would be expected for a highly outcrossed species with high levels of interpopulation gene flow (cf. *G. scapigera*: Rossetto *et al.*, 1995).

Taken together with the allozyme studies of *G. macleayana* populations mentioned above, these findings raise some interesting questions. The RAPD data suggest that populations of adult plants have been strongly interconnected by gene flow. In contrast, the allozyme studies revealed high levels of selfing among seed families in several populations, with highly restricted gene flow indicated in at least one site. Such a pattern of matings, if it persisted, would be expected to lead to differentiation among populations. Thus, perhaps the alteration to the realised mating systems is a relatively recent event, or perhaps a soil-stored seed bank buffers against such differentiation or selection removes inbred offspring.

Although restricted to a relatively small geographic range (McGillivray, 1993; Briggs & Leigh, 1996), the separated populations of *G. macleayana* we have studied are likely to be natural patches, of varying size, both on road verges and in relatively undisturbed bushland. In contrast, a closely related species *G. caleyi* (Makinson, 1999) is an endangered species which occurs in small population fragments adjacent to urban areas in the Sydney region (Auld & Scott, 1996). It is restricted to a particular habitat type: lateritic ridges in a dissected sandstone landscape. About 85% of the habitat has been lost to urban development. Consequently, *G. caleyi* occurs in about 21 patches of varying size, some of which must have been separated for a very long time (different ridge tops) and others have been recently dissected.

Preliminary AFLP results (T. Llorens, unpublished data) suggest that large genetic differences may exist among adult plants occurring on neighbouring ridge tops (Duffy's Forest vs. Terrey's Green vs. Belrose Quarry vs.

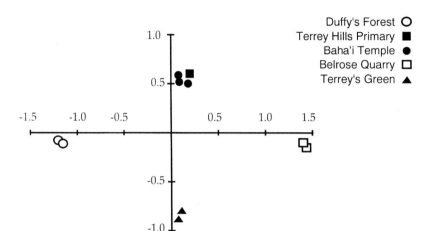

Fig. 15.5. Multidimensional scaling plot based on AFLP variation among plants within and between five *Grevillea caleyi* sites. Each point represents an individual plant. Points close together represent plants with a high degree of similarity in their AFLP profiles (T. Llorens, unpublished data).

Baha'i Temple sites; Fig. 15.5), whereas plants in fragments within a ridge top show minimal genetic differentiation (individual points within each site in Fig. 15.5). It therefore appears that the fragments within each ridge top have been more connected and will be expected to diverge genetically if they are now isolated.

There are several questions that follow from the findings of these various studies. (1) Have there been dramatic changes in the mating systems since the establishment of the current adult plants? (2) Is there a large amount of temporal variability (as well as the spatial variability already observed) in realised mating systems? (3) Combined with variability in the mating system, does a long-lived seed bank (Edwards & Whelan, 1995) accumulate genetic variability over many years, thus buffering the effects of high levels of selfing in a particular year (see Cabin, 1996)?

FUTURE STUDIES

The studies outlined in the above review reveal a number of challenges ahead. However, to explore many of these issues effectively, genetic markers that are highly variable and codominant are needed. Perhaps the most promising molecular marker is the microsatellite (Litt & Luty, 1989; Tautz, 1989; Weber & May, 1989). Of nine microsatellite loci we have identified so far in *Grevillea macleayana*, seven have proved to be polymor-

Table 15.4. Microsatellite primers developed for *Grevillea macleayana* (Gm2, Gm10, etc.) and the results of tests of their transferability to eight other *Grevillea* species

Species	Gm2	Gm10	Gm12	Gm13	Gm15	Gm18	Gm25	Gm29	Gm37
G. acanthifolia	•	•		•	•		•	•	•
G. anethifolia								•	
G. caleyi	•	•	•	•	•		•	•	•
G. floribunda									
G. glabella									
G. iaspicula	•		•	•	•	•	•	•	•
G. longifolia	•	•	•	•	•		•	•	
G. sphacelata				•	•				

phic (England *et al.*, 1999). Although there can be a considerable development time for the use of microsatellite markers in a new system, there is some potential for transferability of primers across species. This has not been explored to date in the family Proteaceae. In *Grevillea*, six out of eight other species we have tested give PCR products with one or more of the nine primers from *G. macleayana*. Eight primers generate PCR products in *G. caleyi* (Table 15.4). Utilisation of these markers will allow us to focus on both the processes that maintain heterozygosity and genetic variation (e.g. the role of a soil-stored seed bank), and the poorly understood consequences for plant fitness of variation in the source and quality of pollen transferred within and between population fragments.

ACKNOWLEDGMENTS

We thank Andrew Young and Geoff Clarke for organising the symposium and inviting this contribution. Two anonymous reviewers made comments which improved the manuscript. The Australian Research Council provided funding for projects run by RJW and DJA, an Australian Postgraduate Award to FB and an Australian Postgraduate Award (Industry) to TL. The Australasian Pollination Ecologists' Society provided TL with a student grant. The Australian Flora Foundation and the University of Wollongong provided research funding for some of the studies described in this paper. We also acknowledge the financial and other assistance of the New South Wales National Parks and Wildlife Service, especially for work on *Grevillea caleyi,* and the long-standing co-operation of the staff of Bouderee National Park (Jervis Bay) and the Department of Defence (Naval Bombing Range, Beecroft Peninsula, Jervis Bay). This is contribution number 186 from the Ecology and Genetics Group at the University of Wollongong.

16

Genetic and demographic influences on population persistence: gene flow and genetic rescue in *Silene alba*

CHRISTOPHER M. RICHARDS

ABSTRACT

While migration of individuals has been shown to increase the persistence of small isolated populations, the role that pollen-mediated gene flow may have on population viability is not known empirically. In this study, I investigate the consequences of limited gene flow in newly established colonies within a regional metapopulation of *Silene alba* (Caryophyllaceae) (Antonovics *et al.*, 1994). Given that colonisation events are frequent, but that they often fail, I focused on factors that influence colonisation success. In particular, I examined the extent of inbreeding and gene flow in isolated colonies. In addition to these empirical approaches, a simulation model was used to examine the interaction of inbreeding and demographic stochasticity on population persistence and metapopulation spatial structure. The chapter reviews theoretical and empirical studies pertinent to the viability of new established colonies and uses work related to the *S. alba* system as a case study.

OVERVIEW: SPATIAL STRUCTURE, GENETICS DEMOGRAPHY AND POPULATION ESTABLISHMENT

It is generally accepted that most plant species are not uniformly distributed over space, but are often divided into subpopulations that are interconnected to varying degrees through dispersal of pollen and seeds. Spatial subdivision, local extinction and colonisation have long been recognised as important features of natural populations by both ecologists addressing the broad issues of distribution, abundance and community structure of species and geneticists interested in localised adaptation and evolution. These two distinct approaches to a common population structure are in part a legacy of biogeographic theory advanced by MacArthur & Wilson (1967) and early work of Wright on genetic structure and adaptation (1931). Currently, research in population subdivision encompasses both of these approaches (often in a mutually exclusive fashion), yet recently there has emerged a greater emphasis on the interplay of numerical and gene frequency dynamics especially as it relates to applied conservation projects (Schemske *et al.*, 1994; Soulé & Mills, 1998).

The general acceptance of metapopulations has provided, to some degree, a conceptual framework within which to evaluate genetic and demographic interactions (reviewed by Hanski & Simberloff, 1997; and see Thrall *et al.*, Chapter 5, this volume). In these systems of interacting populations, regional stability is maintained in spite of high local population turnover. The establishment of new populations, therefore, is critical in maintaining this stable dynamic (Ebenhard, 1991). Each colonisation episode represents a progression of events, starting with dispersal and initial survival at the new site, to growth where the colony itself can act as the exporter of propagules. While establishment of new populations is characteristic of all species, the factors that affect this process are likely to vary widely among plant taxa (Lewontin, 1965; Rabinowitz, 1981). In spite of this diversity, there are many common elements in the establishment process.

In the 34 years since the symposium volume on *The genetics of colonising species* (Baker & Stebbins, 1965), much information has been collected on the characteristics associated with colonisation in plants (Harper, 1977; Brown & Marshall, 1981; Burdon & Marshall, 1981; Jain, 1983; Parsons, 1983; Rice & Jain, 1985; Barrett & Richardson, 1986; Gray *et al.*, 1987; Polans & Allard, 1989; Barrett & Husband, 1990; Sun & Corke, 1992). There have been important generalisations concerning the diversity of colonising strategies related mainly to mating system, phenotypic plasticity and reproductive allocation. Comparison of these shared attributes provides an evolutionary context within which to view the ecological and genetic qualities most likely to be important in colonisation success. However, in contrast to these studies which describe the genetic and life-history characteristics likely to be associated with the evolution of weediness or invasiveness, there have been few experimental studies that examine directly the genetic and demographic factors that affect the colonisation processes (Barrett & Husband, 1990).

It is reasonable to generalise that most colonising events are typified by small population size and some degree of isolation from neighbouring populations. In plants, when long-range dispersal is involved, seeds may be deposited in habitats that are geographically isolated and ecologically stressful. Hence, survival and growth of the founding cohort is often acted on by extreme selective forces. These may include both biotic interactions with competitive species and pathogens as well as the physical or abiotic qualities of the environment. In addition, the genetic attributes of the founding cohort may interact with the environmental conditions in ways that are not readily predicted. Thus, the conditions that typify population establishment are relevant to many broad fields of biological research including ecology, evolutionary biology and conservation biology.

Interacting influences of genetics and demography

How might genetic and demographic approaches be reconciled and how would more inclusive approaches lead to better understanding of population viability? Clearly, theoretical and empirical studies of both genetics and demography are relevant to the study of population viability, and their interactions need further exploration (Pimm *et al.*, 1989; Nunney & Campbell, 1993). The general arena where genetics and demography directly interact is the population size. The relation between the number of individuals and the genes they contain is, however, often confounded in most empirical studies because in natural systems there is often a substantial difference between the local census count (N) and the genetic effective population size (N_e). In fact, there exists no fixed relationship between N and N_e (Nunney & Elam, 1994). This pattern is especially evident in clonal organisms. The relationship between N and N_e within a population depends primarily upon the variance in lifetime reproductive success of individuals and clones. The ratio of N to N_e is reduced when there are deviations from equal sex ratio, periodic fluctuations in breeding size, changes in the frequency of self-incompatibility alleles or mating types and reproductive inequalities among individuals. These factors are largely influenced by demographic and environmental stochasticities and are particularly strong in small populations (Richter-Dyn & Goel, 1972; DeMauro, 1993; Byers, 1995). In addition, the genetic effective size of a population can be affected by among-population processes such as the source of colonisers and the frequency of subsequent gene flow or immigration – dispersal patterns that are likely to be influenced by spatial isolation.

Gene flow among founding populations, and between founding and source populations, may also greatly affect colonisation success by changing the effective population size. Gene flow influences the extent to which genetic changes in local populations are independent. The most prominent evolutionary consequence of gene flow is that it tends to homogenise population structure, acting against drift and diversifying selection. Even in continuous populations, limited gene flow distance can result in genetic differentiation (Wright, 1943). When N_e is small, it may magnify the influence of genetic and demographic stochasticity. In the absence of gene flow or migration in these small colonies, the probability that any two copies of a gene are derived from a common ancestor increases substantially with time, thus reducing within-population variation and increasing the inbreeding coefficient. Gene flow into small isolated colonies may act to increase the effective population size and prevent the increase in inbreeding depression, yet the precise relationship between inbreeding depression and gene flow in natural populations has yet to be explored experimentally or theoretically (Ellstrand & Elam, 1993).

Broad, integrative models that take into consideration the various types of stochasticities that affect the viability of a founding cohort have generally been conceptual or heuristic in nature. The first to use this approach was the model by Gilpin & Soulé (1986) who described how these numerous stochastic effects might interact and, more importantly, how their interactions may create self-reinforcing feedback loops or 'extinction vortices' that could have drastic effects on population survival (Soulé & Mills, 1998). Theoretical and empirical studies aimed at establishing a firm link between measures of genetic diversity and demographic attributes of populations have only recently emerged. In addition, recognition that these factors could interact in complex ways has resulted in modelling efforts focused on population extinction probabilities (Halley & Manasse, 1993; Mills & Smouse, 1994; Tanaka, 1997; Thrall et al., 1998). Recent work in this area promises to move the discourse away from a hierarchical perspective and instead to shift the focus on the range of possible interactions (Nunney & Campbell, 1993; Mills, 1996).

While it is clear that the long-term evolutionary potential of a population is influenced by genetic variation, its short-term impact on extinction is less clear (Franklin, 1980; Frankel & Soulé, 1981). There are several general areas where reduced genetic variation may have distinct short-term consequences including disease susceptibility, reduced growth rate due to genetic incompatibility and reductions in population fitness due to the expression of inbreeding depression. While many authors have suggested that genetic factors may only come into play long after a population has undergone significant demographic decline, species that regularly undergo colonisation and extinction events may be susceptible to genetic effects long before they undergo widespread demographic decline. Theory suggests that small population sizes have reduced genetic variation, yet the effects of population size and genetic variation (as measured in neutral characters) is complex and variable (Antonovics & Via, 1988; Karron, 1989; van Treuren *et al.*, 1991; Ouborg & van Treuren, 1994) (but see Meagher *et al.*, 1978). There are few studies that empirically examine the joint effects of population size and genetic variability on extinction, perhaps because it requires the ability to experimentally decouple genetic and demographic effects.

The most studied consequence of reduced genetic variation relates to the expression of inbreeding depression. Inbreeding depression has been implicated in extinction but the factors influencing the rate of inbreeding depression are often confounded with demographic attributes of declining populations. A generalised decrease in viability, survivorship and fecundity correlates with increased homozygosity. In angiosperms, inbreeding depression has been documented in many species (Charlesworth & Charlesworth, 1987; Dudash, 1990; Molina-Freaner & Jain, 1993; reviewed by Waser, 1993; Willis, 1993; Nason & Ellstrand, 1995; Husband & Schemske, 1996). It is reasonable to expect that without new genetic input through migration (gene flow), these reductions may negatively affect population survival, growth rate and subsequent export of propagules.

Empirical studies of genetics and demography

Small and isolated populations may, for a number of reasons, be particularly prone to reduced population viability. Recently, there has been a number of studies that have investigated the interaction of genetic and demographic factors in population viability in natural populations. Studies on the perennial herb *Silene regia* have shown significant reductions in seedling germination in small populations throughout its range and have indicated that this reduction may in fact have a genetic basis (Menges, 1991*a*; Menges & Dolan, 1998). These studies were some of the first to examine whether small population size and the resulting effects of inbreeding depression may have detrimental effects on the demography of a historically widespread plant made rare by habitat fragmentation. Similarly, seed size, seed germination and regrowth following herbivory were all reduced in small populations of *Ipomopsis aggregata* (Heschel & Paige, 1995). In addition, these fitness reductions could be substantially reduced by hand-pollination with unrelated pollen drawn from a different population. In an extensive survey of demographic and environmental conditions, Fischer & Matthies (1997, 1998) showed that individuals of *Gentianella germanica* drawn from smaller populations and grown in a common garden had lower mean fitness than those drawn from larger populations. They concluded that seed production and offspring fitness in small populations are strongly reduced by genetic effects and have negative consequences for population dynamics in the field. Newman & Pilson (1997) undertook an elegant experimental approach to understand the effects of effective population size and extinction rates in the annual *Clarkia pulchella*. Their experimental design allowed them to establish populations that were demographically equivalent but differed in their effective population size. This allowed a direct evaluation of how reductions in genetic variation due to drift and inbreeding in small populations affected population survival probabilities. The results from these studies showed that populations established with a high effective population size had a 9-fold multiplicative fitness advantage over those populations established with a low effective population size. These data show that reduced genetic variation can significantly increase the short-term population extinction rate above that expected by demographic stochasticity alone. Detailed studies of the genetic and phenotypic variation in two endangered plant species in the Netherlands, *Salvia pratensis* and *Scabiosa columbaria*, were conducted over several years to assess how the depletion of genetic variation impacts the demographic attributes of small fragmented populations (Ouborg *et al.*, 1991; van Treuren *et al.*, 1991; van Treuren, 1993; Ouborg & van Treuren, 1994). Their findings show that these two species are in genetic decline and that the amount of genetic load has not been purged in small populations. They underscore that genetic erosion of viability leaves smaller populations particularly prone to demographic stochastic extinction. In addition to these studies, the work by Oostermeijer and colleagues (1996) on the endangered *Gentiana pneumonanthe* have focused on both demographically and genetically influenced extinction probabilities (see Chapter 18, this volume).

CASE STUDY: INBREEDING AND GENE FLOW IN *SILENE ALBA*

Overview

While the previously mentioned empirical studies have examined the genetic and demographic contribution to extinction in rare plant species, the research presented in this case study focuses on metapopulation dynamics and gene flow in a common, weedy species (*Silene alba*) as a model system. The intention of the work was to develop a general framework for how the relative effects of genetics and demography might influence population establishment and, in turn, affect the regional spatial structure of a metapopulation. A great deal of the data from these studies were derived from collaborations with Janis Antonovics, David McCauley and Peter Thrall.

Given that colonisation events are frequent, but that they often fail, the studies described below focus on the factors that influence colonisation success. The overall study is partitioned into three main sections. The first part addresses whether there is appreciable genetic load in this species and whether this results in the expression of inbreeding depression in greenhouse crosses and in naturally colonised populations in the field. The second part of the study examines how the timing and magnitude of pollen-mediated gene flow into experimental colonies varies with population size and isolation. And the third part uses data from the first two studies and demographic data collected in the field to parameterise a simulation model that examines how gene flow and inbreeding depression affects meta-population spatial structure.

Silene alba as a model system

Over the past decade, our interests have focused on various aspects of the population biology of the plant *Silene alba*. This is a short-lived dioecious perennial herb, originally introduced from Europe in the mid-1800s, that has since spread throughout eastern North America and Canada (McNeill, 1977). In Giles County in south-eastern Virginia, where we have studied population dynamics and genetic structure of this system in nature, *S. alba* is found primarily along roadsides, which makes it easier to do demographic studies.

The main focus of understanding the regional dynamics of the *Silene* metapopulation has been a large-scale longitudinal survey of hundreds of natural populations of *S. alba*. This survey has taken place annually since 1988 within an area 25 × 25 km in Giles County, Virginia, where the plant is common along roadsides (Antonovics *et al.*, 1994). Within this area, populations are defined as the number of individuals in a 40-m segment of roadside or 'psilon' (*sensu* Thrall & Antonovics, 1995). Our definition of what constitutes a population was based on reasons of practicality as well as biology. Firstly, it is generally difficult to delineate population boundaries in nature except in some situations (e.g. aquatic species in systems of lakes). Secondly, 40-m segments can easily be relocated using a car odometer. Finally, seed and pollen dispersal are limited such that the scale of 40 m encompasses at least one, but not many ecological and genetic neighbourhoods (Alexander, 1990; Antonovics *et al.*, 1994).

Fertilised females produce small seeds (similar to mustard seed) in fruits containing about 300 seeds. Seeds of this species possess no obvious adaptation for dispersal and it is thought that most of the long-distance seed dispersal events are the result of human activity (road grading, mowing, etc.). Seedlings emerge in early spring, flower continuously through-out the late spring and summer and can overwinter as rosettes. Seeds typically mature by mid to late summer. The species is easily grown in the greenhouse and flowers in about six weeks under long day length; it is dioecious and easily crossed. The white flowers of both sexes open before dusk and remain open through the night, closing in late morning. The species is pollinated by a variety of generalist nocturnal and diurnal insect pollinators. In studies of pollinator foraging behaviour and fungal spore dispersal of the floral smut pathogen of *S. alba* (*Ustilago violacea*), Altizer *et al.* (1998) found that diurnal visitors were mainly bees, lepidopterans and flies, whereas nocturnal visitors were mainly noctuid, geometrid and sphingid moths. Studies using fluorescent dye markers indicated that nocturnal visitors disperse dye to more flowers than diurnal visitors (Shykoff & Bucheli, 1995). Although differences in dispersal ability are most likely due to differences in pollinator foraging behaviour, there was no direct comparison of how these different suites of pollinators moved pollen over distances of more than several metres.

The census comprises a total of 7500 segments; between 400 and 500 are occupied at any one time. The census results for the first nine years indicate a high rate of population turnover, with a substantial rate of colonisation and extinction (mean colonisation and extinction rate = 0.20) (Antonovics *et al.*, 1994; Thrall & Antonovics, 1995; C. M. Richards, unpublished data). Data from this 10-year study show that the initial size of newly colonised populations is typically five individuals or fewer. Within these colonising groups, there is no consistent sex-ratio bias (C. M. Richards, unpublished data). Most newly colonised sites are located within 80 m of an established population, although long-distance dispersal events can occur. Extinction rates appear to be 'size dependent', where the smallest newly colonised populations have a higher extinction rate than larger ones (Antonovics *et al.*, 1994). In addition, populations that are more isolated tend to have higher extinction rates (Antonovics *et al.*, 1994).

The genetic structure of the *Silene* metapopulation was first estimated from allele frequencies calculated from 12 persistent populations located within the metapopulation census area (all within 10 km of Mountain Lake). Two classes of genetic markers were used. Firstly, seven polymorphic allozyme loci were used to estimate the local geographic partitioning of nuclear genetic variance. From these data F_{ST} was estimated as 0.12, suggesting moderate genetic structure at the local level. In addition, the F_{ST} estimated from the distribution of four chloroplast DNA (cpDNA) haplotypes was 0.67, clearly much greater than that seen with the allozymes

(McCauley, 1994). These differences in the F_{ST} estimates were in part due to the higher level of gene flow expected for nuclear genes relative to chloroplast genes, since the former can also move between populations in pollen (McCauley, 1994; McCaulay *et al.*, 1995). In contrast, only a small amount of nuclear genetic structure ($F_{ST} < 0.05$) was detected among patches separated by 10s of metres, indicating near-panmixis at that spatial scale, though a high degree of genetic structure could still be detected using cpDNA markers (McCauley *et al.*, 1996). In addition, the results showed that F_{ST} for newly founded populations was much greater than for older established populations. McCauley (1994, 1997b) has hypothesised that the difference between genetic structure estimates based on nuclear and cpDNA markers underscores the importance of pollen movement relative to seed movement for gene flow, and the pattern supports the idea that most colonisation events occur as a result of a small number of founders (McCauley *et al.*, 1995).

In systems where colonisation/extinction processes are important, most local populations will be founded by small numbers of individuals. These conditions set the stage for subsequent inbreeding in two ways. When the founding event involves few unrelated individuals, their progeny produced in the first generation would be entirely outcrossed as a result of dioecy in this species, but subsequent inbreeding would arise in the following generations if there is limited gene flow. If the colonising cohort is composed of related individuals (kin structure founding), progeny in the first generation would be the product of full-sib mating. If inbreeding depression could arise immediately following colonisation when the population size is small, it may be particularly susceptible to chance extinction due to reduced effective population size (Newman & Pilson, 1997).

Inbreeding studies

The idea that immigrants from a nearby population can prevent the extinction of small populations has come to be known as the 'rescue effect' (Brown & Kodric-Brown, 1977). However, the pathways of immigration for individuals are not always identical to the pathways for the immigration of genes. In plants, immigration of genes can occur by pollen flow, and assuming pollen is not limiting seed set, this may have no direct effect on population growth through seed production. Yet where populations have become subdivided into local breeding groups, gene flow can mitigate the loss of genetic diversity brought about by genetic drift and inbreeding. Immigrants can therefore have two effects; they may increase the numerical abundance directly or indirectly because they can reduce the levels of inbreeding. I use the term 'genetic rescue' to describe the increase (over and above the purely demographic effects) in the probability of a population's survival due to immigration of genes from another population.

The necessary preconditions for genetic rescue within a metapopulation framework include the presence of inbreeding depression in naturally colonised local populations and the amelioration of this condition by pollen-mediated gene flow. To investigate the consequences of biparental inbreeding, studies were conducted using seeds collected from the metapopulation study site (Richards, 2000). Progeny from crosses made among individuals derived from seed collected from widely separated locations within the metapopulation served as a base population from which the inbred lines were derived. Experimental inbred lines were established over a three-year period, during which crosses between full and half sibs were conducted. Inbred lines were produced with inbreeding coefficients of 0.0 (outcrossed), 0.125 (half-sib mated), 0.25 (full-sib mated) and 0.375 (second-generation sib mated) determined by pedigree. These inbreeding coefficients represent reasonable values of inbreeding likely to develop within a few years after colonisation without gene flow. After the seeds from these crosses matured, a series of germination trials was initiated. Two random samples of 50 seeds were taken from each of two seed capsules for each level of inbreeding within each line. Each sample of seed from each capsule was germinated in flats with standard greenhouse research soil mix on different days over a two-month period. Seedling emergence after three weeks was scored as percentage germination.

Results from the greenhouse inbreeding depression studies showed that *S. alba* was strongly affected by repeated biparental inbreeding. The negative relationship between the inbreeding coefficient (F) and germination rate appeared largely linear. Sib-mated progeny showed up to a 60% reduction in germination compared to outcrossed seeds (Fig. 16.1). Seeds that failed to germinate were not viable using tetrazolium staining. Results of logistic regression showed that the level of inbreeding, maternal family and their interaction were all significant indicating that families differed in their response to inbreeding (Table 16.1). Ten lines were not included in the highest levels of inbreeding because they failed to flower. These data were not included in the regression. Additional studies are now under way to assess inbreeding depression in later stages of development. Results have shown that pollen viability in progeny derived from sib mating is roughly half of that derived from outcrossing (C. M. Richards, unpublished data).

In order to compare the seed germination rate within the controlled greenhouse trials to those in newly colonised populations, a series of cross-

Table 16.1. Analysis of logistic regression of inbreeding and maternal family on germination success

Source	df	Likelihood ratio χ^2	$P > \chi^2$
Inbreeding coefficient	13	3377.06	0.001
Maternal family	9	110.75	0.004
Inbreeding coefficient × family	27	290.27	<0.001

The inbreeding coefficient was used as a fixed effect and the maternal family was used as a random effect. Analysis was run on 10 maternal families that could be bred to produce four levels of inbreeding. The R^2 for the whole model was 0.33. (From Richards, 2000.)

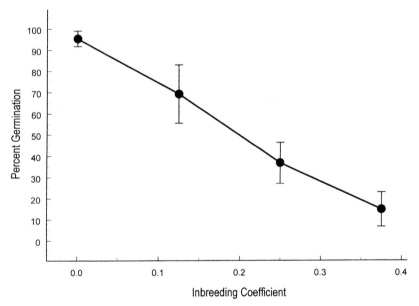

Fig. 16.1. Relationship between the germination rate and the level of inbreeding in seeds of *Silene alba* produced from greenhouse crosses (see text). Points represent the mean (± SD) germination rate for different families used in the crosses. (From Richards, 2000.)

es was initiated using field-collected progeny. These crosses were intended to investigate the degree of inbreeding in newly colonised populations (Richards, 2000). The crossing programme within and among sites was accomplished using a three-level design (Fig. 16.2). The first level crossed individuals from within a family, resulting in progeny with inbreeding co-efficients of at least 0.125 (half-sib relatedness). The second level crossed individuals between widely separated sites. These first two crosses served as

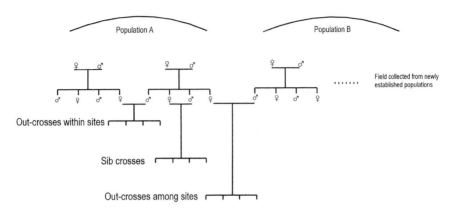

Fig. 16.2. Design of crosses of greenhouse-raised inbred *Silene alba* with field-collected seeds. Maternal sibships were collected from 10 isolated and two central newly established populations. Progeny from these maternal families were subjected to a three-level crossing design. (From Richards, 2000.)

controls (inbred and outcrossed progeny) with which to evaluate crosses made among families within a site. The crossing hierarchy was replicated within and among families collected in two location classes (isolated and central colonies).

The effects of within-family crosses on seed germination differed dramatically between central and isolated populations (Fig. 16.3). Crosses conducted within and among newly founded populations showed that rather high levels of inbreeding depression occur within the first or second generation after founding within crosses among individuals in isolated populations (where gene flow is unlikely) but little or no inbreeding depression occurs in crosses among individuals within a population located near to large established populations that may probably experience significant gene flow. In isolated populations, crosses among maternal families resulted in germination rates statistically similar to full-sib mating, indicating that populations are often composed of related individuals. The same type of crosses in central populations, however, showed germination rates consistent with outcrossing, suggesting that these populations are more often composed of unrelated individuals. Within-family crosses in isolated populations displayed a level of inbreeding depression on a par with full-sib mating (~ 60% reduction in germination rate) whereas those same crosses done in central populations showed inbreeding depression equivalent to that found in half-sib-mated progeny (~ 40% reduction in germination rate). The difference in within-family crosses suggest that central populations may have a more genetically variable pollen pool than isolated popula-

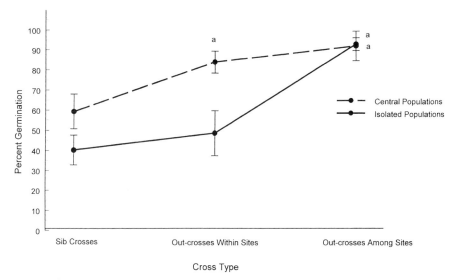

Fig. 16.3. Percentage germination rate among sites within a cross type. Isolated populations were founded at least 250 m away from the nearest established population. Central populations were founded no less than 40 m but no more than 80 m away from an established population. Each point represents the mean (±SD) germination rate from each class of population. Superscripts denote points that are statistically similar using a Tukey–Kramer HSD (Honestly Significant Difference) means comparison. (From Richards, 2000.)

tions. This may be because the local pollen donors in central populations are either unrelated or that there is a high level of gene flow coming from the nearby established populations.

The crossing experiments agree with the genetic structure data and confirm that isolated populations are inbred and are likely to be founded by related individuals. Values of F_{ST} taken from the metapopulation among recently established populations ($F_{ST} = 0.20$) correspond very closely to estimated values for F based on germination success of progeny from the among-family, within-site crosses. If colonisation (seed movement) is rare and migration (pollen movement) is common then the potential for inbreeding to occur soon after establishment is offset by genetic rescue through pollen-mediated gene flow. Crosses among isolated sites restored germination rates to levels statistically equivalent to outcrossed lines. No rescue effect was seen in these same crosses among central colonies.

Gene flow studies

Given that recurrent founding events contribute substantially to the observed genetic structure, the next part of the study focused on how gene flow mitigates the degree of genetic differentiation (Richards et al., 1999a). The movement of genes in plants is mediated by the dispersal capabilities of both seeds and pollen, gene flow agents whose distances and modes of dispersal are often quite different (Crawford, 1984; Fenster, 1991a; Ennos, 1994; McCauley, 1994, 1997b). In insect-pollinated plant species, the degree of patch isolation may be largely controlled by ecological factors influencing pollinator foraging behaviour (Levin & Kerster, 1969; reviewed by Handel, 1983). These might include the physical distance between populations, the size and geometry of the populations, the density of the floral display (Rathke, 1983; Kunin, 1993, 1997a, b; Morris, 1993; Karron et al., 1995b) and the interaction of these factors (e.g. how far and how large the nearest neighbouring source population is relative to the target population). All these factors could vary with population history.

Initial experiments were designed to measure the range of pollen dispersal from a single pollen source to a series of artificial all-female populations that varied in size and distance from the source. These pollen-dispersal experiments helped define the spatial scale at which subsequent gene flow studies were conducted. The second set of experiments measured the rate of gene flow by pollen into experimental target populations that varied in size and degree of isolation. We used a linear array of eight artificially constructed populations that contained both sexes and where alternate populations were homozygous for different allozyme alleles. The populations in these arrays were separated by either 80 m or 20 m. By harvesting the seed produced in these patches and monitoring the flowering abundance at the time of fertilisation, it was possible to obtain data on the ecological context of each gene flow event. These experiments were also used to evaluate factors that influenced seed production including population size, sex ratio and degree of isolation from neighbouring pollen sources.

In the pollen-dispersal experiments, over 85% of pollination occurred within the first 40 m from the pollen source. Using target size (two or 10 plants) and distance class (log transformed) as the main effects, and the number of fruits per female flower (arcsine transformed) as the dependent variable, the analysis of variance showed a highly significant effect of distance ($df = 6$, $P = 0.001$). There were no significant effects of patch size (number of plants) and no distance × size interaction.

The results from the gene flow arrays showed a clear difference in gene flow and fruit set between populations separated by 20 m and 80 m (Fig. 16.4–16.6). Gene exchange among populations separated by 20 m approached panmixis (subunits appeared more as patches within an interbreeding population) whereas populations separated by 80 m showed

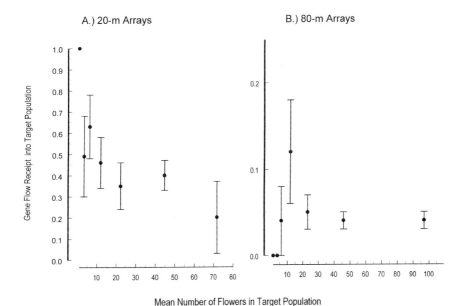

Fig. 16.4. Gene flow receipt into a target population as a function of target population size. (A) 20-m arrays; (B) 80-m arrays. Points represent means (± SE) of experimental gene flow records pooled within seven size classes of target size. Gene flow receipt measures the fraction of seed within a target size class sired from pollen-mediated gene flow into the target population. Note differences in the scale of the *y*-axis. (From Richards *et al.*, 1999*a*.)

substantially less gene flow and were clearly not panmictic. While isolation strongly influenced the rate of gene flow overall, there was a large degree of variation in gene flow into populations of similar isolation. This variance could be explained in part by the flowering abundance of the nearest source and target populations at the time of the gene flow event. Our experimental data suggest that factors which influenced gene flow at one spatial scale may not act at another. Specifically, the rates of gene flow were shown to be differentially affected by the target size and relative size at 20 m and 80 m (Fig. 16.4 and 16.5). This is most likely a consequence of pollinator foraging behaviour.

In general, pollinator visitation increases with higher floral density. In addition, pollinators visiting larger patches visit more flowers within that patch than they do in small populations. Although pollinator visitation is higher, mean flight distances are much shorter, which acts to increase the effective local pollen pool and may, in turn, dilute the migrant pollen. In small populations, where floral density and number is low, pollinators are

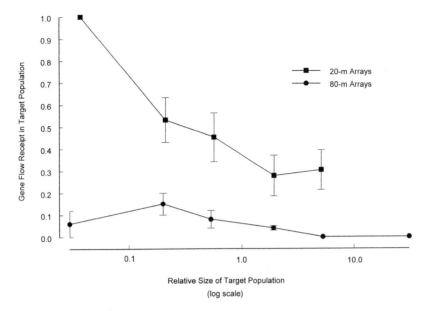

Fig. 16.5. Gene flow receipt into target population as a function of relative population size (number of flowers in target population / number of flowers in source population). Gene flow receipt measures the fraction of seed within a target size class sired from pollen-mediated gene flow into the target population. Points represent means (± SE) of experimental gene flow records pooled within six classes of relative size. The largest relative size class in the 20-m arrays set no seed and is therefore omitted. (From Richards *et al.*, 1999*a*.)

not retained to the degree they are in large populations (flight distances tend to be longer) and migrant pollen constitutes a higher relative fraction of the total pollen pool (Handel, 1983). This relationship appears to occur in our 20-m arrays, where there was a clear negative relationship between flower abundance and gene flow, though no apparent linear relationship exists among populations spaced at 80 m (Fig. 16.4A and B).

These data confirm that large populations have higher pollinator visitation (higher fruit set) and lower gene flow receipt at both spatial scales. While the smallest populations in our study showed the highest levels of gene flow in the 20-m arrays, gene flow in the smallest populations within the 80-m arrays fell to zero (Fig. 16.4). Fruit set data from these small populations show that small isolated populations are more pollen-limited than small populations that are near to other patches (Fig. 16.6). This trend has been confirmed in other species where there is a demographic cost to being isolated (Sih & Baltus, 1987; Jennersten, 1988; Groom, 1998). Gene flow into these isolated populations was maximal at some intermediate size

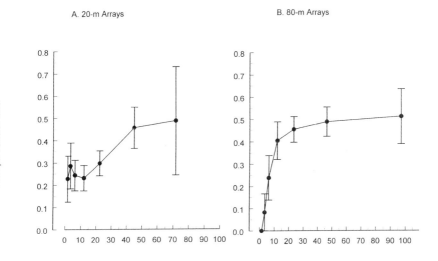

Fig. 16.6. Fruit production per female flower as a function of population size (number of flowers) for (A) 20-m and (B) 80-m arrays within the experimental gene flow populations. (From Richards *et al.*, 1999*a*.)

where pollinator visitation rates are non-zero but retention was possibly low.

Interactions

The previous studies illustrate some of the important genetic ecological features that shape the process of population establishment in *Silene alba*. While the dichotomy of genetic and demographic factors influencing population extinction serves as a useful heuristic device, it does not adequately allow for investigations aimed at their interaction. The major finding from these studies is that genetic connectedness can have a very direct demographic consequence.

Might inbreeding within newly established populations threaten the stability of entire metapopulations? In simple metapopulation models the proportion of suitable sites that are occupied depends on the relative frequencies of extinction and colonisation (Hanski & Gilpin, 1991*b*). If colonisation rates are not sufficient to offset extinction, the entire system will collapse. More realistically, colonisation could be considered to consist of two stages (Thrall *et al.*, 1998). The first stage requires that colonists make their way to an empty patch of suitable habitat. This depends largely on the rate at which surrounding demes export dispersing individuals. Once a site is occupied, the second stage of colonisation is achieved when the newly founded population grows to be large enough to itself export colonists. Only until this second phase is completed does the colonisation event increase the long-term stability of the metapopulation. The transition between these stages is directly affected by demographic and environmental uncertainty including Allee effects associated with increased isolation. In addition, inbreeding depression could reduce the probability of effective colonisation by reducing the population growth rate in the early generations of population establishment. It is possible that this local effect could cascade across an entire metapopulation. If inbreeding depression decreases the probability of successful colonisation, this should result in lowered occupancy rates which will, in turn, limit the number of emigrants from the remaining established populations. Not only would this reduce the frequency of successful colonisation events, but any reduction in the size of the colonising propagule would further reduce the rate of successful transition from stage one to stage two of colonisation, since the smaller the number of founders the greater the potential for inbreeding depression.

The empirical studies discussed earlier indicate that in *Silene alba* (1) there is a substantial amount of genetic structure at the among-population level, (2) levels of inbreeding can potentially be quite high in small and/or isolated populations and (3) there is potential for gene flow among populations to 'rescue' inbred populations from such effects to a significant extent. These results further suggest the possibility that changes in dispersal (e.g. through loss of a pollinator) could result in decreased potential for genetic rescue. This may result in local extinction, causing cascading effects leading to a radical change in spatial distribution of local demes within the regional metapopulation.

To investigate these issues further, I embarked on a collaborative study with Thrall, McCauley & Antonovics (1998) using a spatially explicit simulation model of sets of local populations that incorporate both within-population dynamics and among-population dispersal, migration and extinction processes. Although the model incorporates diploid genetics, for simplicity we assumed that all genotypes were equivalent (with the same birth and death rates) and focused on population-level processes. The goal was to illustrate the interplay between dispersal, gene flow and population-level effects of inbreeding using a phenomenological approach, rather than to make precise predictions about the dynamics of *Silene alba*. However, for realism we parameterise the model using data from prior field experiments (Thrall & Jarosz, 1994*a*, *b*; Richards *et al.*, 1999*a*), the roadside censuses

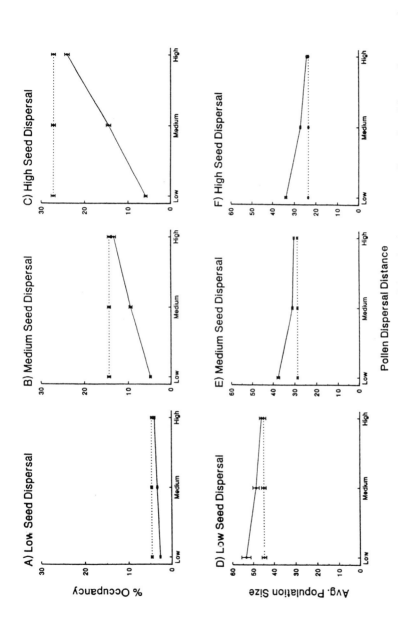

Fig. 16.7. Relationship between percentage occupancy, average population size and the scale of migration (pollen and seed dispersal) in a simulated metapopulation. Within each level of seed dispersal (low, medium, high), pollen dispersal was simulated at one of three levels (low to high). Graphs (A)–(C) show how percentage occupancy changes with increasing seed and pollen dispersal; graphs (D)–(F) show changes in average population size. For each graph, the dotted lines indicate values without inbreeding depression for comparative purposes. Data shown are the mean and standard errors of 100 random runs. (From Thrall et al., 1998.)

(Antonovics *et al.*, 1994; Thrall & Antonovics, 1995) and experimental studies of inbreeding depression and genetic rescue (Richards, 1997, 2000).

When inbreeding was included in the simulation, the percentage occupancy in the metapopulation changed dramatically for different levels of pollen and seed dispersal (Thrall *et al.*, 1998). As the pollen dispersal was reduced, the metapopulation's spatial structure collapsed from a collection of ephemeral local populations into several large, persistent core populations. When genetic rescue through pollen flow was included, percentage occupancy increased with the scale of pollen dispersal (Fig. 16.7A–C). Regardless of seed dispersal, however, at the highest level of pollen dispersal, percentage occupancy was nearly equal to that seen with no inbreeding effects. In addition, it was observed that during the simulated collapse of the metapopulation, measurable statistics that could realistically be collected by a population ecologist studying the system might not indicate that a widespread decline was occurring. For instance, when pollen dispersal was reduced and the metapopulation coalesced into several core populations, the mean population size increased (Fig. 16.7D–F). Also, the apparent colonisation rate increased (since much of the suitable habitat is unoccupied) as the core populations continued to send out large numbers of propagules. A census taken during this point of decline (without any historical census data) might appear to indicate large healthy populations with a high colonisation rate.

CONCLUSIONS

Overall, the experimental, descriptive and simulation studies show that the interacting effects of migration, demography and inbreeding depression can not only have consequences at the level of single newly colonised populations, but may also cascade to produce effects across the metapopulation as a whole. Several important points emerge. Firstly, genetic and demographic factors interact and result in dynamics that would not be predicted if one were to examine either factor individually. The fact that genetic factors may influence the demography of a common weed shows that hierarchies of relative importance (genetics vs. demography) may not be straightforward or even informative. Secondly, the simulation studies suggest that when inbreeding depression is included, subtle changes in pollinator availability can have significant effects on metapopulation structure. Detection of such cascading effects is likely to be difficult from short-term cross-sectional studies, regardless of whether these studies encompass few or many populations. These data clearly demonstrate the importance of genetic connectedness and the consequences that gene flow can have on the demographic attributes of newly founded populations. Indeed, maintaining pollinator diversity may prove to be an essential element in sound plant-conservation policies (Kearns *et al.*, 1998). Thirdly, these data underscore how the spatial and ecological context of patches influences the factors important to persistence. Landscape-level variation in the rates of migration and gene flow are important determinants of how populations respond to fragmentation and should be incorporated more often into metapopulation studies (Wiens, 1997). Specifically, models that examine dispersal and gene flow as fixed entities may have to be modified. Finally, these studies suggest the need for further work on empirical and theoretical studies to understand the consequences of inbreeding and immigration on local and regional dynamics.

ACKNOWLEDGMENTS

Thanks are due to Janis Antonovics, David McCauley and Peter Thrall for many instructive discussions, and advice concerning all aspects of this work. Also I would like to thank Matt Olson, Deb Marr and Jim Leebens-Mack for helpful improvements to the manuscript. This work was supported by a National Science Foundation grant (DEB-9610496) to D. E. McCauley and a travel award from Vanderbilt University to CMR to attend the symposium in Australia.

17

Fragmentation in Central American dry forests: genetic impacts on *Swietenia humilis* (Meliaceae)

GEMMA M. WHITE & DAVID H. BOSHIER

ABSTRACT

Human disturbance is widespread within tropical forests, with large tracts of once-continuous forest reduced in many cases to highly fragmented stands of trees. Formulation of management strategies for the conservation of tree species within such landscapes requires an understanding of the consequences of fragmentation for the levels and spatial distribution of genetic diversity and the extent of genetic connectivity between stands. The study investigated the genetic impacts of fragmentation in a population of the neotropical tree *Swietenia humilis* in a seasonally dry forest in Honduras. Microsatellite markers showed high levels of genetic variation over 10 loci, with little genetic differentiation across the sampled range. Comparisons of pollen flow within continuous forest to that within and between the fragments showed that although near-neighbour interactions predominate throughout, the proportion of long-distance pollen flow increased with a decrease in the population size of the fragments. At the extreme of isolation, a remnant tree in pasture, pollen flow up to 4.5 km was detected, with the proportion of pollen flow governed by the size of the pollen source rather than distance. Pollen donors were highly variable between trees, and influenced by both the frequency and the flowering phenology of surrounding trees, while seed production increased with an increased degree of disturbance. Fragmentation did not result in the reproductive isolation of the stands, but rather resulted in increased levels of long-distance pollen flow, possibly leading to restoration of rare alleles lost in the original fragmentation. For *S. humilis,* and other tropical tree species with similar mating systems, density and distribution, remnant stands appear to have long-term conservation value, with lone trees in pastures also acting as focal points for pollinator movement, and providing stepping-stones between forest fragments.

INTRODUCTION

Clearance of forest by humans has left the landscape severely altered, with much forest now persisting as what are often seen as 'islands' in a sea of agricultural land. The dry forests of Central America and Mexico are amongst the world's most threatened habitats ($<2\%$ remain, with 0.1% in protected reserves) (Janzen, 1986a), and typify this situation. Contrary to the views of HMS *Sulphur*'s botanist (Fig. 17.1), they consist of a mosaic of floristically rich habitats which, under human influence, have become highly fragmented and localised. The same forests are the source of many tree genera which are used, or show potential for use, both locally and worldwide for timber and as multipurpose trees (e.g. *Gliricidia, Leucaena, Prosopis*), while several other important timber species (e.g. *Astronium graveolens, Bombacopsis quinata*) are almost commercially extinct and *Swietenia humilis* is listed in Appendix II of the Convention on International Trade in Endangered Species (CITES). With general limitations on the resources to directly support conservation, incorporation of genetic conservation criteria into forest- and farm-management practices will offer, in many cases, the best prospects for achieving conservation goals. Remnant forest patches and trees in farming systems are therefore increasingly favoured for the realisation of conservation objectives *circa situm* (Kanowski & Boshier, 1997). Effective management and conservation of the natural resources within such modified environments requires an understanding of the impacts of disturbance on the species present and their interactions. Accordingly, in recent years, an increasing amount of research has focused on understanding the impacts of forest fragmentation on tropical forest environments (e.g. papers in Laurance & Bierregaard, 1997). Until recently, however, there has been little empirical evidence regarding the genetic viability of fragmented populations of tropical tree species, and the genetic value of such fragments for conservation has been the subject of speculation.

It has been postulated that only large tracts of undisturbed forest have potential conservation importance (Lande & Barrowclough, 1987; Soulé, 1987), and that the fragmentation of populations will disrupt pollinator patterns (Levin & Kerster, 1974; Levin, 1978), leading to their genetic isolation, reduction in population sizes and genetic diversity (Young et al., 1996). Curtailment of gene flow may lead to increased levels of inbreeding and subsequent losses in genetic variation (Gilpin & Soulé, 1986). Conversely, fragmentation may lead to increased levels of gene flow between fragments and hence the maintenance, or increase of, genetic diversity

Fig. 17.1. 'Our visits to the gulfs of Nicoya and Fonseca were not productive, indeed the sameness of an unbroken but dreary and profitless forest was nowhere more forcibly felt.' *The botany of the voyage of HMS* Sulphur, *1836–42.*

within populations (Hamrick, 1992; Young *et al.*, 1996). Persistence of a group of fragmented patches as a metapopulation of interacting stands would deserve consideration for their long-term conservation and management as functional, though highly modified, communities (Nason & Hamrick, 1997). Studies have revealed that levels of genetic variation are high for most tropical tree species and the majority of the variation resides within rather than among populations, suggesting extensive gene flow (e.g. Hamrick & Loveless, 1989). But the reliance of most tropical tree species on animal vectors for cross-pollination means that the extent of gene flow is largely determined by the foraging behaviour of pollinators and seed-dispersal agents. Changes in behaviour as a consequence of fragmentation could strongly affect the patterns of gene flow within remnant tree populations. Consequently some tropical tree species may be particularly vulnerable to landscape fragmentation due to both their low densities and the potential disruption to pollen and seed vector associations.

Assessment of the potential for sustainable management of the genetic resources of tree species, within such modified and fragmented environments, requires direct assessment of the effect of increased spatial isolation and reduction in population sizes on the levels and distribution of genetic diversity and on the dynamics of gene flow. Lack of variability in biochemical markers has hindered the ability to directly characterise the dynamics of gene exchange within populations. However, the development of highly variable microsatellite markers and recognition of the importance of determining the genetic consequences of fragmentation have recently led to a surge of research in this area (Chase *et al.*, 1996*a, b*; Dow & Ashley, 1996; Morgante *et al.*, 1996; Dawson *et al.*, 1997; Aldrich & Hamrick, 1998; Nason *et al.*, 1998; Dayanandan *et al.*, 1999). This chapter provides a synthesis of a study that addressed these concerns in *Swietenia humilis* (Meliaceae), using a combination of field and laboratory studies involving the development of microsatellite markers and whose results are reported in more detail in White *et al.* (1999). Specifically, the objectives were to: (1) evaluate the levels and spatial distribution of genetic variation within a fragmented population of *S. humilis*, (2) determine the patterns of gene flow, via pollen, within a fragmented population of *S. humilis* compared to that within a continuous 'control' subpopulation, (3) evaluate how flowering phenology influences mating patterns and (4) evaluate seed production under increasing degrees of disturbance.

STUDY SPECIES: BIOLOGY AND CONSERVATION STATUS

Swietenia humilis is a medium-sized deciduous tree confined mainly to the Pacific coast of Central America and Mexico. It is traditionally of less commercial significance for timber than the other two species within the genus, *S. macrophylla* (big-leaf mahogany) and *S. mahagoni* (West Indian or small-leaf mahogany)(Styles, 1981), although its heavy timber is used and marketed within local economies. *S. humilis* has a reasonably wide ecological range and is found in seasonally dry tropical forest and savanna, through a range of 0 to 1200 m altitude (Styles, 1981). The species is monoecious, with flowers that appear perfect but have either vestigial anthers or ovaries (Lee, 1967; Fig. 17.2) and under controlled pollination is shown to be self-incompatible (D. H. Boshier, unpublished data). The faintly sweet-scented small white flowers are borne as dense terminal clusters on large highly branched inflorescences, and are visited by small bees, butterflies and other insects (Styles, 1972; D. H. Boshier, unpublished data). Trees flower generally for about a month during the dry season, although the precise period varies throughout the natural distribution. The fruit, which take almost a year to mature, are erect strongly wooded ovoid capsules containing winged seeds approximately 8 cm long (Styles, 1972). The capsules ripen in Central America between January and March, with the valves dropping off as they dry, leaving the seeds exposed to dispersal by the wind.

Populations of *S. humilis* have been reduced and fragmented over much

Fig. 17.2. External and cross-section views of male and female flowers in *Swietenia humilis*, showing (A) and (B) vestigial anthers, (C) and (D) vestigial ovary.

of its range, leaving small forest remnants and scattered, isolated individuals among agricultural crops, in pastures, and along the borders of fields. It therefore provides an ideal species in which to examine levels of genetic diversity and gene flow within such fragmented populations and so assess the potential of these fragments for the conservation of *S. humilis*. Susceptibility to shoot-tip-borer attack (*Hypsipyla grandella*) has limited its use in plantations (Newton *et al.*, 1996), and the concern for its conservation has been such for it to become, along with *S. mahagoni*, the first hardwood trees of commercial significance to be listed under CITES.

STUDY SITE

The study was carried out mainly in the Punta Ratón region of the Honduran Pacific alluvial coastal plains, by the Gulf of Fonseca (Fig. 17.1), and lying within the native range of *Swietenia humilis*. Fragments of mainly secondary forest of varying size are left on a number of small hills. The plains and estuaries surrounding the hills are principally occupied by cattle pasture, shrimp farms, sugar-cane, cantaloups and water-melons, although some trees (mainly *Albizia niopoides, A. saman, Bombacopsis quinata, Ceiba pentandra, Enterolobium cyclocarpum, S. humilis, Tabebuia neochrysantha* and *T. rosea*) are also maintained in pasture. The study area includes the largest remnant of dry forest in the Punta Ratón region (Cerro Las Tablas, total area approximately 500 ha), with a predominance of the above species and *Bursera simaruba, Cochlospermum vitifolium* and *Lysiloma* spp. Aerial photography from 1954 (31 January 1954, Instituto Geografico Hondureño) shows the hillside to be covered predominantly by low regeneration following evident clearance at some time prior to this date. Some remnant mature trees of *S. humilis* are visible within this regeneration, as are occasional pockets of what was possibly the original dry forest cover on the steepest slopes. Recent aerial photography shows that more minor clearing of the regeneration for cultivation has occurred in small patches from time to time. Currently human intervention of this type is still evident, as is the cutting and sawing on site of *Bombacopsis quinata* and *S. humilis* trees. Ageing by ring counts of disks cut from felled *S. humilis* trees showed mature trees ranging from approximately 20 to 75 years old.

The study sites were selected on the basis of size and degree of separation of fragments, with all *S. humilis* trees located, surveyed and mapped. The area comprised: two remnant stands of secondary dry deciduous forest principally confined to hillside areas, Jiote ($n = 22$) and Butus/Jicarito ($n = 44$); one stand of remnant trees in pasture, Tablas Plains ($n = 7$); and an isolated tree, 501, located at the most westerly point of *S. humilis* distribution within the area (Fig. 17.3). A plot of 97 trees, Las Tablas, situated within an approximately 68-ha stand of the Cerro Las Tablas forest, was used as a control of continuous forest. The *S. humilis* trees in the other fragments are similarly remnants of previous forest cover and some newer regeneration. The 1954 aerial photographs, however, show the surrounding plains to have been forested at that stage, with extensive deforestation likely to have followed the agrarian reforms and cattle expansion of the early 1960s, such that the Tablas Plains site is exclusively made up of remnant trees, as is tree 501.

GENETIC DIVERSITY AND GENE FLOW

Plant material: collection and evaluation

Leaf material was collected from all mapped trees, and stored (methodology in White *et al.*, 1999). Bulked progeny arrays from each of 17 trees selected across the control and fragments (five in Las Tablas, two in Tablas Plains, three in Jiote, six in Butus/Jicarito and tree 501) were collected and stored (methodology in White *et al.*, 1999). Total genomic DNA was extracted from dried leaves or germinated seedling leaves, with genetic diversity assessed for 10 microsatellite loci isolated from an enriched genomic library.

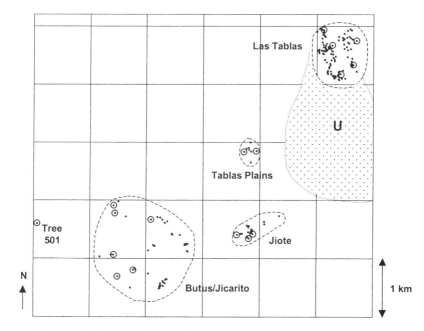

Fig. 17.3. The locations of the *Swietenia humilis* trees sampled at Punta Ratón, Choluteca, Honduras. Each tree is represented by a dot; the 17 trees selected for the progeny analysis are circled. The location of the unsampled trees (U) adjacent to the Las Tablas site is shown by the spotted enclosed area.

Isolation, PCR conditions, electrophoresis and scoring details are described in White & Powell (1997). To study gene flow 30 seeds per tree were genotyped for four of the microsatellite loci, selected for their high levels of allelic polymorphism and relatively high frequency of rare alleles distributed throughout the study area: *mac38* (26 alleles), *mac45* (11 alleles), *mac49* (21 alleles) and *mac58* (13 alleles). The paternity of seeds within individual capsules was also studied for four or five capsules in four trees across the population. Inheritance of the loci was confirmed by evaluating progeny produced under controlled pollination.

Data analysis

Estimates of genetic diversity were made using the POPGENE program (Yeh *et al.*, 1997). Deviations from Hardy–Weinberg expectations, genotypic linkage equilibria and the inbreeding estimator F_{IS} were determined using GENEPOP version 3.1b (Rousset & Raymond, 1997). Genetic differentiation under the stepwise model of microsatellite evolution

(Slatkin, 1995) was assessed by an estimator of R_{ST}, called ρ, using the RSTCALC 2.2 program (Goodman, 1997). Indirect estimates of the number of migrants per generation (N_m) between fragments were calculated by $N_m = 1/4[(1/\rho) - 1]$. Slatkin's original derivation assumes populations of equal size, with all loci having equivalent variances, which in natural populations is generally not the case. When data sets contain loci with widely differing variances, loci with low variances contribute little to the final R_{ST}, even when they show a high degree of differentiation. RSTCALC deals with both sources of bias, by initially globally standardising the data set to take into account loci that may have widely differing variances. Bootstrapping (Effron, 1979; Van Dongen, 1995) was used to assign 95% confidence intervals to ρ and N_m, with significance testing following Lynch & Crease (1990). Within-population genetic structure was evaluated using co-ancestry (f_{ij}) autocorrelation analysis within the control subpopulation Las Tablas, using the programs f_{ij}Anal and BS-f_{ij} (see Loiselle et al., 1995).

Direct measures of pollen flow were made using a fractional paternity programme (Devlin et al., 1988). A constraint of sampling was an area of unsampled trees adjacent to the Las Tablas site, giving an area of unknown potential pollen donors between Las Tablas and the remnant stands (Fig. 17.3, area U), so that pollen-donor distances could only be accurately characterised up to this area. The frequency of progeny arising from pollen donors potentially within the area U was summed for each fragment and plotted as coming from a distance equal to or greater than the maximum specified limit. Maximum distance limits for pollen-flow characterisation were 0.6 km for Las Tablas, 0.9 km for Tablas Plains, 1.5 km for Jiote, 3.6 km for Butus/Jicarito and 4.5 km for tree 501. The exclusion power of the microsatellite markers gave 45% of parent–offspring matches with 100% certainty of paternal assignment. The frequency of the definite paternal donors was determined for 300-m distance categories from the maternal tree, calculated as the number of definites against the number of definites + the number of progeny in that array for which the pollen source could not have been the father. The data were plotted, together with the frequency of trees within each distance category, generating direct pollen-flow distance frequency histograms for each fragment of forest (Fig. 17.4). A subset of possible fathers was listed for the remaining progeny from which the minimum distance of pollen flow was inferred. This latter method underestimates the true distance of gene exchange, although the distance frequency curves from both minimum and direct data show similar patterns of gene flow within and between the fragments. In both data sets, there is also the possibility that external gametes that mimic local ones

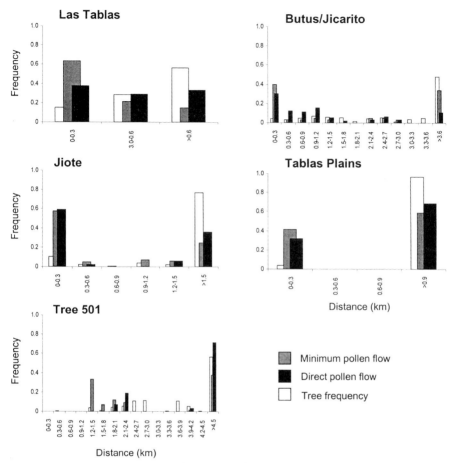

Fig. 17.4. Distance frequency histograms of pollen flow to selected *Swietenia humilis* trees in each forest fragment, Punta Ratón, Choluteca, Honduras. The frequency of paternal donors every 300 m was calculated from the pollen-flow distance data as estimated from the paternity-exclusion analysis. The histograms are a representation of the proportion of mapped trees within each distance category.

are misidentified, such that progeny identified as possible with a given set of fathers may in fact have been sired from outside that same set (Adams & Birkes, 1991). Therefore the pollen flow identified may be an underestimate of the true patterns within such a population.

Genetic diversity
All 10 loci were highly polymorphic (3–26 alleles per locus) with high levels of genetic variation, typical of a self-incompatible species, in all fragments

Table 17.1. Number of low-frequency alleles (<0.05, 10 microsatellite loci) for *Swietenia humilis* in the Las Tablas control forest and their presence in neighbouring forest fragments

	Control: Las Tablas	Fragments Butus/Jicarito	Jiote	Tablas Plains
Size of population $(n)^a$	98	44	22	7
Total low-frequency alleles	54	29	19	8
Percentage	100	53.7	35.2	14.8

^{a}Population size given as number of trees.

(mean $H_e = 0.548$). Intrafragment genetic diversity was high, with the mean number of alleles per locus (A) ranging from 4.2 (Tablas Plains) to 9.5 (Las Tablas) and the proportion of polymorphic loci from 90% to 100%. Initial effects of fragmentation were seen through the loss of low-frequency alleles and the number of alleles per locus declining in conjunction with reductions in fragment population size. The overall percentage of rare alleles (frequency <0.05) across the 10 loci varies within each fragment. Butus/Jicarito, the largest fragment, has 53.7% of the rare alleles found in Las Tablas, while Tablas Plains, the smallest, has only 14.8% (Table 17.1). The level of heterozygosity detected in the Tablas Plains fragment ($H_o = 0.472$, $H_e = 0.509$) was slightly less, although not significantly so in comparison to the other larger stands ($H_o = 0.490$, $H_e = 0.543$). Heterozygosity is most affected by middle-frequency alleles, whereas it is the rare alleles that small populations are most likely to lose immediately, from a reduction in the representation of the original gene pool (Taggart *et al.*, 1990). Indeed, Tablas Plains has lost across the 10 loci nearly all of the rare alleles found in the control forest, Las Tablas, and the proportions of rare alleles in Jiote and Butus/Jicarito were considerably reduced in comparison to those found in the control (Table 17.1). The size of such a founder effect will depend on the extent of forest loss, the degree to which it is random and its coincidence with any fine-scale genetic structure. An immediate loss in heterozygosity would however only be expected if population sizes were very reduced.

A high proportion of loci (0.4) also showed significant departures from Hardy–Weinberg equilibrium with associated significant levels of F_{IS}, although there were inconsistencies between fragments in the loci showing departures. Deviations from Hardy–Weinberg expectations, due to non-random mating or inbreeding within the fragments, might be expected across all loci. Selective advantage of certain loci, linked with local adaptation, may influence the dominance of particular alleles within the population, and/or the presence of null alleles may contribute to significant high levels of homozygosity. The latter is, however, unlikely owing to the non-detection of any null alleles in loci exhibiting high overall F_{IS} values in the segregating progeny of a controlled cross. Rather, the presence of local genetic structure over short distances within a stand (see below) may contribute to higher levels of homozygosity, and hence significant positive F_{IS} values (Wahlund effect) at particular loci. Any local genetic structure may be accentuated as a result of both tree loss at fragmentation and post-fragmentation regeneration within the stands.

Genetic differentiation and measures of gene flow

As for other studies of genetic variation in tropical tree species made over similar distances to those in this study (e.g. Loveless, 1992; Schierenbeck *et al.*, 1997), most variation was found within fragments (96.8%), giving high indirect measures of gene flow ($N_m = 8.93$). Multilocus pairwise estimates of fragment differentiation showed no correlation between geographical and genetic distance, suggesting an interbreeding network of trees, with levels of gene exchange sufficient to prevent their genetic differentiation by drift or selection. However, the history of land utilisation in this area indicates that a large proportion of the trees used in this study would have been present prior to fragmentation. The distribution of genetic variation detected is therefore largely a reflection of the previous more continuous forest, rather than being indicative of the impact of more recent disturbance. Therefore, current levels of genetic variation provide little indication of the population's ability to maintain genetic variation in the long term. Measures of current gene flow within and among the subpopulations are required to determine the long-term effects of fragmentation, and to assess whether the immediate consequence of rare-allele loss through reduced population size will be followed by the more extensive theorised genetic depletion.

Contrasting patterns of pollen flow were evident both within and between the fragments. Common to each fragment was a high frequency of pollen flow from within the first 300 m of the maternal tree, indicating the predominance of near-neighbour mating. However, a large proportion of pollen donors were from outside of each fragment and located across the sampled area, indicating extensive gene flow over this spatial scale. Increases in the proportion of long-distance pollen flow were paralleled by a reduction in fragment size (number of trees: Table 17.1). So Las Tablas had the highest percentage of near-neighbour interactions (64% pollen flow from within the mapped area of forest; 0–0.6 km), compared to the lowest percentage in the smallest fragment Tablas Plains (31% intrafragment pollen flow, the remainder from > 0.9 km). Although within-fragment pollination predominates in the larger fragments (Jiote 62% and Butus/Jicarito 53%) the remainder comes from the largest pollen source, rather than the nearest fragment, with 24% and 34% of pollen detected from distances greater than 1.5 km and 3.6 km respectively. The isolated tree 501 showed 100% external pollen sources, in agreement with the evidence of self-incompatibility from controlled pollinations, with the proportion of pollen again governed by the size of the pollen source rather than distance. While the nearest flowering trees were 1.2 km away, pollen flow was greater, with only 26% of the pollen coming from the nearest fragment (Butus/Jicarito at 2 km), and over 70% from the Las Tablas/unsampled areas, over 4.5 km away. The increase in spatial isolation appears to have promoted long-distance gene flow between it and the fragments, in contrast with the predictions that spatially isolated trees are more likely to deviate from random mating and receive pollen from fewer donors (Murawski & Hamrick, 1991).

Seed dispersal and genetic structure

Localised or correlated seed dispersal within populations of highly outcrossing tree species has been predicted to cause high levels of genetic structure (e.g. Loveless & Hamrick, 1984; Furnier et al., 1987). Co-ancestry analysis showed a significant level of spatial genetic relatedness in the first 50 m in Las Tablas (Fig. 17.5), although the value is much less than that expected for either full or half sibs, suggesting some mixing of seed (Loiselle et al., 1995). This was in strong agreement with measures of seed dispersal in S. humilis from another site in Honduras, where the majority of seed fell out within the first 50 m, although some was traced up to distances of 300 m (Fig. 17.5). The large distances of pollen flow within Las Tablas suggests that seed dispersal (or localised adaptation) is the main factor dictating the very localised genetic structure, rather than isolation by distance as a consequence of restricted gene flow (Wright, 1943, 1946, 1951). The level of genetic structure in tree populations often disappears or decreases from juveniles to adults as selection or random thinning occurs in the seedling and sapling stages (Hamrick et al., 1993; Loiselle et al., 1995). Persistent structure in adult populations, as here, suggests that biparental inbreeding may play a major role in determining the microevolutionary dynamics in a natural population (Schnabel et al., 1998). The spatial struc-

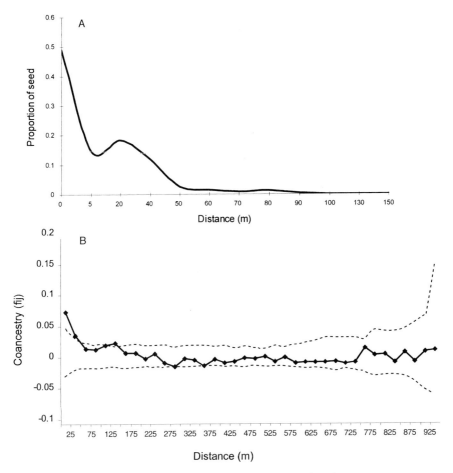

Fig. 17.5. (A) Seed dispersal from four trees of *Swietenia humilis* at Comayagua, Honduras. (B) Correlogram of estimated co-ancestry (f_{ij}: Loiselle *et al.*, 1995) for pairs of *S. humilis* within 25-m distance classes at Las Tablas, Punta Ratón, Choluteca, Honduras. Dashed lines represent upper and lower 95% confidence limits around zero relationship.

ture within this stand is, however, at a much smaller scale (50 m) than most pollen flow (within first 350 m), indicating half-sib maternal relatedness as the cause, rather than persistent biparental inbreeding within the area of genetic structure. Given the increased levels of pollen flow measured here, the extent to which any localised genetic structure within the fragments is maintained will depend on the effects of habitat disturbance on the dynamics of the wind-dispersed seed.

A)

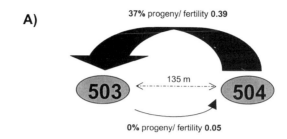

37% progeny/ fertility **0.39**

503 $\xleftarrow{\text{135 m}}$ 504

0% progeny/ fertility **0.05**

B) Flowering period

Male

7.4 days (se 3.9)

31.9 days (se 5.6)

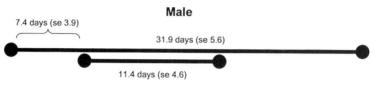

11.4 days (se 4.6)

Female

Fig. 17.6. Schematic representations of (A) the relationship of neighbouring trees 503 and 504 with respect to the percentage of progeny fathered and fertility likelihoods of each tree to the other, and (B) the overlapping flowering period of the male and female flowers of *Swietenia humilis* (se: standard error).

FLOWERING, FRUITING AND INDIVIDUAL TREE MATING PATTERNS

Study of flowering and fruiting phenology in *Swietenia humilis* was revealing in terms of (1) the restrictions it places on mating between trees and (2) the extent of reproduction under disturbance. Within the Las Tablas control, whilst near-neighbour interactions predominated in two of the sampled trees, a lower frequency of near-neighbour pollen flow was observed in the other three trees, even though there were possible donor trees at closer distances. Field observations made in the year of seed set (D. H. Boshier, unpublished data) showed a high proportion of neighbouring trees to be asynchronous in flowering with respect to the seed tree. Together with a low frequency of neighbours, this gives some explanation as to the patterns of pollen flow to the three maternal trees where near-neighbour interactions were low. Interactions are further complicated by temporal differences in the appearance of male and female flowers within individual trees, and illustrated by biased mating between trees 503 and 504. These two trees are nearest neighbours, separated by only 135 m, with tree 504 the definite father of 37% of the progeny sampled from 503; conversely 503 is not a father of any of the progeny sampled from 504 (Fig. 17.6A). Male

Table 17.2. Percentage of *Swietenia humilis* trees at each of four sites producing seed over four fruiting seasons (years)

Site[a]	0 Years	1 Year	2 Years	3 Years	4 Years
Lomas Barbudal Reserve	25.0	31.3	28.1	12.5	3.1
Las Tablas	9.4	18.8	12.5	31.3	16.7
Punta Ratón	16.7	13.9	30.6	22.2	28.1
Recursos Naturales	0.0	5.6	5.6	11.1	77.8

[a]Sites are ranked for increasing degree of disturbance: Lomas Barbudal, a relatively undisturbed reserve; Las Tablas, forest with remnant trees and secondary regeneration; Punta Ratón, remnant trees and secondary regeneration in small fragments and pasture; Recursos Naturales, planted trees by roadside.

flowers appear for a period of approximately 32 days, while female flowering is limited to a period of approximately 11 days (D. H. Boshier, unpublished data; Fig. 17.6B). Tree 504 flowered ahead of 503, such that its period of female flowers would probably have finished before 503 started to produce male flowers, and would therefore not have received pollen from 503. Conversely, 504 would still have an extended period of male flowering that overlapped with 503's initial period of female flowering, providing a large pollen source at a minimum distance interval. The short female flowering period in *S. humilis* would appear to reduce considerably the number of potential pollen sources open to any one tree.

Aldrich & Hamrick (1998) reported that in a disturbed pasture the largest individuals of *Symphonia globulifera* had the highest fecundity and dominated mating. A shift in the foraging behaviour of the pollinators was recorded, apparently influenced by an increase in resources (flowers/nectar) in the disturbed population. Such a dominance of large individuals was not observed here in *Swietenia humilis*, with no relationship between the size (diameter at breast height) and the fertility likelihood of paternal trees. There was however a marked increase in the number of trees producing seed with an increase in disturbance (Table 17.2). Such an increase of floral resources may promote pollen flow between the disturbed fragments, attracting pollinators and resulting in the extensive network of gene flow seen at Punta Ratón. Extensive pollen flow and multiple paternity was evident within the individual capsules examined, with the source of pollen donors varying between capsules and trees (Table 17.3). Pollen donors within capsules were from both intra- and interfragment sources, with the maternal tree's position within a fragment influential. Tree 639, located on the edge of the Las Tablas study plot, had paternal donors with the highest fertility likelihoods external to the fragment in four out of five capsules.

Table 17.3. Fertility likelihoods from paternity analysis of progeny within five capsules from each of two *Swietenia humilis* trees in the Las Tablas forest; paternal donors of fertility likelihood >0.1 (highlighted in bold) are listed for each capsule, with their distance (km) from the maternal tree, and their respective fertilities in the other capsules of the same tree

Paternal donor	Distance (km)	Capsule/fertility 1	2	3	4	5
Tree 609[a]						
607	0.06	0.018	**0.129**	0.019	—	0.008
597	0.29	—	**0.168**	—	—	**0.100**
1216	0.55	—	0.042	**0.339**	0.079	**0.180**
624	0.79	—	0.041	**0.312**	0.070	**0.108**
1234	0.85	**0.400**	0.064	—	**0.125**	0.039
518	4.29	0.029	**0.127**	0.059	**0.203**	0.078
514	4.35	**0.114**	0.073	—	0.045	0.007
506	4.45	**0.132**	0.024	—	—	0.008
1416	4.88	—	0.001	0.027	—	**0.118**
Tree 639[a]						
634	0.21	**0.189**	—	—	—	—
638	0.07	**0.290**	—	—	—	—
1216	0.59	**0.260**	—	—	—	—
538	3.75	—	—	—	**0.209**	—
546	3.76	—	—	0.004	**0.157**	0.011
518	4.50	—	0.066	**0.161**	**0.104**	**0.168**
527	4.56	—	0.004	—	**0.185**	0.013
515	4.54	**0.110**	—	—	—	—
514	4.58	—	**0.347**	**0.308**	0.011	**0.167**
506	4.96	—	0.084	0.017	0.032	**0.114**

[a] Tree 609 is located in the middle of the plot and 639 on the edge.

Tree 609, located in the middle of the Las Tablas study plot, had neighbouring trees within Las Tablas feature as at least one of the fathers with high fertility likelihood in each capsule examined. Tree 552, in the Tablas Plains fragment, had two capsules (4 and 5) where two common paternal donors (trees 554 and 555, both within 90 m of 552) accounted for approximately 70% of the fertility across all potential fathers, but which did not even feature as potential donors in the other three capsules.

CONCLUSIONS

This study indicates that for *Swietenia humilis*, at the degree of separation studied, fragmentation does not impose a genetic barrier between remnants, but alters and increases levels of interfragment gene flow, facilitated by the ability of pollinators (small bees and butterflies) to move between the spatially isolated stands. The fragments showed reduced genetic diversity through the loss of lower-frequency alleles, although the enhanced levels of long-distance pollen flow into the smaller fragments promoting genetic mixing across this spatial scale could potentially restore, maintain or even increase levels of genetic variation within the subpopulations of *S. humilis*. The subsequent development from the initial loss of alleles to a reduction in heterozygosity levels is dependent on the degree of genetic isolation and the pressures of genetic drift and selection on a small population (Young *et al.*, 1996). Higher rates of pollen flow into small 'island' populations (distances < 1 km) than into continuous and large 'island' populations have also been shown for the tropical tree *Spondias mombin* (Nason & Hamrick, 1997), with similar fragmentation effects seen in the temperate tree *Acer saccharum* (Foré *et al.*, 1992; Young *et al.*, 1993; Ballel *et al.*, 1994; Young & Merriam, 1994). Conversely, a positive relationship was detected between remnant population size and heterozygosity in *Eucalyptus albens*, but small remnant stands were genetically impoverished only beyond a certain isolation distance from larger populations (Prober & Brown, 1994).

Clearly a variety of factors can influence the degree of relatedness that results from mating in trees. Observed mating patterns are likely to result from interactions between a number rather than any one particular factor, but any increases in inbreeding subsequent to fragmentation will be particularly dependent on the strength of the species' self-incompatibility mechanism. Only one example of selfing was observed in all the investigated trees and was not associated with any spatial isolation. Tree 536 (Jiote fragment) showed 30% of its progeny to be from selfing, although there was still a large proportion of long-distance pollen flow represented in the remaining progeny. Any increased, abnormal levels of inbreeding may be unimportant from an evolutionary viewpoint, with selfed individuals selected against during regeneration, although they may be critical in terms of the levels of diversity sampled for *ex-situ* conservation, tree breeding or plantation programmes.

Different patterns of pollinator foraging, seen by variable patterns of pollen flow, may depend on the spatial structure of the subpopulation. Whilst there was a tendency for mating with nearest neighbours, flowering phenology and the frequency of neighbouring trees at different distances clearly influenced mating patterns. Populations of *Spondias mombin* and *Turpinia occidentalis*, with clumping of reproductive trees, showed a predominance of near-neighbour interactions, with an increase in pollinator distance mirrored by a decrease in density of conspecific flowerers (Stacy *et al.*, 1996). If pollinator flight distances are density- or number-dependent in *Swietenia humilis*, reductions in the number of flowering trees in the fragments may promote interfragment pollinator flight, and hence pollen flow between the smaller stands. The capacity of some insect pollinators to move long distances has been shown (Webb & Bawa, 1983), although the potential to move between stands is dependent on their behavioural response to habitat fragmentation. Indeed, changes in pollinator assemblages in fragments could strongly affect patterns of gene flow and genetic variation within remnant tree populations. Some bat species have been shown to move preferentially down forest tracks and pathways (e.g. Estrada *et al.*, 1993). Whereas some bees have been shown to be restricted in their movements (Powell & Powell, 1987), others move across agricultural landscapes up to 4 km between forest fragments (e.g. Raw, 1989), and the genus *Ficus* with its species-specific wasp pollinator may apparently form extensive metapopulations in fragmented landscapes (Nason *et al.*, 1998). The small generalist insect pollinators of *S. humilis* are shown in this study to move across the large interfragment distances. The distance of pollinator movement cannot, however, be specifically inferred, as carry-over, due to secondary pollinator movements, is also likely to be a contributing factor to long-distance pollen flow (Adams, 1992). With a range of non-specialist pollinators the dynamics of pollen flow in *S. humilis* are probably much less susceptible to the effects of habitat disturbance than tree species with more specialist pollinator relationships. However, it is likely that all pollinators will have a distance threshold beyond which they will not move, but this being pollinator- and habitat-specific indicates how fragmentation effects will be variable and complex.

With pollination occurring over considerably greater distances and more frequently than has previously been considered, some tropical tree species may be more adaptable to fragmentation than previously thought. The description of many remnant trees as 'isolated' or 'living dead' (Janzen, 1986*b*) may be more a human perception than a true reflection of actual gene flow patterns. For some species, individuals and populations of forest species, connectivity may in fact be high across landscapes with little forest cover. It is more realistic to view remnant patches not as islands, but rather as existing within a patchwork of land uses which differ in their capacity to provide habitat or permit movement for any organism. With adequate gene flow and seed production, some remnant forest patches and trees may be important contributors to connectivity and conservation more generally. The study shows clearly the danger of extrapolating from data on pollen flow in undisturbed forests to draw conclusions about the impacts under fragmentation. Other demographic and in particular social factors not considered here may however ultimately determine the future viability of particular fragmented populations. In the Punta Ratón population, over the four years after the trees were mapped, 11.6% were felled, bringing into question the population's continuation, despite its clear viability from a genetic perspective.

ACKNOWLEDGEMENTS

Many organisations and people helped with the field and laboratory work related to this study. We thank CONSEFORH/COHDEFOR for facilitating the fieldwork in Honduras, Sea Farms S.A., Secretaria de Recursos Naturales (Comayagua), Srs Cornelio and Hernan Corrales, Sr Hernan Videl and Sr Enrique Weddle for access to their land; José Dimas Rodriguez and Modesto Castillo for their help in seed and leaf collection; Allan Booth for his assistance in both the DNA extractions and running the gels; and Wayne Powell for his advice. This paper is an output from research projects (Projects R5729, R6080, R6516, Forestry Research Programme) funded by the Department for International Development of the United Kingdom; however, the Department for International Development can accept no responsibility for any information provided or views expressed.

18

Population viability analysis of the rare *Gentiana pneumonanthe*: the importance of genetics, demography and reproductive biology

J. GERARD B. OOSTERMEIJER

ABSTRACT

The relative importance of demography, genetics and reproductive biology for the viability of small populations was assessed for the rare perennial herb *Gentiana pneumonanthe*. Based on a previously published analysis, a stochastic matrix projection model was constructed to simulate population dynamics during heathland succession after management-induced disturbance (sod-cutting). Demographic data were used mainly from one population that has been censused from the invasive to the regressive stage of population succession. In earlier studies, I established the relationships between population size, selfing rate and inbreeding depression, as well as between population size and reproductive success. Three variants of the basic, demographic model were made, incorporating (1) the relationship between population size and reproductive success (RE variant), (2) the relationships between population size, selfing rate and inbreeding depression (ID variant) and (3) both of the above relationships (RE + ID variant). Reduced reproductive success did not have a significant effect on population performance. Inbreeding depression, however, resulted in significantly lower peak population sizes, a more rapid extinction and a lower growth rate. The latter effects were even stronger in the RE + ID model. In this variant, the time to extinction, T_E, of small populations decreased by six years relative to the basic model. Significant differences in T_E between the basic model and this variant persisted up to an initial population size of 250 plants. Further simulations with various time intervals between sod-cutting again showed that the RE variant differed very little from the basic model. The ID variant differed only for initial population sizes of $\leqslant 50$ individuals. Populations simulated in the RE + ID variant, however, went extinct more often, even when similar populations in the basic model were growing rapidly. This depended on the time interval between disturbances (sod-cutting) and was mainly present in populations with an initial size ⩽ 100 reproductive individuals. Small populations suffering from reduced reproductive success and inbreeding depression needed shorter time intervals between disturbances in order to grow than large populations and populations that did not experience these problems. I conclude that small populations firstly depend on suitable ecological conditions to enable successful regeneration. Inbreeding, by altering values of demographic transitions, has a significant effect on the regeneration capacity of small populations. This effect is increased by an additional reduced reproductive success, which by itself does not affect viability very much. The results clearly demonstrate that small populations cannot be treated in the same way as larger populations, but need more cautious and frequent management. In special cases, even reinforcement may be needed to counteract inbreeding effects.

INTRODUCTION

A matter of much concern to the conservation biologist, and to all those involved in the conservation of rare and threatened species, is the viability of small remnant populations. An important question in this respect, frequently posed by reserve managers, is how large the minimum viable population (MVP) has to be. Several aspects are important for the viability of small populations. Firstly, they are more susceptible to *environmental stochasticity* (Menges, 1990, 1991b). Secondly, *demographic stochasticity* may become important, especially in very small populations (<50 individuals: Menges, 1991b). A third effect associated with small effective population size is the loss of genetic variation through drift and inbreeding. Drift is a random process, leading to the fixation of alleles (either advantageous or deleterious ones) by sampling errors (Barrett & Kohn, 1991). Inbreeding occurs because the probability of matings between close relatives increases if population size gets smaller (Lande & Schemske, 1985; Charlesworth & Charlesworth, 1987). For self-compatible, animal-pollinated plants, however, inbreeding may also result from higher selfing rates in small populations. These may be a consequence of enforced *autogamy* (because pollinators are scarce, or because of the Allee effect: Lamont et al., 1993) or more frequent *geitonogamy* (because visitation behaviour of pollinators changes at low plant densities). Inbreeding increases homozygosity, leading to the expression of deleterious recessive alleles which reduce offspring fitness (inbreeding depression: Charlesworth & Charlesworth, 1987; Barrett & Kohn, 1991). In plants, any increase in autogamy and geitonogamy in small populations may not only result in inbreeding depression, but may also lower fecundity, because the quantity or quality of the pollen reaching the stigma may be insufficient for normal production of viable seeds (Casper, 1984; Galen *et al.*, 1985; Karron, 1987, 1989; Lyons *et al.*, 1989; Cruzan, 1990).

During the last decades, conservation biologists have attempted to test the effects of small population size on demography, genetics and reproductive biology, stimulated by the papers of Soulé (1980, 1986*b*, 1987). Several studies have been performed to investigate the relationship between population size and genetic variation (Levin *et al.*, 1979; Ledig & Conkle, 1983; Moran & Hopper, 1983; McClenaghan & Beauchamp, 1986; Karron *et al.*, 1988; Billington, 1991; van Treuren *et al.*, 1991; Prober & Brown, 1994; Raijmann *et al.*, 1994), inbreeding depression (Ouborg & van Treuren, 1994; van Treuren *et al.*, 1994), offspring fitness (Ledig & Conkle, 1983; Menges, 1991*a*; Oostermeijer *et al.*, 1994*b*, 1995), environmental and demographic stochasticity (Menges, 1990, 1991*b*) and reproductive success (Jennersten, 1988; Kwak & Jennersten, 1991; Kwak *et al.*, 1991; Aizen & Feinsinger, 1994*b*; Oostermeijer *et al.*, 1998; Young & Brown, 1998). Only rarely, however, have attempts been made to incorporate all of these relationships in an assessment of the viability of small populations in a single species. This is unfortunate, because there still is debate on the relative importance of demography vs. genetics for the conservation of small populations or rare species (Lande, 1988; Young *et al.*, 1996). Demographers suggest that the minimum population size necessary to buffer environmental stochasticity is many times greater than the size needed to counteract the negative effects of genetic drift and inbreeding (Lande, 1988). On the other hand, geneticists argue that populations may comprise thousands of individuals to buffer environmental stochasticity, but may nevertheless go extinct if the variability among them is too small to escape conditions to which none of these individuals is adapted (Menges, 1991*b*; Ellstrand & Elam, 1993; Nunney & Campbell, 1993; Schemske *et al.*, 1994).

It is extremely difficult to estimate the relative importance of demography, genetics and reproductive biology for population viability. Only rarely will there be sufficient data to make an assessment of any of these aspects separately on short- and long-term viability, let alone data to test all of these effects simultaneously for a single species. Simulation models have sometimes been used to project certain environmental conditions (e.g. management strategies) to the future, to investigate the effects of increasing levels of stochasticity on the probabilities of extinction (Menges, 1991*b*, 1992), or to simulate certain genetic processes in (small) populations (Charlesworth, 1991; van Treuren, 1993; Ouborg, 1993*b*).

For the species presented in this paper, *Gentiana pneumonanthe* L., data have been gathered on demography (Oostermeijer *et al.*, 1994*b*, 1996), genetics (Raijmann *et al.*, 1994; Oostermeijer *et al.*, 1994*b*, 1995) and reproductive biology (Petanidou *et al.*, 1995; Oostermeijer *et al.*, 1998). This enables development of a detailed simulation model, based on data from a six-year demographic study (1987–93), in which empirically determined effects of population size can be incorporated either directly (via simulations with the basic model using a range of starting population sizes) or via the studied relationships between population size and inbreeding depression, offspring fitness and reproductive success. By comparison of models with and without these relationships, the relative importance of each for the probabilities of extinction or regeneration of small, remnant populations can be assessed. Some introductory information on the demographic behaviour of *Gentiana pneumonanthe* is vital for a good understanding of the first version of the simulation model that will be presented in this paper.

Gentiana pneumonanthe (Fig. 18.1) may exhibit three successional stages in its demography, which are strongly related to local vegetation structure (Oostermeijer *et al.*, 1994*a*, 1996). In very open vegetation types, at the start of (secondary) succession, populations are 'invasive', and can be characterised by relatively high finite rates of increase ($\lambda > 1$), high numbers of seedlings and juveniles and a relatively low proportion of vegetative and reproductive adults. These populations are generally observed in recently sod-cut heathlands (Oostermeijer *et al.*, 1994*a*, 1996). The second, 'stable' population stage is found in half-open vegetation types, such as grasslands which are annually mown or heathland vegetation that stays open by winter inundations. These populations are characterised by a stable growth rate ($\lambda \approx 1$) and by the presence of low numbers and densities of seedlings and juveniles and a relatively high proportion of adult plants. The third, and last, 'regressive' stage is characterised by the absence of seedlings and juveniles and a very low density of adult (often mainly reproductive) individuals. The growth rate of these populations indicates a steady, albeit slow, decline ($\lambda < 1$).

Throughout Europe, *Gentiana pneumonanthe* is most commonly found in declining 'regressive' populations in heathland vegetation or abandoned hay meadows dominated by grasses (mainly *Molinia caerulea*). This habitat can only be conserved and managed by setting back the succession by means of cutting and removing the top (15–25 cm) layer of the entire vegetation, litter layer and top soil. This type of land use was very common before 1950, when farmers mixed these hand-cut sods with sheep dung in special stables and spread the mixture as fertiliser over the arable lands near the villages. *Gentiana pneumonanthe*, as well as other heathland species with a

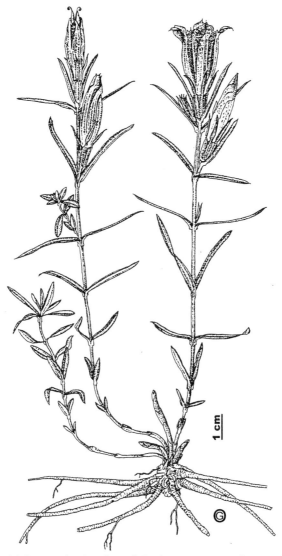

Fig. 18.1. Adult, reproductive plant of *Gentiana pneumonanthe*. Drawing by the author.

similar life strategy, obviously thrived on this type of land use, and was reported as a common species before 1950.

The main questions of this paper are (1) whether small, regressive populations of *Gentiana pneumonanthe* are able to regenerate to a large, viable population after suitable ecological restoration, and (2) to what extent the

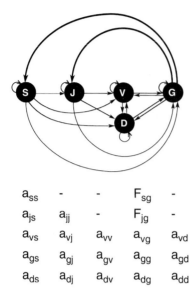

$$
\begin{array}{ccccc}
a_{ss} & - & - & F_{sg} & - \\
a_{js} & a_{jj} & - & F_{jg} & - \\
a_{vs} & a_{vj} & a_{vv} & a_{vg} & a_{vd} \\
a_{gs} & a_{gj} & a_{gv} & a_{gg} & a_{gd} \\
a_{ds} & a_{dj} & a_{dv} & a_{dg} & a_{dd}
\end{array}
$$

Fig. 18.2. Life cycle and basic projection matrix of *Gentiana pneumonanthe*: S, seedling; J, juvenile; V, vegetative; G, generative; D, dormant. The thin arrows indicate transition probabilities (i.e. values between 0 and 1) and the thick ones fecundity transitions (numbers of seedlings or juveniles per generative adult, which can be > 1). At the bottom, the basic projection matrix that was derived from the life-cycle diagram is shown, along with the coding used for the different transitions (*a* for transition probabilities, *F* for fecundity).

inbreeding depression and reduced reproductive success observed in such populations affect this regenerative capacity.

METHODS

Demographic monitoring and construction of matrix models

Since 1987, six populations of *Gentiana pneumonanthe* have been demographically monitored in a total of 10 fixed permanent quadrats (Oostermeijer *et al.*, 1996). In each quadrat, all individual plants (seedlings, juveniles, vegetative and reproductive adults) (see Oostermeijer *et al.*, 1994*a*) were accurately mapped annually in September. In this way, the fate of each individual plant could be monitored. On the basis of the life-history diagram (Fig. 18.2), the numbers of individuals going from one life stage to another could be determined for every year-to-year transition. These counts allow the construction of stage-based transition matrices (Lefkovitch, 1965; Caswell, 1989), representing the probabilities of all possible transitions between life stages occurring between two censusing dates (in this case one year). Detailed analyses of the 35 matrices that have been constructed for the six populations of *Gentiana pneumonanthe* for the period 1987–93 have been presented elsewhere (Oostermeijer *et al.*, 1996). These matrix analyses form an important basis of the simulation model described in the present paper.

Construction of the simulation model

Basic demographic data

Because the main environment of *Gentiana pneumonanthe* is successional (Oostermeijer *et al.*, 1994*b*, 1996), a demographic model has to simulate this. This implies that the transition values of the stochastic matrix model could not be randomly selected from given distributions, but that they had to change with the time since disturbance, as succession of the surrounding vegetation progresses.

In order to obtain information on the relationship between time since disturbance and the values of the transition elements in the projection matrix, demographic data were used from the populations in sod-cut heathlands, as these represent the relationships with secondary succession most strongly. The main population that was used for this purpose was population Terschelling, in which sods were cut in 1981. This population was monitored demographically from 1987 up to 1993. In 1987, the vegetation at this site was quite open (average total cover 25%), and gradually closed by natural succession to a total cover of 98% in 1993.

The period from 1987 to 1993 allowed the construction of six transition matrices from the available data (Oostermeijer *et al.*, 1996). Since the succession in the censusing plot was set back completely by sod-cutting in 1981, the monitoring period represents six to 12 years of succession. Figure 18.3 shows the strong changes in mortality that occurred in this population. In the beginning of the censusing period (six years after sod-cutting), the finite rate of population increase in this population was still relatively high ($\lambda = 1.23$ and 1.59 in the first two censusing years, respectively) but sharply declined to values below 1.0 ($\lambda = 0.62–0.82$) after 1990 (nine years after sod-cutting). This is correlated with the increasing mortality, but mainly of the seedling, juvenile and vegetative stages (Fig. 18.3).

Statistical basis of the matrix elements

The relationships between the various transition elements in the matrices and time since sod-cutting were studied by means of regression analyses on the Terschelling data. Significant regression equations were used instead of

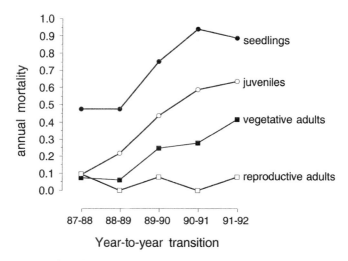

Fig. 18.3. Changes through time in the mortality of different life stages in population Terschelling, which was sod-cut in 1981, and demographically monitored from 1987 to 1993.

the 'normal' transition probabilities in the replacement functions of the various stages in the matrix projection model. For the transition probabilities that were not significantly related to time since sod-cutting, the average values and variances obtained from the data of all heathland matrices (Oostermeijer et al., 1996) were used in the replacement functions. The results of the regression analyses are presented in Table 18.1.

Basic simulation model

The simulation model was set up in RAMAS/stage 1.4 (Ferson, 1991). With the help of this computer program, the relationships between matrix elements and time could easily be incorporated into the model. In RAMAS/ stage, driver variables can be specified as statistical distributions (e.g. normal or log normal) with a given mean and variance. Drivers can for example be chosen to represent transitions between life stages or regression coefficients that determine the rate at which these transitions vary with time. Each (time) step of the simulation, the driver variables are selected by means of Monte Carlo methods from their specified distributions, which allows the introduction of different (observed or simulated) levels of environmental stochasticity in the model. The problem of independence of matrix elements that this method introduces was avoided by the introduction of time into the model, which automatically led to a correlation among the time- (or succession-) dependent transitions, as was observed in the original matrices (Oostermeijer *et al.*, 1996).

Relationship between population size and reproductive success
In earlier studies on reproduction (Oostermeijer *et al.*, 1992, 1998) in a range of populations of different sizes, we found a significant relationship between individual fecundity (viable seed production per flower) and the number of reproductive individuals in the population. This relationship was incorporated as a logarithmic regression (see Table 18.1) into the 'Reproduction' (RE) variant of the basic model, to assess the effects of low seed production in small populations on population dynamics.

The number of flowers per reproductive individual (N_{fl}) showed a significant non-linear relationship with time since sod-cutting (Table 18.1). This relationship is logical, as the number of flowers of the newly recruited reproductive plants is initially low and gradually increases as plants age, but decreases again (presumably as a consequence of increased competition) when the vegetation closes by succession. Using the number of seeds produced in year $t-1$ (the product of flower number per plant and seed production per flower) and the number of seedlings and juveniles in year t, the seedling and juvenile recruitment (SR and JR) from $t-1$ to t could be calculated, assuming that there is no seed bank of importance (Oostermeijer *et al.*, 1994*a*, 1996). These recruitment values were also used in regression analyses to investigate their relationship with time since sod-cutting (Table 18.1).

The numbers of newly recruited seedlings and juveniles at a given time step were calculated from the product of the number of reproductive plants, the number of flowers per plant, the number of viable seeds per flower and the recruitment rate. In the basic model, which does not incorporate any relationship between population size and reproductive success, the number of viable seeds per flower was selected from a normal distribution with a mean of 350 and a variance of 3000 (data from Oostermeijer *et al.*, 1998).

Relationships between population size, selfing rate and inbreeding depression
In an earlier study, the selfing rate showed a significant relationship with the number of reproductive individuals in the population (Raijmann *et al.*, 1994). This relationship was used in an 'Inbreeding depression' (ID) variant of the simulation model to determine which fraction of the offspring experienced inbreeding depression. The effects of 100% inbreeding on several of the demographic transitions were based on the results of a transplant experiment with inbred and outbred offspring described in

Table 18.1. Formulation of the stochastic matrix projection model for *Gentiana pneumonanthe* in a successional environment after sod-cutting management[a]

(1) Transitions[b] not related to time after sod-cutting (lognormal distributions), variances[c] in brackets

Transition	Mean value		Transition	Mean value	
a_{ss}	0.003 (0.2)	(seed to seed)	a_{vd}	0.832 (1.9)	(dormant to vegetative)
a_{js}	0.067 (0.8)	(seed to juvenile)	a_{gd}	0.097 (0.9)	(dormant to generative)
a_{jj}	0.005 (0.3)	(juvenile to juvenile)	a_{dd}	0.072 (0.5)	(dormant to dormant)
a_{dv}	0.032 (6.3)	(vegetative to dormant)	mortality G	0.023 (1.8)	(mortality generatives)
a_{dg}	0.025 (2.3)	(generative to dormant)			

(2) Transitions[b] related to time since sod-cutting, T (variances[c] in brackets)

Regression equations	R^2	$F_{[1,4]}$	P	
$a_{vs} = 0.355 \ (10.3) - 0.0197 \ (1.2) \cdot T$	0.811	17.169	0.014	(seed to vegetative)
$a_{vj} = 0.710 \ (109) - 0.10 \ (70) \cdot T$	0.872	27.170	0.006	(juvenile to vegetative)
$a_{gj} = 0.077 \ (16) - 0.0076 \ (0.2) \cdot T$	0.808	16.809	0.015	(juvenile to generative)
$a_{dj} = 0.054 \ (1.8) - 0.005 \ (0.2) \cdot T$	0.601	6.033	0.070	(juvenile to dormant)
$a_{vv} = 0.907 \ (120) - 0.06 \ (14) \cdot T$	0.823	18.607	0.013	(vegetative to vegetative)
$a_{gv} = 0.30 \ (140) - 0.022 \ (2) \cdot T$	0.744	11.599	0.027	(vegetative to generative)
$a_{gg} = 0.90 \ (200) - 0.025 \ (16) \cdot T$	0.555	4.995	0.089	(generative to generative)
$a_{vg} = 1 - (\text{mortality } G) - (a_{gg}) - (a_{dg})$	—	—	—	(generative to vegetative)
$SR = 0.0155 \ (0.4) - 0.00156 \ (0.005) \cdot T$	0.746	11.647	0.027	(seedling recruitment)
$JR = 0.0168 \ (0.4) - 0.00162 \ (0.005) \cdot T$	0.656	7.636	0.051	(juvenile recruitment)
$N_{fl} = 0.7614 + 0.278 \ (1) \cdot T - 0.021 \ (1) \cdot T^2$	0.967	96.779	≤ 0.0001	(number of flowers/plant)

(3) Other parameters in the basic model and in the different model variants (variances[c] between brackets)

Number of seeds/flower: F, normal distribution, mean = 350 (3000)

Number of seeds/flower in relation to population size: $F = 162.5\,(407) - 23.5\,(70.5) \cdot \ln(G)$

Selfing rate in relation to population size: $s = 0.898\,(0.008) - 0.148\,(0.002) \cdot \ln(G)$

Inbreeding depression in relation to selfing rate: $ID = 1 - (0.40 \cdot s)$

(BA & ID, RE + ID)
(RE & RE + ID)
(ID & RE + ID)
(ID & RE + ID)

(4) Replacement functions for the different stages in the model

$S_{t+1} = [S]_t \cdot a_{ss} + [G]_t \cdot [N_{fl}] \cdot [F] \cdot [SR]^d$ *seedlings*

$J_{t+1} = [S]_t \cdot a_{js} + [J]_t \cdot a_{jj} + [G]_t \cdot [N_{fl}] \cdot [F] \cdot [JR]^d$ *juveniles*

$V_{t+1} = [S]_t \cdot a_{vs} + [J]_t \cdot a_{vj} + [V]_t \cdot a_{vv} + [G]_t \cdot a_{vg} + [D]_t \cdot a_{vd}^{\,d}$ *vegetative adults*

$G_{t+1} = [S]_t \cdot a_{gs} + [J]_t \cdot a_{gj} + [V]_t \cdot a_{gv} + [G]_t \cdot a_{gg} + [D]_t \cdot a_{gd}^{\,d}$ *generative adults*

$D_{t+1} = [J]_t \cdot a_{dj} + [V]_t \cdot a_{dv} + [G]_t \cdot a_{dg} + [D]_t \cdot a_{dd}$ *dormant plants*

[a] BA = basic model, based on demographic data only, ID = variant of BA incorporating inbreeding depression, RE = variant of BA incorporating declined reproductive success, and RE + ID = variant of BA with both inbreeding depression and declined reproductive success

[b] Transition subscripts are to be read as follows: a_{ij} means the transition from j to i (i.e. what stage i *gets* from j).

[c] All variances have to be multiplied by 10^{-5}, except those given in section (3).

[d] In ID variant, a_{js}, a_{vs}, a_{vj}, a_{vv}, a_{vg}, a_{gj} and a_{gv} are multiplied by [ID]; see section (3).

Oostermeijer (1996*b*). From this experiment, we established a maximal inbreeding depression (that is, if the selfing rate was 1) of approximately 30%–40%, that applied to the transitions P_{js}, P_{vs}, P_{vj}, P_{vv}, P_{gj}, and P_{gv}. Since the model is intended to simulate regeneration capacity over a relatively short time interval (50 years), it was assumed that too few generations are involved for purging to decrease the rate of inbreeding depression (van Treuren, 1993).

In the third variant of the basic model, both the relationship between population size and reproductive success and between population size, selfing rate and inbreeding depression were incorporated. In the following, this variant will be referred to as the RE + ID variant.

Simulations with the basic model and the model variants

In order to assess the relative effects of initial population size on regeneration capacity and extinction probability in a newly sod-cut heathland area, various kinds of simulations were made with the basic model and the model variants described in the above. Firstly, the model was run for 50 years with 100 iterations, simulating a single sod-cutting event, which in all cases resulted in an initial increase in population size and a subsequent decline to extinction. Extinction always occurred within 50 years. The models were run with initial population sizes of 10, 25, 50, 75, 100, 250, 500, 1000 and 5000 individuals. Initial population stage structure was derived from the typical 'regressive' structure observed in many small remnant populations (Oostermeijer et al., 1994a). This meant that 50%–75% of the initial population consisted of reproductive individuals. The remaining plants were either vegetative adults or dormant plants (Oostermeijer et al., 1992, 1996), although the latter stage was always scarce (1%–10% of the initial size). It was assumed that no seedlings or juveniles were present.

Of each run, peak size (maximum number of reproductive plants attained during the simulation) and time until extinction (defined as the moment when the number of reproductive plants was < 1) were determined. Peak size and average time to extinction were compared for different initial population sizes and for the basic model and the RE, ID and RE + ID variants. In addition, this type of simulation was also run with a 10-fold increase in the variances which are incorporated into the basic model and its variants. This increase was not applied to the drivers for the various regression coefficients, because this removed the succession trend from the simulations.

Secondly, the models were run with a variable time interval between disturbance of the population by sod-cutting. The purpose of these simula-

Table 18.2. Criteria used for the effects of sod-cutting
on modelled populations of *Gentiana pneumonanthe* of
different sizes

Number of generatives		Other stages
Remaining	Spared	Spared
< 100	100%	20%
100–500	75%	10%
500–1000	50%	5%
> 1000	25%	1%

tions was to determine at which disturbance frequencies populations increased or decreased within a period of 50 years. The average value and variance of λ was determined for each series of simulations and statistically compared between different population sizes and model variants by means of analysis of variance. In the simulations, the effect of sod-cutting on the standing population was made dependent on the number of reproductive adults that was present at the moment of sod-cutting in order to simulate a management that recognises the importance of reproductive plants and population size. Because the limited seed dispersal causes a clustering of individuals, the fraction of the other life stages that is spared during cutting depended on the proportion of reproductives that was saved. The criteria that were used to determine the effects of sod-cutting on population size and structure are given in Table 18.2.

Significant differences between simulation results (peak size, time to extinction and population trajectories) were determined using 95% confidence intervals around the means, as provided by RAMAS/stage. The confidence intervals were based on 100 replicate simulations (iterations) of the model.

RESULTS

Peak population size and mean time to extinction after a single sod-cutting event

Peak population size increased logarithmically with the initial number of individuals in all model variants. At small population sizes, there were considerable differences in peak size between the basic model and its variants (Fig. 18.4).

At initial sizes 10 and 25, peak size differed significantly between all four model variants. In all simulations, the highest size was achieved in the

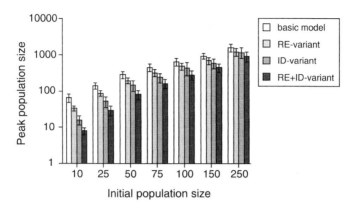

Fig. 18.4. Average peak population sizes derived from simulations with the basic model and the three variants for different initial population sizes. Simulations were made of a single sod-cutting event, which led to an initial increase to a peak, and a subsequent decline of the population to extinction within 50 years in all simulations. Error bars represent 95% confidence intervals, based on 100 model iterations.

basic model, and the lowest in the RE + ID variant. The average peak size of the basic model remained significantly different from that of the RE + ID variant until the initial size was 250. The basic model and the RE variant did not differ when initial sizes were larger than 25 plants. The ID variant differed significantly from the basic model until the initial size > 100. At population sizes > 250, no differences between the variants were observed any longer (Fig. 18.4).

The mean time to extinction, T_E, also differed considerably between the different model variants, but mainly at small population sizes (Fig. 18.5). The differences between the basic model and the RE variant were mostly small and within the range of the 95% confidence interval, and totally absent when the initial size was > 100. The same was true for the ID and the RE + ID variants, which showed an equal T_E when the initial size was larger than 100 plants. In comparison, the mean T_E of the basic model and that of the ID and RE + ID variants was significantly different up to an initial size of 500 individuals (Fig. 18.5). At very small initial population size (10 individuals), the RE + ID variant reduced T_E by six years in comparison with the basic model. At initial size > 100 individuals, mean T_E still differed by four years.

A 10-fold increase in the variances in the model, simulating higher environmental stochasticity, did not affect the outcome of the simulations very much, besides increasing the variances around the means of the iterations.

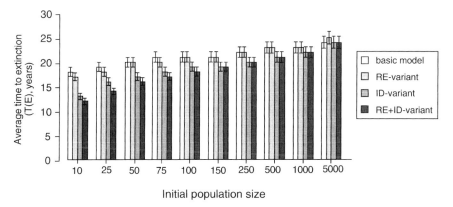

Fig. 18.5. Mean time to extinction in single sod-cutting simulations with the basic model and its three variants. The 95% confidence interval around T_E, based on 100 model iterations, is 1–2 years in all cases.

Mean peak sizes and extinction times were not affected significantly (data not shown).

Dynamics of small populations at different sod-cutting intervals

When the four model variants were compared at a sod-cutting interval of 10 years, it appeared that at a very small initial size of 10 individuals (Fig. 18.6a), there was a highly significant difference in population performance ($F_{[3,\,153]} = 10.469$, $P < 0.0001$). Although the basic model and the RE and ID variants resulted in growing populations, the RE + ID variant caused a decline to extinction in 17 years (even before the second attempt at sod-cutting). The trajectories of the models clearly showed that there is only a minor difference between the basic model and the RE variant, while the ID variant (initially) grows at a slower rate. No significant difference was found between the average λ of the basic model and that of the RE variant (*t*-test, $P = 0.423$, $\lambda = 1.108$ and $\lambda = 1.09$ respectively), but the difference with the ID variant ($\lambda = 1.05$) was significant ($t = 2.31$, df = 90, $P = 0.023$). The average λ of the RE + ID variant was 0.940, which was significantly different from all three other models ($P < 0.001$).

When the initial population size was increased to 25 individuals (Fig. 18.6b), there was still a significant difference in the finite rate of increase between the four model variants ($F_{[3,180]} = 10.116$, $P < 0.0001$). However, the basic model and the RE and ID variants no longer exhibited significantly different growth rates ($P > 0.20$ in all cases). The RE + ID variant still had a negative growth rate ($\lambda = 0.992$), though, which was significantly

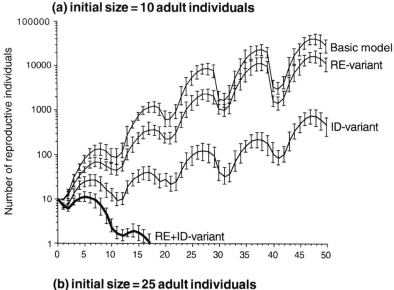

(a) initial size = 10 adult individuals

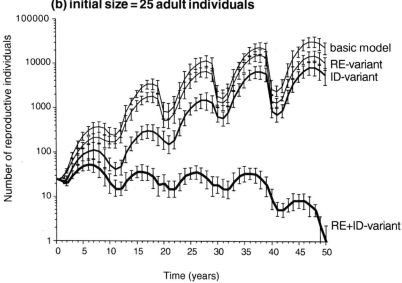

(b) initial size = 25 adult individuals

Fig. 18.6. Results of stochastic simulations with a sod-cutting interval of 10 years for each of the four model variants (see text) with an initial size of (a) 10 adult individuals and (b) 25 adult individuals. Error bars represent 95% confidence intervals around the mean, based on 100 model iterations.

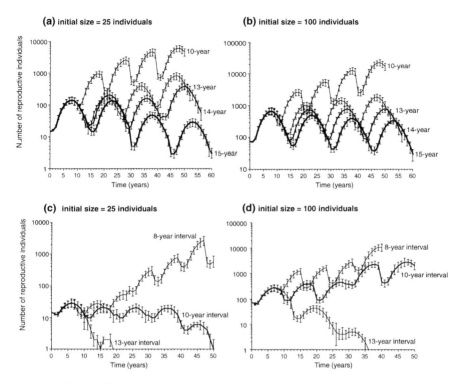

Fig. 18.7. Results of stochastic simulations with (a) and (b) the basic model, and (c) and (d) the variant incorporating data on both reduced reproductive success and inbreeding depression (RE + ID variant) with different sod-cutting intervals and an initial size of (a) and (c) 25 and (b) and (d) 100 individuals. Error bars represent 95% confidence intervals around the mean, based on 100 model iterations.

different from the other model variants ($P < 0.001$ for all comparisons). This led to extinction of the population in approximately 50 years.

Because the differences in growth between the basic model and the RE and ID variants were not very pronounced, except at extremely small population sizes, the effects of different sod-cutting frequencies on population dynamics are presented only for the basic model and the RE + ID variant (Fig. 18.7a–d). It can be seen that at an initial size of 25 individuals, the basic model resulted in growing populations when sod-cutting was performed at least every 14 years. When the vegetation was disturbed less frequently, the simulated populations went extinct within 100 years (Fig. 18.7a). An increase in the initial population size to 100 plants did not seem to make much of a difference in this model (Fig. 18.7b). Simulations with the RE + ID variant, however, showed that sod-cutting must take place at much shorter intervals (every 8–9 years or less) in order to prevent small populations from going extinct. When the initial size is 25 individuals, sod-cutting every 10 years already results in extinction, as we have seen before, and the time to extinction only decreases when cutting takes place at longer intervals (for example 13 years: Fig. 18.7c). When the initial population size was larger (100 plants), the time interval between sod-cutting could also be shorter (10–11 years).

DISCUSSION

Comparison with previous studies

Using empirical data on genetical, demographical and reproductive effects of small population size, it was possible to make an assessment of their relative importance. To this date, not many studies have gathered sufficient data to enable a thorough assessment of the importance of genetics for the probability of extinction. Ouborg (1993b) made such an attempt, but found only a slight to moderate effect of inbreeding depression on the probability of extinction and then mostly in the largest of his study populations. He attributed this mainly to a relatively low level of environmental stochasticity in this larger population, enabling a higher expression of inbreeding depression as compared to the other three populations which exhibited higher environmental stochasticity. However, it is more likely that the higher impact of inbreeding in his largest population is a consequence of the fact that only there a significant recruitment of new (inbred) individuals was observed, while the others were almost completely regressive. This shows that an assessment of the importance of inbreeding depression is best done by using demographic data of one or more large, genetically viable populations and imposing the effects of inbreeding in simulations using data from these populations. If the actual dynamics of small remnant populations are projected and experimental data on inbreeding depression are then included in the model, the inbreeding effects are most likely overestimated, because they have already affected the demography of the population studied. In the current paper, a stochastic simulation model was used that was based on data from a population that was monitored from the invasive to the regressive stage of development in a vegetation succession series (Oostermeijer et al., 1996).

A similar approach was followed in a study on Banksia cuneata by Burgman & Lamont (1993), who modelled population dynamics in response to different fire frequencies and incorporated inbreeding effects. They found only a minor effect of inbreeding in comparison with the importance of stochastic fluctuations in the annual precipitation, which dominantly influenced seedling recruitment after a fire. However, inbreeding effects were not based on data from actual field experiments with the same species, which means that still no reliable conclusions can be made about the importance of inbreeding in natural *Banksia* populations.

Although the model presented here does not pretend to simulate the actual dynamics of populations of *Gentiana pneumonanthe* in a successional environment, it is about as close as one can get. Still, caution must be taken before the results from the model (minimal viable population sizes and optimal sod-cutting intervals) are directly used as management guidelines. The successional trends are still mainly based on data from one population, although field observations in other sod-cut populations support the data from this now nearly extinct population. The vegetation around the study plot was cut in 1991, so regeneration could have taken place, but probably the population had become too small and no successful recruitment occurred. Based on another modelling study on *Gentiana pneumonanthe*, Chapman *et al.* (1989) found that a disturbance of the vegetation (in their populations mainly by heathland fires) every 13 years would lead to a stable or growing population. This is in line with the results presented here, which show that normal, large populations grow only when the sod-cutting interval is not longer than 13 to 14 years. This suggests that the model simulates the actual dynamics rather well, and is suitable for an assessment of the effects of inbreeding depression and loss of reproductive success.

Conclusions from the current model version

Two important conclusions can be made from the (preliminary) simulations performed with the model. Firstly, it seems that reproductive success is by itself a less important factor for the regeneration capacity of small populations, while in comparison, inbreeding has a stronger effect on population performance. On the other hand, the levels of inbreeding depression we modelled (30%–40%) also appeared to be not severe enough to prevent successful regeneration of small populations after ecological restoration. It is very interesting, however, that the combined effects of reduced reproductive success and inbreeding depression clearly make small populations more vulnerable to extinction than demographic and environmental stochasticity alone. The simulations with different initial (i.e. remnant) population sizes show that the effects of low seed production and inbreeding play an important role only in rather small populations, in which ⩽ 100 reproductive individuals remained. One explanation for the strong additive effect of reduced seed production on the regeneration capacity may be a smaller opportunity for selection of the fittest, less inbred individuals from a limited number of offspring.

A second important conclusion is that despite all the negative consequences of inbreeding and low seed production, small populations could be regenerated to a growing, large population by means of an optimal management regime. It is also clear, however, that small populations need more cautious and intensive management than larger populations. For large populations, sod-cutting at intervals of up to 14 years yielded growing populations. For small populations, the maximum time interval between two sod-cutting events to maintain $\lambda > 1$ is 9–11 years, depending on their size. It is important to realise that in the model, the management already recognised the importance of reproductive individuals for population regeneration, as well as the importance of population size. This meant that as many reproductive plants as possible were spared from sod-cutting in small populations. This is, however, not a common practice in nature management. Hence, the effects of less favourable management on the survival probability of small populations are likely to be even more severe than in these simulations.

The maximum level of inbreeding depression that was incorporated into the simulation model did not change over time. Theoretically, this is not correct, since under the partial-dominance hypothesis (which best explains the inbreeding depression in *Gentiana pneumonanthe*: Oostermeijer *et al.*, 1995), it may be expected that the genetic load is reduced by selection after several generations of inbreeding (Schemske & Lande, 1985). In the simulation model, a maximum inbreeding depression of 30%–40% was used, and the actual level expressed in the population was determined by the proportion of self-fertilisation, which was related to population size (Raijmann *et al.*, 1994). From the few studies available, it may be deduced that in many plant species, especially perennials, the small remnant populations have not been small long enough to have experienced several generations of inbreeding (Ouborg & van Treuren, 1994; van Treuren *et al.*, 1994). Modelling studies by van Treuren (1993) have shown that even under rather strong selection, many generations of inbreeding (>25) are needed to significantly reduce genetic load. In perennial plant species, which have long generation times, it is not likely that populations have recruited this many generations since the time that the ecological conditions became worse.

It should be also realised that population size is not the only important factor reducing reproductive success. Especially the composition of the surrounding vegetation may also play an important role (Oostermeijer *et al.*, 1998). In vegetation with few other flowering, insect-pollinated plant species, facilitation for pollination no longer occurs (Oostermeijer *et al.*, 1998). This results in low seed set and higher selfing rates, even when the population is large. On the other hand, small populations surrounded by co-flowering entomophilic plant species may suffer less from a loss of reproductive success than small populations on sites dominated by grasses.

The simulation model presented here only provides information on the short-term effects of small population size and the associated inbreeding depression. On a longer time scale, even if a population is growing because of careful and optimal management, the loss of genetic variation that has occurred through inbreeding and genetic drift may make it more vulnerable to environmental stochasticity, especially if catastrophes are involved (Fisher, 1930; Beardmore, 1983; Menges, 1991*b*, 1992). Such long-term effects have not been considered here, because it is not easy to incorporate them into a demographic model. If management enables a small population to regenerate rapidly to a large size, however, the overall losses of genetic variation may be kept at a limited level (Barrett & Kohn, 1991).

In *Gentiana pneumonanthe*, and probably in other long-lived perennials of open, successional habitats, loss of genetic variation during bottleneck population sizes may be reduced by the fact that the reproductive individuals are able to persist for a relatively long period in closed vegetation (Oostermeijer *et al.*, 1994*a*). Because there is such a strong relationship between heterozygosity and various fitness-related traits (Oostermeijer *et al.*, 1995), it can be expected that there is selection against homozygotes. This in turn may result in a higher frequency of heterozygotes among the reproductive individuals, which are the founders of the new, regenerating population (Schaal & Levin, 1976; Wolff & Haeck, 1990; Lesica & Allendorf, 1992; Tonsor *et al.*, 1993). Allozyme electrophoresis of samples of flowering and vegetative adults from field populations of *Gentiana pneumonanthe* indicate that the reproductive plants are indeed relatively more heterozygous than the vegetative plants (Oostermeijer, 1996*a*). The importance of the conservation of genetic variation in surviving heterozygotes has not yet been sufficiently studied, although the phenomenon has been put forward as an argument to conserve small populations (Lesica & Allendorf, 1992).

ACKNOWLEDGMENTS

The author would sincerely like to thank all of those who have contributed in any way to the model presented in this chapter. Special thanks are due to Hans den Nijs, Sheila Luijten and Eric Menges for their particularly valuable role in the study that led to this paper.

19

Genetic erosion, restricted mating and reduced viability in fragmented populations of the endangered grassland herb
Rutidosis leptorrhynchoides

ANDREW G. YOUNG, ANTHONY H. D. BROWN, BRIAN G. MURRAY, PETER H. THRALL & CATHY H. MILLER

ABSTRACT

The endangered herb *Rutidosis leptorrhynchoides* from south-eastern Australia has had its population numbers and sizes severely reduced by habitat loss and fragmentation. Assessment of allozyme variation in 18 diploid and five tetraploid remnant populations shows a strong positive relationship between the logarithm of population size and allelic richness for diploid populations, but not for tetraploids, which generally maintain higher diversity. Diploid populations smaller than 200 plants also show reduced *S* allele richness at the self-incompatibility locus, which constrains mate availability. Reduced mate numbers are associated with increased variance in male fitness in small isolated populations as reflected by high correlated paternity and increased deviation in allele frequencies between adult and pollen gene pools. Such changes reduce effective population sizes in small populations below their census size, further exposing them to effects of genetic drift.

Stochastic matrix projection models based on two years of demographic data from five populations show that population persistence is related in a general way with both higher genetic diversity and lower correlated paternity. Variation in population performance is primarily due to differences in the magnitude of the variance in transition probabilities rather than their mean values, which are generally not significantly different.

Measurement of genetic variation and fixation indices suggests that small re-established populations suffer the same constraints as small remnant populations. Conservation strategies for *R. leptorrhynchoides* should focus on maintaining population sizes above 200 plants to avoid demographic effects of *S* allele erosion and future re-establishment efforts should source seed broadly for the same reasons. Populations currently smaller than 200 plants and isolated by more than 5 km have low expected persistence times.

However it is possible that their prospects could be improved if they are supplemented with new S alleles.

INTRODUCTION

Habitat loss, fragmentation and degradation are pervasive threatening processes for many indigenous plant species in Australia, as in much of the rest of the world. Large-scale conversion of land for grazing and agriculture over the past two centuries in Australia has dramatically reduced the extent of native ecosystems, often limiting previously extensive species' ranges and reducing the number and size of plant populations. Currently, approximately 5000 species of Australian plants are listed as rare or endangered, representing 23% of the flora (Briggs & Leigh, 1996). Many species, covering a range of life forms from herbs to trees, now persist primarily in habitat fragments or remnants. If such species are to be managed effectively for conservation, an understanding of the factors that limit the viability of fragmented populations is required.

Many deterministic threats are associated with habitat loss and fragmentation (see Holsinger, Chapter 4, this volume). In the Australian context, land clearing has been associated with changes in the dynamics of disturbance due to fire (Gill & Williams, 1996), increased weed invasion (Scarlett & Parsons, 1990) and widespread soil salination due to raised water tables (Neil & Fogarty, 1991), all of which can impact on the viability of populations occupying remnant habitat. However, of the 103 plants species currently classified as critically endangered in Australia, about half have population sizes small enough (<100 plants) that genetic erosion and inbreeding might also be expected to play a role in determining their fate through effects on evolutionary potential, mate availability and fitness. Thus removal of deterministic threats alone may not be enough to prevent extinction.

Here we review the results of recent genetic and demographic research on remnant and re-established populations of the endangered grassland herb *Rutidosis leptorrhynchoides* (Asteraceae), a species that has had its population structure radically altered by loss and fragmentation of its grassland habitat. This species is listed as endangered under the Commonwealth Endangered Species Act 1992 as well as under local State and Territory Acts. It is currently the subject of a formal Recovery Plan being developed by the New South Wales National Parks and Wildlife Service (Briggs *et al.*, 1998). Part of this planning process is determining the viability and conservation value of the remaining populations. Another focus for species management is on planning for population re-establishment (Cropper, 1993).

The research presented here combines the use of genetic markers, controlled pollination experiments, cytological studies, field monitoring of demography and population simulation modelling to examine how population size and isolation affect genetic diversity and mating system parameters, and the implications for fecundity, demography and viability. Several questions are of particular interest:

1. Do smaller populations exhibit reduced genetic diversity?
2. Do smaller populations have elevated levels of inbreeding, either through selfing or mating among relatives?
3. Are there demographic differences between small and large populations that relate to genetic differences and do these affect viability?
4. Is there significant interpopulation genetic structure that should be taken into account when sourcing plants or seed for re-establishment purposes?

RUTIDOSIS LEPTORRHYNCHOIDES

Rutidosis leptorrhynchoides is a multi-stemmed, herbaceous, perennial daisy indigenous to the temperate grasslands of south-eastern Australia. Though it is unlikely that *R. leptorrhynchoides* has ever been very common, herbarium records show that there has been a significant reduction in the number and size of populations since 1874 (Fig. 19.1). This decline has paralleled that of its habitat. The temperate grasslands of the south-east are Australia's most endangered ecological community, having been reduced to less than 0.5% of their original estimated 2 million ha extent since European settlement (Kirkpatrick *et al.*, 1995).

More recently, invasion of grassland remnants by exotic weeds and reduction in fire frequency have also played a role in reducing the number and size of *R. leptorrhynchoides* populations (Scarlett & Parsons, 1990). *R. leptorrhynchoides* is currently known from 24 populations split into two geographic groups separated by about 500 km, consisting of 16 northern populations in the Australian Capital Territory and New South Wales and eight southern populations in Victoria. Populations range in size from five up to approximately 100 000 flowering plants. However, about half of the populations contain fewer than 200 flowering plants.

Rutidosis leptorrhynchoides is pollinated by a range of flying insects. It is known to be primarily self-incompatible, having a single-locus sporophytic incompatibility system (Young *et al.*, 1999), which is characteristic of mem-

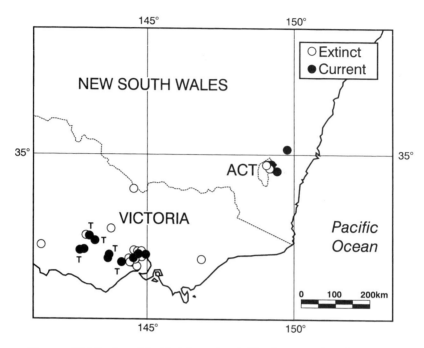

Fig. 19.1. Distribution of *Rutidosis leptorrhynchoides* 1874–1999 based on
herbarium records. Those marked T are tetraploid populations. Reproduced from
Young *et al.* (1999).

bers of the Asteraceae. The seeds of *R. leptorrhynchoides* are wind-dispersed,
but dispersal distances are generally less than 0.5 m (Morgan, 1995*a*) as
plants rarely reach greater than 40 cm in height. *R. leptorrhynchoides*
flowers profusely from November until March, shedding seed from February onwards. Seedlings emerge in late autumn (April–May), generally after
the first major rains of the season. There is no long-term soil-stored seed
bank (Morgan, 1995*b*).

GENETIC EROSION

One of the major predictions of population genetics theory is that when
populations are reduced in size, they will lose allelic richness, and to a
lesser degree heterozygosity, owing initially to founder effects and then
subsequently to genetic drift (Barrett & Kohn, 1991; Sherwin & Moritz,
Chapter 2, this volume). For natural populations of plants the relationship
between population size and genetic diversity is well established empirically (e.g. Billington, 1991). Significant losses of diversity in small popula-

tions created by relatively recent habitat fragmentation have also been observed for several plant species (van Treuren *et al.*, 1991; Prober & Brown, 1994; see Young *et al.*, 1996 for a review).

In the long term, genetic erosion means that populations will have little genetic variation for selection to act on, compromising their evolutionary potential (Frankel *et al.*, 1995). In the short term, reduced diversity at genes of major effect, such as loci controlling self-incompatibility or conferring disease resistance, will exert a direct influence on individual fitness and through this on population demography. For example DeMauro (1993) showed that lack of seed set in small remnant populations of the endangered lakeside daisy (*Hymenoxys acaulis* var. *glabra*) was the result of remaining plants sharing the same incompatibility genotype.

Allozyme variation

To determine whether small or isolated remnant populations of *R. leptorrhynchoides* have suffered genetic erosion, allelic richness and heterozygosity were assessed at nine allozyme loci for 22 of the 24 known populations using horizontal starch gel electrophoresis [see Young *et al.* (1999) for details of methods]. In each population allozyme variation was assayed in 35 seed sampled evenly from as many plants as possible up to a maximum of 35. Measures of genetic variation were then regressed against reproductive population size measured as number of flowering plants and three different measures of isolation: (1) distance to nearest population, (2) distance to nearest larger population and (3) average distance to all other populations within a geographical region.

Although *R. leptorrhynchoides* has previously been reported to be uniformly diploid (Leeton & Fripp, 1991), allozyme analyses showed that the five western Victorian populations are made up of tetraploid plants. This was confirmed by chromosome counts that revealed plants in these populations to be almost exclusively $2n = 44$ rather than $2n = 22$ as found in the two eastern Victorian populations that were surveyed, and in populations in the north (Figs. 19.1 and 19.2). The frequent formation of quadrivalents during meiotic metaphase I suggests that plants are autotetraploids (B. G. Murray & A. G. Young, unpublished data). This is supported by segregation patterns from controlled crosses between putative tetraploids which are in line with expectations for a tetrasomic model of inheritance (Brown & Young, in press).

Allelic richness for diploid populations was strongly positively correlated with the log of reproductive population size (Fig. 19.3a). Below 5000 plants, some loss of alleles was evident and below about 200 plants these

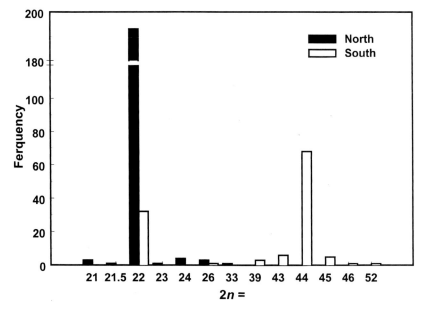

Fig. 19.2. Chromosome numbers for *Rutidosis leptorrhynchoides* in 15 northern and eight southern populations based on mitotic root-tip counts. B. G. Murray & A. G. Young unpublished data.

losses are severe, with populations having only about one-half to two-thirds the number of alleles found in the larger populations. Analysing rare ($q < 0.1$) and common ($q > 0.1$) alleles separately (Fig. 19.3b) shows that most of the genetic erosion in small populations is due to the elimination of low-frequency alleles, as is expected as a result of random sampling due to founder effects and genetic drift. There was no effect of any of the three isolation measures on allelic richness. In contrast to the situation in diploids, allelic richness in the five tetraploid populations is apparently unrelated to population size. These populations consistently have a greater number of alleles for a given population size than their diploid counterparts (Fig. 19.3a).

To allow direct comparison of heterozygosity between diploids and tetraploids, partial heterozygotes were scored following Bever & Felber (1992): $AAAA = 0$, $AAAB = 0.5$, $AABB = 0.66$, $AABC = 0.83$, $ABCD = 1$ and expected heterozygosity was calculated assuming chromosomal segregation with no double reduction following Geringer (1949). Neither observed nor expected heterozygosity was related to population size or isolation for either diploids or tetraploids (data not shown), but again, tetraploid heterozygosity (mean $H_o = 0.34$, SE $= 0.04$; $H_e = 0.36$, SE $= 0.04$) was consistently

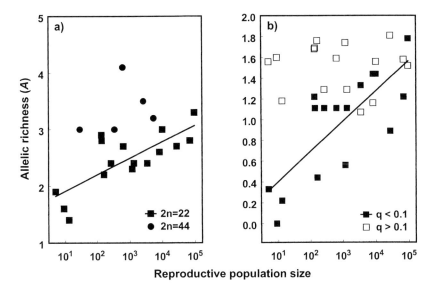

Fig. 19.3. Allelic richness at nine allozyme loci (*Aat-1, Aat-2, Aat-3, Adh-1, Gpi-2, Mdr-1, Pgm-1, Pgm-2, Pgm-3*) in *Rutidosis leptorrhynchoides* populations: (a) diploid vs. tetraploid populations, $R^2 = 0.56$, $P < 0.001$; (b) diploid populations: common ($q > 0.1$) vs. rare ($q < 0.1$) alleles, $R^2 = 0.59$, $P < 0.001$. Data from Young *et al.* (1999) and Brown & Young (in press).

higher than that observed in diploids (mean $H_o = 0.22$, $SE = 0.07$; $H_e = 0.25$, $SE = 0.08$).

These results show that low genetic variation may limit the long-term conservation value of remnant populations of diploid *R. leptorrhynchoides*, in particular the half of the current populations that have fewer than 200 reproductive plants and for which genetic losses are significant. However, the consistently higher levels of allelic richness and heterozygosity in the tetraploid populations suggest that minimum threshold population sizes for maintenance of genetic variation are lower for these than for diploids. For example, even the smallest tetraploid population of 28 plants at Wickcliffe exhibits a greater number of alleles ($A - 3.0$, $SE = 0.3$) and higher heterozygosity ($H_o = 0.43$, $SE = 0.10$) than the average for diploids.

Loss of self-incompatibility alleles

Multi-allelic sporophytic self-incompatibility systems, such as that found in *Rutidosis leptorrhynchoides*, are effective at preventing inbreeding through self-fertilisation in large plant populations because frequency-dependent selection operates to maintain high numbers of *S* alleles at fairly even frequencies (Richman & Kohn, 1996). However, if the number of S alleles becomes low, reductions in the possible number of compatible genotypes will limit mate availability, reducing effective population size, and possibly resulting in lower fecundity (Imrie *et al.*, 1972; Vekemans *et al.*, 1998). This represents a direct link between genetic erosion and demographic performance. Though previous studies have found S allele richness to be high even in quite small populations (Wright, 1964), the strong relationship between population size and numbers of allozyme alleles suggests that loss of S allele richness in small *R. leptorrhynchoides* populations is quite possible.

To test this hypothesis, controlled crosses were used to estimate the relative numbers of S alleles present in each of four diploid populations of different sizes: West Block (five plants), Captains Flat (161 plants), Queanbeyan (10 000 plants) and Stirling Ridge (70 000 plants). For each of the three larger populations single plants were grown from 12 open-pollinated families and crossed in a full diallel design (12 × 12), while for West Block single plants grown from open-pollinated seed from each of the five remaining plants were crossed giving a 5 × 5 diallel. Reciprocal crosses were conducted by bagging flower heads just prior to opening to exclude pollinators, and then hand-pollinating them by rubbing heads together three or four times over the following seven days. Seed set was assessed four to six weeks after pollination.

Minimum numbers of S alleles were estimated as twice the number of plants in the largest self-compatible group for the three larger populations. For the small West Block population putative S allele genotypes were worked out for all individuals based on their compatibility reactions with each other plant in the crossing array. Mate availability in all populations was calculated as the proportion of the other plants within the sample with which a plant could mate as either a female or a male.

Estimates of S allele richness in the population samples ranged from 16 in the largest population of 70 000 plants at Stirling Ridge down to three alleles in the smallest population of five plants at West Block. Regression analysis showed a strong positive relationship between the logarithm of reproductive population size and the number of S alleles in a sample ($R^2 = 0.92$, $P = 0.03$), closely paralleling the data for allozymes ($R^2 = 0.81$, $P = 0.06$) (Fig. 19.4a). Both male and female components of mate availability were also reduced as population size decreased, and this was linearly related to S allele richness ($R^2 = 0.86$, $P = 0.05$ for both) (Fig. 19.4b).

The significant loss of S alleles with reduced population size suggests that frequency-dependent selection has not protected the S locus from gen-

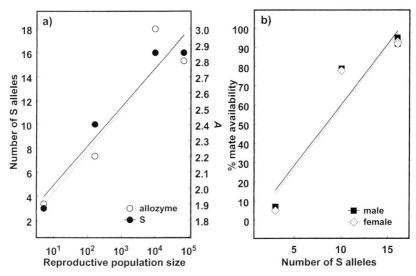

Fig. 19.4. (a) *S* allele richness at the incompatibility locus based on diallel crosses and at nine allozyme loci (from Fig. 19.2) in relation to population size, and (b) male and female components of mate availability in relation to *S* richness in four diploid populations of *Rutidosis leptorrhynchoides*.

etic erosion. This loss of *S* allele richness, and the correlated reduction in mate availability, have several implications for small *R. leptorrhynchoides* populations. Firstly, genetically, reduced *S* allele richness lowers the effective population size, further exposing populations to genetic drift. Using the observed relationships between population size, *S* alleles and mate availability, we predict that mate availability will drop below 50% in diploid populations smaller than approximately 60 reproductive plants. Secondly, demographically, low mate availability may reduce population-level seed set while increasing interplant variance in fecundity, as any remaining individuals with unique *S* genotypes experience increased fitness. Thirdly, severe mate limitation will also favour self-compatible plants that occur at low frequencies (< 15%: Young *et al.*, 1999), shifting the mating system of small populations away from predominant cross-fertilisation towards mixed mating. Such bottleneck-induced dissolution of self-incompatibility has been observed previously for the daisy *Aster furcatus* by Reinartz & Les (1994). The likelihood of this occurring depends critically on the interacting effects of mate limitation, pollen competition and inbreeding depression. The mechanisms underlying the shift to self-compatibility (i.e. mutation to a null allele, reversal of dominance relationships in the style and stigma or mutual weakening) (De Nettancourt, 1977), will also be important in promoting or limiting any spread of self-compatibility.

The expectations for effects of S allele erosion on small populations of tetraploids are not as clear-cut. Though the frequency of interplant incompatibility may be greater owing to the greater potential for matching a single allele between two $4x$ individuals, higher allelic richness in tetraploid populations, as evidenced by the allozyme results, might mitigate this effect to some degree. Data comparing seed set from controlled crosses among half-sib and unrelated plants in diploids and tetraploids indicate that tetraploids show about a 30% reduction in mate availability over diploids for a given level of relatedness (Young *et al.*, 1999). Based on this, genetic and demographic effects of erosion at the S locus are probably more pronounced in tetraploid populations. Intriguingly, Morgan (1999) has identified strong relationships between population size and seed set over several years for populations of *R. leptorrhynchoides* in Victoria, which are mostly tetraploid.

INBREEDING

The second main genetic threat to short-term viability of small populations is reduced fitness owing to increased selfing or mating among relatives (biparental inbreeding). Inbreeding depression is likely to be particularly severe for species like *R. leptorrhynchoides* that are predominantly outcrossers and so maintain high genetic load (Fenster & Dudash, 1994). Shifts towards inbreeding following fragmentation may result directly from the reduced number of potential mates in small populations, or indirectly from reduced pollinator activity or changed pollinator behaviour. Effects of both population size and plant density on rates of cross-fertilisation (outcrossing rates) have been observed for several herbaceous species (e.g. Karron *et al.*, 1995a; Raijmann *et al.*, 1994) and in some instances increased selfing in small populations has been correlated with low progeny fitness (Oostermeijer *et al.*, 1994a; Buza *et al.*, 2000).

In the case of *R. leptorrhynchoides* there are three possible changes in the mating system that could occur in small mate-limited populations: (1) increased selfing, due to breakdown of the self-incompatibility system and selection for self-compatible plants; (2) increased biparental inbreeding due to an increased proportion of matings among relatives, which is possible in self-incompatible species because dominance hierarchies allow some cross-fertilisation between individuals that share only a single S allele; and (3) increased correlation of outcrossed paternity (i.e. generation of paternal bottlenecks) due to dominance of the pollen pool by a few plants with unique S genotypes.

To examine whether reduced population size and isolation have affected mating events in *R. leptorrhynchoides*, allozyme markers were used to estimate multilocus (t_m) and single-locus (t_s) outcrossing rates along with the correlation of outcrossed paternity (r_p) (Ritland, 1989) in nine diploid populations in three size classes: large (1000s), medium (100s) and small (5), and two isolation classes: $< 2\,km$ and $> 5\,km$ (Table 19.1). All estimates were based on samples of 15 seed bulked from several flower heads for each of 15 mothers (225 total seed per population), and eight allozyme loci, except for the smallest population at West Block for which seed were collected from all five remaining plants (Young & Brown, 1999).

Multilocus outcrossing rates were uniformly high regardless of population size or isolation indicating that cross-fertilisation is maintained even in very small populations (Table 19.1). Differences between multilocus and single-locus outcrossing rate estimates ($t_m - t_s$), which are a measure of biparental inbreeding, were not significantly different from zero indicating that mating among relatives is limited. This general lack of inbreeding of either sort is supported by the low fixation coefficients observed in all of the populations examined (Table 19.1). Based on this, and the maintenance of heterozygosity in small populations, inbreeding depression is unlikely currently to represent a major threat to the viability of small remnant *R. leptorrhynhoides* populations.

In contrast to the apparent stability of rates of cross-fertilisation, patterns of paternity were affected by both population size and isolation, with small populations and medium-sized populations isolated by $> 5\,km$ having significantly higher production of full-sib progeny within open-pollinated families (Fig. 19.5). Two out of three of these populations (West Block and Captains Flat) also showed an order of magnitude increase in divergence between adult and pollen pool allele frequencies (Table 19.1). Combined, these two results mean that relatively few males are dominating the pollen pool. Such reduced male diversity in the pollen pool suggests that loss of *S* alleles has led to skewed male fitness distributions and reduced effective population sizes. While an increase in the production of full-sib progeny in the current generation is not in itself inbreeding, it does predispose small or isolated populations to the possibility of future biparental inbreeding, as well as to more severe limitation of mates in subsequent generations, especially as seed dispersal is limited.

POPULATION VIABILITY ANALYSES

The results presented so far show that significant changes in both genetic diversity and mating events occur in small isolated populations of *Rutidosis*

Table 19.1. Mating system parameters for *Rutidosis leptorrhynchoides* populations based on eight allozyme loci: Aat-1, Aat-2, Aat-3, Gpi-2, Mdr-1, Pgm-1, Pgm-2, Pgm-3 (standard errors in brackets)

Population	Number of flowering plants	Isolation	Maternal fixation coefficient (F)	Multilocus outcrossing rate (t_m)	Biparental inbreeding ($t_m - t_s$)	F_K[a]
Large						
Stirling Ridge	70 000	< 2 km	0.007	0.84 (0.06)[b]	0.02 (0.03)	0.004
Queanbeyan	10 000	< 2 km	0.013	0.92 (0.06)	0.01 (0.04)	0.008
Mean				0.88		0.006
Goulburn	95 200	> 5 km	0.011	0.96 (0.04)	0.05 (0.01)	0.003
Majura	27 626	> 5 km	0.009	0.93 (0.04)	0.09 (0.02)	0.009
Mean				0.95		0.006
Medium						
Capital Circle	220	< 2 km	0.001	0.86 (0.06)[b]	0.01 (0.03)	0.019
Barton	133	< 2 km	0.003	0.93 (0.06)	0.04 (0.04)	0.005
Mean				0.90		0.012
Captains Flat	161	> 5 km	0.005	0.92 (0.07)	0.04 (0.05)	0.044
St Albans	137	> 5 km	0.003	0.94 (0.06)	0.03 (0.02)	0.006
Mean				0.93		0.025
Small						
West Block	5	< 2 km	0.0	1.0 (0.01)	0.04 (0.01)	0.098

[a] Pollak estimator of allele frequency differentiation between parental and pollen gene pools.
[b] Significant at $P < 0.05$.
(Data from Young & Brown, 1999.)

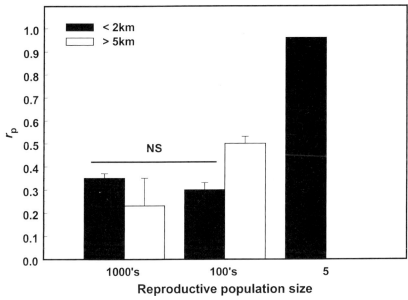

Fig. 19.5. Correlation of outcrossed paternity (r_p) for nine diploid populations of *Rutidosis leptorrhynchoides* in relation to population size and isolation. Based on eight allozyme loci (*Aat-1, Aat-2, Aat-3, Gpi-2, Mdr-1, Pgm-1, Pgm-2, Pgm-3*). NS: not significant at $P < 0.05$. Data from Young & Brown (1999).

leptorrhynchoides and that associated mate limitation provides a possible link between these genetic changes and reduced effective population size and fecundity. To assess the effects of these changes on population viability, multi-year demographic data have been collected from remnant populations that vary in their levels of genetic diversity (*S* alleles/allozymes) and mating system parameters. Here data for five populations, from the first two years of this study, are used to compare population growth rates and to estimate parameter values of stochastic matrix projection models to examine differences in population persistence times.

StA. G. Young & udy populations and data collection
The five populations studied ranged in size from 5 to 70 000 plants and encompassed both < 2 km and > 5 km isolation classes, as well as having different levels of allozyme and *S* allele variation and degrees of correlated paternity (Table 19.2). In each of the three larger populations eight to 16 1-m² study plots were the basis of demographic monitoring, while in the smaller populations at West Block and Jerrabomberra all plants were monitored. Within each plot all plants were tagged and mapped in the

Table 19.2. Demographic transition probabilities averaged over plots and years, and population growth rates for five *Rutidosis leptorrhynchoides* populations

Population	Stirling Ridge	Capital Circle	Captains Flat	Jerrabomberra	West Block	P
Number of flowering plants	70 000	220	161	9	5	
Isolation	< 2 km	< 2 km	> 5 km	< 2 km	< 2 km	
Allelic richness (A)	2.8	2.4	2.2	1.6	1.9	
Correlated paternity (r_P)	0.33	0.32	0.53	—	0.96	
Number of S alleles	16	—	10	—	3	
			Transition probability (SE)			
Seedlings per reproductive plant (f_{rs})	1.29 (0.23)	1.94 (0.82)	2.86 (0.78)	0	0	
Seedling to:						
Seedling (a_{ss})	0.13[a] (0.03)	0.18 (0.04)	0.30[a] (0.07)	0.13 (0.09)	0.50 (0.35)	0.03
Juvenile (a_{js})	0.09 (0.05)	0.03 (0.02)	0.01 (0.01)	0	0	
Vegetative (a_{vs})	0.01 (0.01)	0.01 (0.01)	0.02 (0.02)	0	0	
Survival	0.23	0.22	0.33	0.13	0.50	
Juvenile to:						
Juvenile (a_{jj})	0.43 (0.09)	0.38 (0.12)	0.45 (0.12)	—	—	
Vegetative (a_{vj})	0.06[a] (0.03)	0.25 (0.10)	0.29[a] (0.09)	—	—	0.02
Reproductive (a_{Tj})	0.10 (0.05)	0.03 (0.02)	0	—	—	
Survival	0.59	0.66	0.74	—	—	

Vegetative to:			
Vegetative (a_{vv})	0.56 (0.16)	0.39 (0.09)	0.35 (0.14)	—
Reproductive (a_{rv})	0.30 (0.14)	0.39 (0.11)	0.50 (0.14)	—
Survival	0.86	0.78	0.85	—
Reproductive to:			
Vegetative (a_{vr})	0.04 (0.01)	0.17 (0.11)	0.13 (0.04)	0
Reproductive (a_{rr})	0.92 (0.02)	0.80 (0.11)	0.86 (0.04)	0.95 (0.06)
Survival	0.96	0.97	0.99	0.95
Growth rate (λ)	0.99	0.97	1.03	—

[a] Different at $P < 0.05$.

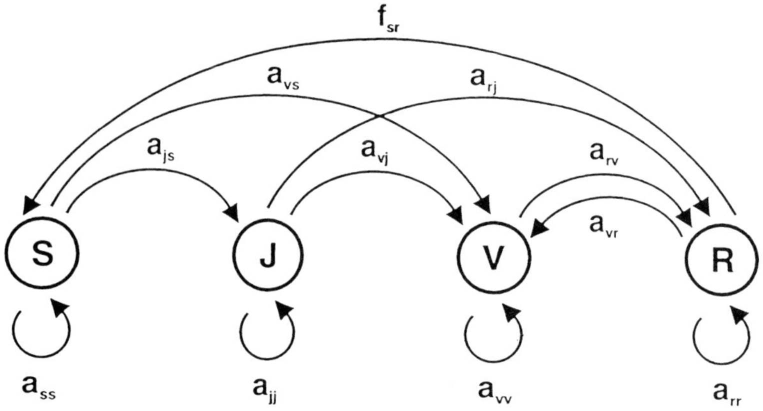

Fig. 19.6. *Rutidosis leptorrhynchoides* population model structure. S, seedling; J, juvenile; V, vegetative; R, reproductive. a_{js} is the transition probability from seedling to juvenile.

spring of 1995 (approximately 750 individuals in total). Plants were divided into four classes. Seedlings (S) were plants < 1 year old or < 3 cm in height. Juveniles (J) were plants > 3 cm in height but single-stemmed and therefore not big enough to be reproductive. Vegetative plants (V) were multi-stemmed but with no flowers. Reproductive plants (R) were multi-stemmed individuals that were flowering.

Plots were monitored four times a year. In late spring (November), survival and growth of all individuals were assessed, along with counts of flowers on all reproductive individuals. In late summer (February) seed set was assessed by bagging three flower heads on each of two plants per plot prior to seed dehiscence. These were collected in March–April and filled seed were counted. Seedling germination was assessed in May and again in July, at which time survival and growth of plants in all classes were reassessed. As *R. leptorryhnchoides* has no long-term seed bank it was assumed that all seedlings emerging during this period came from that year's seed crop.

Population growth rates and elasticity analysis

Monitoring data were collapsed to give a single-year population model of the form shown in Fig. 19.6 and were used to estimate transition probabilities among classes for each plot. Plot-level estimates were averaged over plots and years to produce a mean population transition matrix consisting of 11 possible elements in a 4×4 array. Population matrices were used to calculate population growth rates (λ), as well as to conduct elasticity analyses to quantify the relative importance of each transition probability on growth rate, following Caswell (1989).

Transition matrices and growth rates for all populations are presented in Table 19.2. Several transition probabilities could not be estimated for the two smallest populations as these events were not observed during the two-year monitoring period. As a result it was also impossible to calculate growth rates for these populations. There were only two significant differences among the five populations for transition probabilities. In both cases, seedling to seedling (a_{ss}) and juvenile to vegetative (a_{vj}), probabilities were significantly higher in the isolated Captains Flat population ($n = 161$) than in the large Stirling Ridge population ($n = 70\,000$). Growth rates for all three populations were close to 1.0 suggesting that these populations are performing in a similar fashion and are fairly stable.

Elasticities are based on the eigenvectors of the population transition matrices and give the proportional contribution of each demographic pathway to the overall population growth rate. Values for the three populations examined are presented in Fig. 19.7 and show that the most important transitions are those involving survival of vegetative and reproductive individuals. Seedling recruitment appears to have relatively little impact on growth rate. These results were similar despite differences in population size, isolation, genetic diversity or level of correlated paternity.

Stochastic simulation models

The previous analyses of growth rate are based on mean transition probabilities and so represent a steady-state model of population performance, taking no account of among-plot or among-year demographic variation, a factor which based on the differences in standard errors of parameter estimates (Table 19.2) is likely to be significant. Indeed skewed male and female fitness distributions associated with low S allele diversity will more directly influence the variance of demographic parameters than the mean. This may also be true for any second-order effects due to biparental inbreeding.

To assess the influence of demographic variability on population viability, a stochastic matrix projection model was developed using the same data and model structure as used previously for the steady-state analyses. The variance in the data was explicitly incorporated by sampling transition matrix elements from the possible sets of transition probabilities among plots and years to build synthetic population matrices that were then normalised so that values for each matrix row summed to 1.0. This transition matrix was then used to calculate the change in numbers of seedlings,

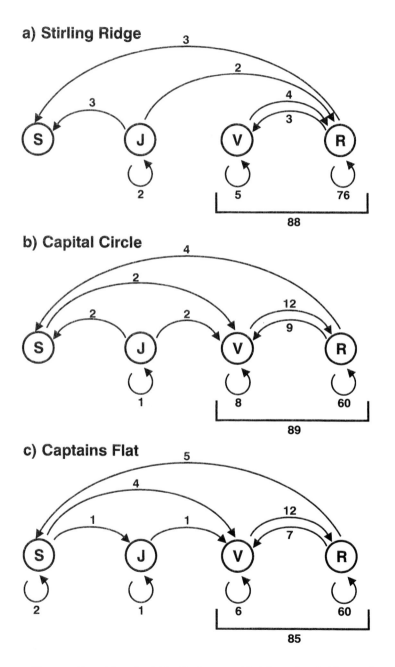

Fig. 19.7. Transition elasticities for three *Rutidosis leptorrhynchoides* populations: Stirling Ridge (*n* = 70 000), Capital Circle (*n* = 220) and Captains Flat (*n* = 161). S, seedling; J, juvenile; V, vegetative; R, reproductive.

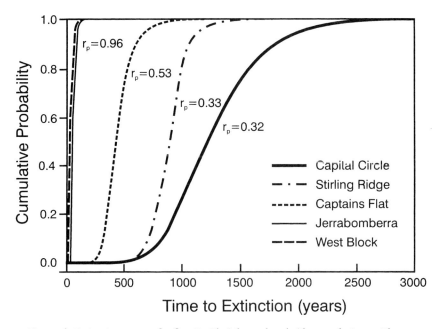

Fig. 19.8. Extinction curves for five *Rutidosis leptorrhynchoides* populations with different genetic parameters (see Table 19.2) based on stochastic matrix projection models.

juveniles and so on from time t to $t+1$. A new transition matrix was then generated and the process was repeated. For each population 2000 replicate runs of the simulation were conducted and cumulative extinction probabilities were calculated, using the 1995 population structures as starting points. For the small West Block and Jerrabomberra populations, missing matrix elements were replaced with elements from the large Stirling Ridge population. This will probably result in overestimates of persistence times for these populations.

Extinction curves for all five populations are presented in Fig. 19.8 and these show, in contrast to the results from the analysis of growth rates based on mean transition matrices, that the different populations have quite different long-term viabilities. Both of the small populations consistently went extinct very quickly, always in less than 150 years. The medium-sized high-isolation population at Captains Flat was the next most likely to go extinct at any point in time, followed by the large Stirling Ridge population and finally the medium-sized low-isolation population at Capital Circle.

These results suggest a general positive relationship between genetic

Table 19.3. Population locations, size and establishment data and genetic parameters for wild and re-established populations of *Rutidosis leptorrhynchoides* in Victoria (standard errors in brackets, * $P < 0.05$)

Population	Location	Wild source	Date established	Population size planted/survived to 1996	Allelic richness (A)	Heterozygosity (H_o)	Fixation coefficient (F)	Number of cytotypes	Number of plants counted
Wild									
Middle Creek	Middle Creek	—	—	610	4.0 (0.89)	0.40 (0.07)	−0.07 (0.01)	2	20
Rokewood	Rokewood	—	—	5419	3.6 (0.40)	0.39 (0.05)	−0.04 (0.01)	5	24
Mean					3.8 (0.20)*	0.39 (0.01)	−0.06 (0.01)	3.5 (1.5)	
Overall					4.4 (0.75)	0.40 (0.06)	−0.05 (0.01)	5	44
Re-established									
MC88	Mooramong	Middle Creek	1988	100/90	3.4 (0.51)	0.35 (0.07)	−0.04 (0.01)	5	17
MC89	Mooramong	Middle Creek	1989	—/300	3.2 (0.49)	0.37 (0.06)	−0.01 (0.02)	2	14
RK87	Mooramong	Rokewood	1987	100/140	3.0 (0.32)	0.40 (0.05)	−0.09 (0.01)	6	16
RK89	Mooramong	Rokewood	1989	76/65	3.4 (0.51)	0.40 (0.06)	−0.05 (0.01)	4	13
MC92	Middle Creek	Middle Creek	1992/93	52/25	3.0 (0.31)	0.36 (0.10)	−0.07 (0.02)	2	11
Mean					3.2 (0.09)*	0.38 (0.01)	−0.05 (0.01)	3.8 (0.8)	
Overall					4.0 (0.55)	0.38 (0.06)	−0.03 (0.01)	9	71

Data from Young & Murray, in press.

diversity, mate limitation and persistence time. It is impossible to escape the fact that the results show correlation rather than direct causality, and that much of the observed effect could be due to reduced population size and increased demographic stochasticity *per se* rather than genetic effects. However, the high viability of the Capital Circle population relative to the low viability of the population at Captains Flat represents a clear contrast between two populations that are similar in size, but that vary in mating behaviour. Nevertheless, it cannot be discounted that the very proximity (< 2 km) to a large population that keeps correlated paternity down at Capital Circle may also be allowing direct demographic rescue by seed immigration.

EVALUATION OF RE-ESTABLISHED POPULATIONS

As with many endangered plant species, re-establishment of new populations has been a feature of recovery actions for *Rutidosis leptorrhynchoides*. Five populations have been re-established in Victoria since 1987, using seed and tubestock derived from two of the western tetraploid populations at Rokewood and Middle Creek (Table 19.3). Four of these populations were established within the Mooramong Nature Reserve, some 30 km from Middle Creek and 40 km from Rokewood. The fifth population, MC92, is located 2.5 km from the wild Middle Creek population.

From the previous results regarding *S* allele erosion and the known cytogenetic complexity of the species in the southern part of its range, it is clear that the persistence of these five populations will depend on their containing enough genetic variation to avoid mate limitation, and being chromosomally uniform enough so as to avoid fitness effects of dysgenesis. However, these populations were established prior to the availability of any genetic information, and numbers of founders were uniformly less than 100 plants.

To assess the conservation value of these re-established populations, allozyme and chromosomal diversity were compared between them and their two wild progenitors. The main questions of interest were: (1) has the re-establishment procedure resulted in a genetic bottleneck that might limit reproduction due to *S* allele effects, and (2) has there been any mixing of tetraploid and diploid cytotypes that could result in infertility?

Seed was collected from up to 24 plants from each of the two wild and five re-established populations. From each of these samples a single seed was germinated and grown for three months at which point its chromosome number was obtained using root-tip squashes following Young &

Murray (in press). From 10 of these families a further five seedlings were germinated and assayed for allozyme variation at five loci (*Aat-1*, *Aat-2*, *Aat-3*, *Gpi-2*, *Mdr-1*) following Young *et al.* (1999).

Polymorphism was uniformly high ($P = 100\%$) in all populations. However, mean allelic richness was significantly lower in the five re-established populations ($A = 3.2$, $SE = 0.09$) than in their two progenitors ($A = 3.8$, $SE = 0.2$), $P < 0.05$ (Table 19.3). With three exceptions alleles in the new populations were a subset of those found in the two wild populations. There were no differences in either heterozygosity or fixation indices between wild populations and their derivatives (Table 19.3).

The most common chromosome number found in all populations was $2n = 44$. Surprisingly, samples from the wild progenitor populations at Rokewood and Middle Creek contained five cytotypes, while the new populations had nine. Not all of this difference could be accounted for by the larger overall sample size from new populations, with the probability of obtaining a difference of four cytotypes due to sampling effects alone being $P < 0.005$. The mean number of cytotypes per population was also slightly higher in re-established populations than wild ones, but not significantly so (Table 19.3).

Both chromosome and allozyme data show that the re-established populations have a different genetic make-up compared to their wild progenitors. The average reduction in allelic richness of approximately 20% in re-established populations represents a fairly dramatic loss of genetic diversity. If there have been parallel losses of *S* alleles, as is expected based on the close relationship between allozyme and *S* allele richness observed in diploid populations, fecundity in re-established populations may well be negatively affected due to increased mate limitation.

In contrast to the predictable loss of allozyme variation, the increase in cytotype diversity in new populations was unexpected. There are several possible origins of these new chromosome numbers. Aneuploids and chromosomal rearrangements could have arisen as a result of increased rates of disjunctional errors at meiosis under increased inbreeding. Several previous studies have implicated increased inbreeding as a factor that can result in reduced chromosomal stability and decreased rates of chiasma formation that lead to production of unbalanced gametes (Rees, 1961; Jones, 1969; Tease & Jones, 1976). Indeed, Parker & Wilby (1989) argue that elevated chromosome polymorphism in a small island population of *Rumex acetosa* is due to just such effects under enforced biparental inbreeding. However, the low fixation coefficients observed here for all re-established populations argue against this as a mechanism in the case of *R. leptorrhynchoides*.

The second possibility is more interesting. The occurrence of triploid $2n = 33$ plants in the new populations in the apparent absence of $2n = 22$ plants raises the possibility that the genetic architecture of the new populations at Mooramong is being influenced by pollen immigration from diploid populations, though the nearest known one of these is some 150 km away. Further support for this hypothesis comes from the presence of two allozyme alleles (*Aat*-2f and *Aat*-2v) in the Mooramong populations that are found nowhere else other than in the south-eastern diploid population at St Albans. Allele frequency and triploid frequency differences between these populations give immigration rate estimates of $N_m = 0.01$–0.17, suggesting that between 1% and 17% of the pollen present at Mooramong has originated from St Albans. However, the very large distances between these populations argue against direct immigration. The other alternatives are that: (1) source material from both diploid and tetraploid populations was in fact mixed at the outset of the re-establishment process, or (2) there is an unknown diploid population of *R. leptorryhchoides* near Mooramong. Whatever its origin, the increased level of dysgenesis in small new populations can only further reduce population viability.

CONCLUSIONS AND MANAGEMENT IMPLICATIONS

The results of this study show that habitat fragmentation has resulted in a significant loss of genetic variation both at neutral allozyme loci and at the self-incompatibility locus in diploid populations of *Rutidosis leptorrhynchoides* of < 200 plants occupying remnant grassland vegetation. The reduction in *S* alleles in these populations has been accompanied by limitation of mate availability. The correlated increase in variance in male fitness evidenced by high correlation of paternity in small populations suggests a direct link between *S* allele erosion, plant fitness and population demography. Indirectly, these effects also reduce effective population size, further exposing small remnant populations to genetic drift and possibly biparental inbreeding in future generations.

The lower persistence times observed for small and medium-sized, genetically depauperate populations relative to large genetically diverse ones suggest that observed genetic and demographic differences translate into reduced population viability. The apparent difference in results between the steady-state and stochastic population models, with steady-state showing persistence of all medium and large populations, and the stochastic model indicating differing extinction probabilities, reinforces the importance of inter-plot and inter-year variation in demographic parameters in determining population viability. This is intriguing as this variance is likely to be strongly affected by the observed genetic changes.

Inbreeding appears of little immediate importance in determining population viability, though future effects of biparental inbreeding cannot be ruled out. More interesting is the possibility that severe mate limitation due to low *S* allele diversity might result in dissolution of the incompatibility system in small populations, via selection for self-compatible plants which occur at low frequencies. Thus it may be that in the long term, human-induced habitat destruction and fragmentation could actually drive an evolutionary shift in the mating system of this species.

Overall these data suggest that populations of less than 200 diploid plants are of limited conservation value, even in the relatively short term. This is of some concern as currently over half of the populations of *R. leptorrhynchoides* fall into this category. Future population management needs to focus on maintaining adult plant numbers so as to preserve *S* allele richness and genotype diversity to avoid mate limitation. The high elasticity values for demographic transitions involving reproductive and vegetative plants also point to the demographic importance of adult survival in influencing population persistence. Thus management strategies that promote adult survival should be favoured. Effective conservation of populations currently smaller than 200 plants is likely to require intervention in terms of introduction of new *S* allele variants. We are currently in the process of identifying genetically compatible plants from the large Stirling Ridge population for reintroduction into the small mate-limited West Block population.

Future establishment of new *R. leptorrhynchoides* populations must encompass adequate *S* allele diversity when sourcing material for planting. Collections should be made from the largest available local population, preferably > 5000 plants. In the absence of such a population, mixing seed from several small source populations is probably appropriate, except where this will result in the mixing of ploidy levels.

How these results translate from diploids to the five tetraploid populations in Victoria is not clear. Effects of *S* allele erosion on mate availability in tetraploid populations are complicated by their ability to retain more genetic diversity than their diploid counterparts. Also, assuming a partial-dominance model of inbreeding depression, tetraploids will be less affected by biparental inbreeding (Husband & Schemske, 1997). Counteracting this, however, is the fact that, for a given level of *S* allele variation, tetraploids suffer more from mate limitation than diploids due to increased probability of matching *S* alleles. While further research is needed to clarify these points, it is already apparent that the biological differences and physical disjunction between diploid and tetraploid populations merit their separate consideration in conservation planning.

ACKNOWLEDGMENTS

This work was funded in part by research grants from the New South Wales National Parks and Wildlife Service and Environment ACT. Melinda Pickup assisted with pollination experiments for estimation of *S* allele richness and Liz Gregory assisted with allozyme analyses.

20

Conclusions and future directions: what do we know about the genetic and demographic effects of habitat fragmentation and where do we go from here?

ANDREW G. YOUNG & GEOFFREY M. CLARKE

There can be little doubt that management of fragmented populations of both animals and plants has become an important element of biological conservation, and will continue to be into the new millennium. Many of the species currently recognised as threatened are restricted to habitat fragments, and the option that was available a century ago, of *in-situ* conservation within relatively pristine environments, is gone. The continuing global trend towards non-sustainable exploitation of natural resources means that more, rather than fewer, species are going to be affected by habitat loss, degradation and fragmentation in the future. This is particularly concerning as the pressure shifts to tropical ecosystems about which we have much less information on the basic biology of many organisms, and so educated guesses as to biological responses are harder to make.

The facts are now, that if many of the world's endangered species are to be preserved at all, it will be either *ex situ* in zoos, aquaria or botanical gardens, or *in situ* in fragmented ecosystems. In the case of *ex-situ* conservation, we can hope to do little more than preserve a snapshot of evolutionary diversity, as maintenance of genetic and demographic processes becomes almost impossible in the 'zoo' context. Avoidance of even common problems, such as loss of stored seed viability through regular sexual regeneration of plant material, is time-consuming and costly and, in any large way, simply impractical. Similarly, the globalisation of breeding programmes among captive-bred populations of animals (an incredibly expensive process), while reducing the immediate problems of inbreeding depression, has limited potential for the long-term maintenance of genetic diversity.

The option of *in-situ* conservation of populations, even in highly fragmented habitats, brings with it the potential for some maintenance of demographic and genetic processes and the prospect of future evolution. Another advantage of this course of action is the maintenance of mutualisms and continued provision of ecosystem services by organisms in habitat fragments. For example Srikwan & Woodruff (Chapter 9) point to the roles played by mammals in seed predation and dispersal of both fruits and mycorrhizal fungi. Ecosystem support roles can also stretch beyond the bounds of the 'natural' system. For instance, it is now well recognised that remnant stands of deep-rooted eucalypt trees play a critical role in maintaining productivity of agricultural systems in Western Australia by keeping water tables low and preventing soil salination (McFarlane *et al.*, 1993).

So how well equipped are we to make informed decisions that will promote the long-term persistence of fragmented habitats and their constituent species? Specifically, can we identify how fragmentation influences demographic and genetic processes and their interactions to determine population viability? Perhaps more importantly, does having this information allow for improved management, or have Lande's (1988) and Caughley's (1994) (often-misinterpreted) calls for greater research integration paid few practical dividends? Though the answers to all of the questions are not entirely clear, it is fair to say that the recent combined application of genetic markers, demographic monitoring and computer simulation modelling, as illustrated by the case studies in this volume, has provided some significant insights into the workings of fragmented populations.

Several authors (e.g. Hedrick & Miller, 1992; Frankham, 1995*b*; Moritz, 1999*b*; Burgman & Possingham, Chapter 6) point to the different uses of genetic information in studying fragmented populations. Firstly, it can be used to identify the unit of conservation concern, that is to define management units by their genetic distinctiveness. A good example of this is the identification of diploid and tetraploid races of the endangered daisy *Rutidosis leptorrhynchoides* (Young *et al.*, Chapter 19), which probably have different genetic, evolutionary and ecological responses to similar disturbance events and should be managed separately. Secondly, genetic markers can provide information about the dynamics of cryptic demographic processes, as shown by Hedrick for salmon (Chapter 7) and Clarke (Chapter 12) in the case of estimating population sizes of the golden sun moth. Thirdly, and by far the most contentious, is in determining whether genetic processes, primarily genetic erosion and inbreeding, actually play a significant role in reducing the viability of small fragmented populations over and above either the influences of external deterministic threats, or the demographic variation inherent in small populations.

Several of the studies in this volume, and much other research reported elsewhere, provide good evidence that reductions in population size that accompany habitat loss and fragmentation are associated with reductions in genetic variation as measured by allelic richness, for a wide range of both animal and plant species, with varying ecologies, reproductive strategies and patterns of abundance. This process of genetic erosion through time has now been explicitly documented by Srikwan & Woodruff (Chapter 9) and appears to represent a general genetic response – a paradigm of habitat-fragmentation effect. In this context, however, the issue is the relationship of this loss, of what are generally low-frequency alleles at neutral marker loci, and population persistence. There are several possible linkages. The first is that the loss of these alleles is symptomatic of what is going on at other more evolutionarily important parts of the genome. In this context, such losses may be detrimental in the long term by reducing the amount of genetic variation upon which selection can act. There are two main objections to this line of argument. Firstly, much of the adaptively significant phenotypic variation in both plants and animals is under polygenic control and such multilocus genetic variation responds rather differently to changes in population size than single-locus markers (Sherwin & Moritz, Chapter 2). The dearth of empirical studies on the effects of habitat fragmentation on quantitative genetic variation is a clear gap in the current research agenda. The second objection is that although significant losses may be observed in single populations, regional variation may not be much reduced, as different populations are likely to lose different suites of alleles, and thus represent complementary sources of gene immigration if moderate levels of gene flow are maintained. If gene flow is limited, however, subsequent losses of small populations for whatever reason may result in extirpation of alleles, reducing overall evolutionary potential. Even if populations are not lost, the increase in regional-level linkage disequilibrium may slow the generation of new multilocus genotypes, impairing adaptive responses to changing selection pressures. Unfortunately, these hypotheses are almost untestable, and though they may be correct, our inability to invalidate them makes them difficult to incorporate as tenets of conservation planning in any meaningful way.

Of considerably more interest are the accumulating data showing that genetic erosion at some genes can have demonstrable short-term impacts on individual fitness and population viability. The best evidence so far for this comes from genetically controlled self-incompatibility systems in plants, where sharing of even a single allele can prevent mating between individuals. Work reported by DeMauro (1993) and Young *et al.* (Chapter 19) clearly demonstrates that small isolated populations of such species experience mate limitation due to reduced S allele diversity. This represents a direct link between genetic and demographic processes that has great potential to influence population viability. A similar link between diversity and fitness is provided by genes conferring disease resistance, such as the major histocompatibility complex (MHC) loci in animals (e.g. Hedrick & Miller, 1994), though in this case the relationship is not quite as direct owing to the need to have both low genetic variation at MHC loci and the presence of a pathogen for which resistance is not available. Regardless of this last caveat, both of these cases represent possible future exceptions to Caughley's (1994) claim that 'no instance of extinction by genetic malfunction has been reported'.

While genetic erosion may represent a long-term threat to many species subject to habitat loss and fragmentation, and place demographic constraints on some self-incompatible plants, the role of inbreeding and associated effects of inbreeding depression in increasing the short-term extinction risk of fragmented populations is rapidly becoming clear. The shift towards greater among-relative breeding in small populations identified by Daniels *et al.* (Chapter 8) for red-cockaded woodpeckers, and its detrimental effect on population persistence is supported by several other studies of animals (e.g. Frankham, 1995a; Dietz *et al.*, Chapter 11), including, recently, at least one invertebrate (Saccheri *et al.*, 1998). The extent of inbreeding depression in wild populations has been heavily debated (see Frankham, 1995a); however a recent review by Crnokrak & Roff (1999) has revealed that inbreeding depression was significantly higher in wild populations than in captive-bred populations and at sufficiently high levels to be 'biologically important under natural conditions'. It is by studying the effects of inbreeding, which are strongly influenced by reproductive strategy, social behaviour and the genetic basis of fitness reduction, that integrated modelling of demographic and genetic processes becomes a powerful analytical tool. Spatially explicit individual-based models facilitate assessment of how inbreeding builds up through time, the possible effects of immigration and the potential for purging of genetic load. However, the limitations imposed by model structure indicated by Burgman & Possingham (Chapter 6) must be kept in mind, as must the fact that models, while representing useful tools for investigation and hypothesis generation, are not magic bullets to put management on a scientific basis.

One of the most interesting recent results is that presented by Oostermeijer (Chapter 18) for *Gentiana pneumonanthe*, showing that effects of inbreeding on population viability are likely to be most important when acting in concert with changes to other processes – in this case reproductive failure. The evidence that either inbreeding or reproductive failure alone have limited demographic effects, but that when combined the two result in significantly reduced population persistence times, is a strong argument for the integration of demographic and genetic approaches to assessing management options for fragmented populations. This result also emphasises the importance of direct effects of fragmentation on reproductive biology, especially with regard to the disruption of mutualisms such as plant–pollinator interactions. Analysis of changes in reproductive biology has shown both positive (Kelly *et al.*, Chapter 14), negative (Lamont *et al.*, 1993) and apparently neutral (Schmidt-Adam *et al.*, in press) effects of fragmentation. This is probably one of the hardest areas to make predictions about, but one that can have strong direct influences on demography and indirect effects on genetic processes.

In summary, there is now good evidence that populations are often subject to significant genetic and demographic changes subsequent to being fragmented and that these interact via a number of direct (e.g. *S* allele erosion and inbreeding depression) and indirect (e.g. reduced evolutionary potential) linkages to affect individual fitness and population viability. The research reported in this volume, and elsewhere in the primary literature, shows that integrated genetic and demographic analysis has proved to be very informative in providing information about the limits to population viability. Brookes's (1997) call for 'A swift divorce' of genetics and conservation can be confidently rejected. Rather, like most good marriages, the value of this integrated approach is in recognising the complementary strengths of the partners instead of focusing on their individual shortcomings or previous experiences of misapplication.

This is not to say that we currently understand all there is to know about the biological effects of habitat fragmentation and the implications for population viability and species persistence. There is still much to be learned. In particular most of the available empirical data are still focused on the genetics and demographics of individual populations, with little emphasis on understanding the dynamics of interpopulation processes and their importance in determining regional species persistence. Although addressing the spatial and temporal dynamics of events such as interpopulation gene flow is difficult, even with current techniques, the value of such information is high. Initial studies such as those by Richards (Chapter 16) and White & Boshier (Chapter 17) show the importance of understanding these events with regard to the possible effects of genetic or demographic rescue of small populations. If such data can be combined into truly integrated genetic and demographic models, then there is potential for making realistic predictions about species' responses to changes in habitat abundance, distribution and quality, and the implications for regional persistence. We will have a management tool. Though it is unlikely that such data will be available for many species, careful choice of target taxa should allow maximum extrapolation of results to other organisms that share life-history characteristics.

References

Abernethy, K. (1994). The establishment of a hybrid zone between red and sika deer (genus *Cervus*). *Molecular Ecology* **3**: 551–562.

Adams, W. T. (1992). Gene dispersal within forest tree populations. *New Forests* **6**: 217–240.

Adams, W. T. & Birkes, D. S. (1991). Estimating mating patterns in forest tree populations. In *Biochemical markers in the population genetics of forest trees*: pp. 157–172, Fineschi, S., Malvolti, M. E., Cannata, F. & Hattemer, H. H. (eds.). The Hague: S.P.B. Academic Publishing.

Addicott, J. F. (1978). The population dynamics of aphids on fireweeds: a comparison of local populations and metapopulations. *Canadian Journal of Zoology* **56**: 2554–2564.

Adler, G. H. & Levins, R. (1994). The island syndrome in rodent populations. *Quarterly Review of Biology* **69**: 473–490.

Aizen, M. A. & Feinsinger, P. (1994a). Habitat fragmentation, native insect pollinators, and feral honeybees in Argentine 'Chaco Serranto'. *Ecological Applications* **4**: 378–392.

Aizen, M. A. & Feinsinger, P. (1994b). Forest fragmentation, pollination, and plant reproduction in a chaco dry forest, Argentina. *Ecology* **75**: 330–351.

Akçakaya, H. R. & Burgman, M. A. (1995). Population viability analysis in theory and practice: a reply to Harcourt. *Conservation Biology* **9**: 705–707.

Akçakaya, H. R. & Raphael, M. G. (1998). Assessing human impact despite uncertainty: viability of the northern spotted owl metapopulation in the northwestern USA. *Biodiversity and Conservation* **7**: 875–894.

Akçakaya, H. R., Burgman, M. A. & Ginzburg, L. R. (1997). *Applied population ecology*. New York: Applied Biomathematics.

Aldrich, P. R. & Hamrick, J. L. (1998). Reproductive dominance of pasture trees in a fragmented tropical forest mosaic. *Science* **281**: 103–105.

Alexander, H. M. (1990). Epidemiology of anther-smut infection of *Silene alba* caused by *Ustilago violacea*: patterns of spore deposition and disease incidence. *Journal of Ecology* **78**: 166–179.

Allee, W. C. (1951). *The social life of animals*. Boston, MA: Beacon.

Allendorf, F. W. (1983). Isolation, gene flow, and genetic differentiation among populations. In *Genetics and conservation: a reference for managing wild animal and plant populations*: pp. 51–55, Schonewald-Cox, C. M., Chambers, S. M.,

MacBryde, B. & Thomas, W. L. (eds.). Menlo Park, CA: Benjamin Cummings.

Allendorf, F. W. (1986). Genetic drift and the loss of alleles versus heterozygosity. *Zoo Biology* 5: 181–190.

Allendorf, F. W. & Leary, R. F. (1986). Heterozygosity and fitness in natural populations of animals. In *Conservation biology: the science of scarcity and diversity*: pp. 57–76, Soulé, M. E. (ed.). Sunderland, MA: Sinauer Associates.

Allendorf, F. W. & Phelps, S. R. (1981). Use of allele frequencies to describe population structure. *Canadian Journal of Fisheries and Aquatic Sciences* 38: 1507–1514.

Allendorf, F. W., Bayles, D., Bottom, D. L., Currens, K. P., Frissell, C. A., Hankin, D., Lichatowich, J. A., Nehlsen, W., Trotter, P. C. & Williams, T. H. (1997). Prioritizing Pacific salmon stocks for conservation. *Conservation Biology* 11: 140–152.

Allstad, D., Curtsinger, J., Abrams, P. & Tilman, D. (1993). POPULUS. DNA@UM NACVX.bitnet: Department of Ecology, Evolution, and Behaviour, University of Minnesota.

Altizer, S. M., Thrall, P. H. & Antonovics, J. (1998). The role of pollinators as vectors for anther-smut infection in *Silene alba*. *American Midland Naturalist* 139: 147–163.

Anderson, W. W. (1968). Further evidence for coadaptation in crosses between geographic populations of *Drosophila pseudoobscura*. *Genetical Research* 12: 317–330.

Andrewartha, H. G. & Birch, L. C. (1954). *The distribution and abundance of animals*. Chicago, IL: University of Chicago Press.

Antonovics, J. & Bradshaw, A. D. (1970). Evolution in closely adjacent populations. VIII. Clinal patterns at a mine boundary. *Heredity* 25: 349–362.

Antonovics, J. & Levin, D. A. (1980). The ecological and genetic consequences of density-dependent regulation in plants. *Annual Review of Ecology and Systematics* 11: 411–452.

Antonovics, J. & Via, S. (1988). The genetic factor in plant distribution and abundance. In *Plant population ecology*: pp. 123–141, Davy, A. J., Hutchings, M. J. & Watkinson, A. R. (eds.). Oxford: Blackwell.

Antonovics, J., Thrall, P. H., Jarosz, A. M. & Stratton, D. (1994). Ecological genetics of metapopulations: the *Silene–Ustilago* plant–pathogen system. In *Ecological genetics*: pp. 146–170, Real, L. A. (ed.). Princeton, NJ: Princeton University Press.

Antonovics, J., Thrall, P. H. & Jarosz, A. M. (1997). Genetics and the spatial ecology of species interactions: the *Silene–Ustilago* system. In *Spatial ecology: the role of space in population dynamics and interspecific interactions*: pp. 158–180, Tilman, D. & Kareiva, P. M. (eds.). Princeton, NJ: Princeton University Press.

Aprion, D. & Zohary, D. (1961). Chlorophyll lethal in natural populations of the orchard grass (*Dactylis glomerata* L.). A case of balanced polymorphism in plants. *Genetics* 46: 393–399.

Armbruster, P., Bradshaw, W. E. & Holzapfel, C. M. (1997). Evolution of the genetic architecture underlying fitness in the pitcher-plant mosquito, *Wyoemyia smithii*. *Evolution* 51: 451–458.

Ashley, M. V. & Dow, B. D. (1994). The use of microsatellite analysis in population biology: background, methods, and potential applications. In *Molecular ecology and evolution: approaches and application*: pp. 185–201, Scherwater, B., Strait, B., Wagner, G. P. & DeSalle, R. (eds.). Basel: Birkhäuser Verlag.

Askins, R. A., Philbrick, M. J. & Sugeno, D. S. (1987). Relationship between the regional abundance of forest and the composition of forest bird communities. *Biological Conservation* **39**: 129–152.

Atkinson, I. A. E. & Cameron, E. K. (1993). Human influence on the terrestrial biota and biotic communities of New Zealand. *Trends in Ecology and Evolution* **8**: 447–451.

Auld, T. D. & Scott, J. A. (1996). Conservation of the endangered plant *Grevillea caleyi* (Proteaceae) in urban fire-prone habitats. In *Back from the brink: refining the threatened species recovery process*: pp. 97–100, Stephens, S. & Maxwell, S. (eds.). Sydney: Surrey Beatty.

Avise, J. C. & Hamrick, J. L. (1996). *Conservation genetics: case histories from nature*. New York: Chapman & Hall.

Ayre, D. J. & Whelan, R. J. (1989). Factors controlling fruit set in hermaphroditic plants and the contribution of studies with the Australian Proteaceae. *Trends in Ecology and Evolution* **4**: 267–272.

Ayre, D. J., Whelan, R. J. & Reid, A. (1994). Unexpectedly high levels of selfing in the Australian shrub *Grevillea barklyana* (Proteaceae). *Heredity* **72**: 168–174.

Baillie, J. & Groombridge, B. (1996). *1996 IUCN red list of threatened animals*. Gland, Switzerland: World Conservation Union.

Baker, A. J. (1991). Evolution of the social system of the golden lion tamarin (*Leontopithecus rosalia*): mating system, group dynamics, and cooperative breeding. PhD thesis: University of Maryland.

Baker, A. J. & Dietz, J. M. (1996). Immigration in wild groups of golden lion tamarins (*Leontopithecus rosalia*). *American Journal of Primatology* **38**: 47–56.

Baker, A. J., Dietz, J. M. & Kleiman, D. G. (1993). Behavioural evidence for monopolization of paternity in multi-male groups of golden lion tamarins. *Animal Behaviour* **46**: 1091–1103.

Baker, H. G. & Stebbins, G. L. (1965). *The genetics of colonizing species*. New York: Academic Press.

Ballel, S. R., Foré, S. A. & Guttman, S. I. (1994). Apparent gene flow and genetic structure of *Acer saccharum* subpopulations in forest fragments. *Canadian Journal of Botany* **72**: 1311–1315.

Ballou, J. D. (1995). Genetic management, inbreeding depression, and outbreeding depression in captive populations. PhD thesis: University of Maryland.

Ballou, J. D. (1997). Ancestral inbreeding only minimally affects inbreeding depression in mammalian populations. *Journal of Heredity* **88**: 169–178.

Ballou, J. D. & Lacy, R. C. (1995). Identifying genetically important individuals for management of genetic variation in pedigreed populations. In *Population management for survival and recovery*: pp. 76–111, Ballou, J. D., Gilpin, M. E. & Foose, T. J. (eds.). New York: Columbia University Press.

Ballou, J. D., Lacy, R. C., Kleiman, D. G., Rylands, A. & Ellis, S. (1998). *Leontopithecus* II: the second population and habitat viability assessment for lion tamarins (*Leontopithecus*). Apple Valley, MN: IUCN/SSC Conservation Breeding Specialist Group.

Banks, M. A., Baldwin, B. A. & Hedgecock, D. (1996). Research on chinook salmon stock structure using microsatellite DNA. *Bulletin of the National Research Institute for Aquaculture*, Supplement **2**: 5–9.

Barlow, B. A., Hawksworth, F. G., Kuijt, J., Polhill, R. M. & Weins, D. (1989). Genera of mistletoes. *Golden Bough* **11**: 1–4.

Barnett, J. L., How, R. A. & Humphreys, W. F. (1977). Small mammal populations in pine and native forests in north-eastern New South Wales. *Australian Wildlife Research* **4**: 233–240.

Barrett, S. C. H. & Charlesworth, D. (1991). Effects of a change in the level of inbreeding on the genetic load. *Nature* **352**: 522–524.

Barrett, S. C. H. & Harder, L. (1996). Ecology and the evolution of plant mating. *Trends in Ecology and Evolution* **11**: 73–79.

Barrett, S. C. H. & Husband, B. C. (1990). Genetics of plant migration and colonization. In *Plant population genetics, breeding and genetic resources*: pp. 254–277, Brown, A. H. D., Clegg, M. T., Kahler, A. L. & Weir, B. S. (eds.). Sunderland, MA: Sinauer Associates.

Barrett, S. C. H. & Kohn, J. (1991). Genetic and evolutionary consequences of small population size in plants: implications for conservation. In *Genetics and conservation of rare plants*: pp. 3–30, Falk, D. A. & Holsinger, K. E. (eds.). New York: Oxford University Press.

Barrett, S. C. H. & Richardson, B. J. (1986). Genetic attributes of invading species. In *Ecology of biological invasions*: pp. 21–33, Groves, R. H. & Burdon, J. J. (eds.). Cambridge: Cambridge University Press.

Barton, N. H. & Turelli, M. (1989). Evolutionary quantitative genetics: how little do we know? *Annual Review of Genetics* **23**: 337–370.

Bascompte, J. & Solé, R. V. (1998). *Modelling spatiotemporal dynamics in ecology.* Heidelberg: Springer-Verlag.

Baur, A. & Baur, B. (1992). Effect of corridor width on animal dispersal: a simulation study. *Global Ecology and Biogeography Letters* **2**: 52–56.

Bayer, R. L. (1991). Patterns of clonal diversity in geographically marginal populations of *Antennana rosea* (Asteraceae: Inuleae) from subarctic Alaska and Yukon territory. *Botanical Gazette* **152**: 486–493.

Beardmore, J. A. (1983). Extinction, survival, and genetic variation. In *Genetics and conservation: a reference for managing wild animal and plant populations*: pp. 125–151, Schonewald-Cox, C. M., Chambers, S. M., MacBryde, B. & Thomas, W. L. (eds.). Menlo Park, CA: Benjamin Cummings.

Beissinger, S. R. & Westphal, M. I. (1998). On the use of demographic models of population viability in endangered species management. *Journal of Wildlife Management* **62**: 821–841.

Bengtsson, J. (1989). Interspecific competition increases local extinction rate in a metapopulation system. *Nature* **340**: 713–715.

Bennett, A. F. (1990). Land use, forest fragmentation and the mammalian fauna at Naringal, south-western Victoria. *Australian Wildlife Research* **17**: 325–347.

Bennett, A. F. (1998). *Linkages in the landscape: the role of corridors and connectivity in wildlife conservation.* Gland, Switzerland: IUCN.

Bensch, S., Hasselquist, D. & von Schantz, T. (1994). Genetic similarity between parents predicts hatching failure: nonincestuous inbreeding in the great reed warbler? *Evolution* **48**: 317–326.

Bever, J. D. & Felber, F. (1992). The theoretical population genetics of autopolyploidy. In *Oxford surveys in evolutionary biology*, vol. 8: pp. 185–217, Antonovics, J. & Futuyma, D. (eds.). New York: Oxford University Press.

Billington, H. L. (1991). Effect of population size on genetic variation in a dioecious conifer. *Conservation Biology* **5**: 115–119.

Blackwell, B. F., Doerr, P. D., Reed, J. M. & Walters, J. R. (1995). Inbreeding rate and effective population size: a comparison of estimates from pedigree analysis and a demographic model. *Biological Conservation* **71**: 299–304.

Blows, M. W. & Hoffmann, A. A. (1993). The genetics of central and marginal populations of *Drosophila serrata*. I. Genetic variation for stress resistance and species borders. *Evolution* **47**: 1255–1270.

Bond, W. & van Wilgen, B. W. (1996). *Fire and plants*. London: Chapman & Hall.

Bouzat, J. L., Cheng, H. H., Lewin, H. A., Westemeier, R. L., Brawn, J. D. & Paige, K. N. (1998*a*). Genetic evaluation of a demographic bottleneck in the greater prairie chicken. *Conservation Biology* **12**: 836–843.

Bouzat, J. L., Lewin, H. A. & Paige, K. N. (1998*b*). The ghost of genetic diversity past: historical DNA analysis of the greater prairie chicken. *American Naturalist* **152**: 1–6.

Bowcock, A. M., Ruiz Linares, A. R., Tomfohrde, J., Minch, E., Kidd, J. R. & Cavalli-Sforza, L. L. (1994). High resolution of human evolutionary trees with polymorphic microsatellites. *Nature* **368**: 455–457.

Box, G. E. P. (1979). Robustness in the strategy of scientific model building. In *Robustness in statistics*: pp. 201–236, Launer, R. L. & Wilkinson, G. N. (eds.). New York: Academic Press.

Boyce, M. S. (1992). Population viability analysis. *Annual Review of Ecology and Systematics* **23**: 481–506.

Boyce, W. M., Hedrick, P. W., Muggli-Cockett, N. E., Kalinowski, S. T., Penedo, M. C. T. & Ramey, R. R. (1997). Genetic variation of major histocompatibility complex and microsatellite loci: a comparison in bighorn sheep. *Genetics* **145**: 421–433.

Bradford, D. F., Tabatabai, F. & Graber, D. M. (1993). Isolation of remaining populations of the native frog, *Rana muscosa*, by introduced fishes in Sequoia and Kings Canyon National Parks, California. *Conservation Biology* **7**: 882–888.

Bradstock, R. A. & Myerscough, P. J. (1981). Fire effects on seed release and the emergence and establishment of seedlings in *Banksia ericifolia* L.f. *Australian Journal of Botany* **29**: 521–531.

Breese, E. L. & Mather, K. (1960). The organization of polygenic activity within a chromosome in *Drosophila*. II. Viability. *Heredity* **14**: 375–400.

Briggs, D. & Walters, S. M. (1997). *Plant variation and evolution*, 3rd edn. Cambridge: Cambridge University Press.

Briggs, J. D. & Leigh, J. H. (1996). *Rare or threatened Australian plants*. Melbourne: CSIRO.

Briggs, J. D., Corrigan, V. T. & Zich, F. A. (1998). *Rutidosis leptorrhynchoides* (Button Wrinklewort) Recovery Plan (unpublished). Sydney: New South Wales National Parks and Wildlife Service.

Briton, J., Nurthen, R. K., Briscoe, D. A. & Frankham, R. (1994). Modelling problems in conservation genetics using *Drosophila*: consequences of harems. *Biological Conservation* **69**: 267–275.

Brnic, D. (1954). Heterosis and the integration of the genotype in geographic populations of *Drosophila pseudoobscura*. *Genetics* **39**: 77–88.

Brockelman, W. Y. & Baimai, V. (1993). *Conservation of biodiversity and protected area management in Thailand*. Mahidol University, Bangkok: World Bank/GEF.

Brockwell, P. J. (1985). The extinction time of a birth, death, and catastrophe process and of a related diffusion model. *Advances in Applied Probability* **17**: 42–52.

Brockwell, P. J. (1986). The extinction time of a general birth and death process with catastrophes. *Journal of Applied Probability* **23**: 851–858.

Brookes, M. (1997). A clean break. *New Scientist* **156**, no. 2106: 64.

Brookes, M. I., Graneau, Y. A., King, P., Rose, O. C., Thomas, C. D. & Mallet, J. L. B. (1997). Genetic analysis of founder bottlenecks in the rare British butterfly *Plebejus argus*. *Conservation Biology* **11**: 648–661.

Brown, A. H. D. (1990). Genetic characterisation of plant mating systems. In *Plant population genetics, breeding and genetic resources*: pp. 145–162, Brown, A. H. D., Clegg, M. T., Kahler, A. L. & Weir, B. S. (eds.). Sunderland, MA: Sinauer Associates.

Brown, A. H. D. & Marshall, D. H. (1981). Evolutionary changes accompanying colonization in plants. In *Evolution today, Proceedings of II International Congress of Systematics and Evolution*: pp. 351–363, Scudder, G. & Reveal, J. (eds.). Pittsburgh, PA: Hunt Institute for Botanical Documentation, Carnegie-Mellon University.

Brown, A. H. D. & Young, A. G. (in press). Genetic diversity in tetraploid populations of the endangered daisy *Rutidosis leptorrhynchoides* and implications for its conservation. *Heredity*.

Brown, A. H. D., Barrett, S. C. H. & Moran, G. F. (1985). Mating system estimation in forest trees: models, methods, and meanings. In *Population genetics in forestry*: pp. 32–49, Gregarious, H. R. (ed.). Berlin: Springer-Verlag.

Brown, A. H. D., Young, A. G., Burdon, J. J., Christidis, L., Clarke, G. M., Coates, D. & Sherwin, W. B. (1997). *Genetic indicators for State of the Environment reporting*. Canberra: State of the Environment Reporting Unit, Environment Australia.

Brown, J. H. & Kodric-Brown, A. (1977). Turnover rates in insular biogeography: effect of immigration and extinction. *Ecology* **58**: 445–449.

Bruford, M. W. & Wayne, R. K. (1993). Microsatellites and their application to population genetic studies. *Current Opinion in Genetics and Development* **3**: 939–943.

Bruford, M. W., Hanotte, O., Brookfield, J. F. Y. & Burke, T. (1992). Single-locus and multilocus DNA fingerprinting. In *Molecular genetic analysis of populations: a practical approach*: pp. 225–269, Hoelzel, A. R. (ed.). Oxford: Oxford University Press.

Bruford, M. W., Cheesman, D. J., Coote, T., Green, H. A., Haines, S. A., O'Ryan, C. & Williams, T. R. (1996). Microsatellites and their application to conservation genetics. In *Molecular genetic approaches in conservation*: pp. 278–297, Smith, T. B. & Wayne, R. K. (eds.). New York: Oxford University Press.

Brunner, H. & Coman, B. (1974). *The identification of mammalian hair*. Melbourne: Inkata Press.

Brussard, P. F. (1984). Geographic patterns and environmental gradients: the central–marginal model in *Drosophila* revisited. *Annual Review of Ecology and Systematics* **15**: 25–64.

Buchanan, F. C., Littlejohn, R. P., Galloway, S. M. & Crawford, A. M. (1993). Microsatellite and associated repetitive elements in the sheep genome. *Mammalian Genome* **4**: 258–264.

Buechner, H. K. (1960). The bighorn sheep of the United States: its past, present and future. *Wildlife Monographs* **4**: 1–174.

Bulmer, M. G. (1985). *The mathematical theory of quantitative genetics*. Oxford:

Clarendon Press.

Burdon, J. J. & Jarosz, A. M. (1991). Host–pathogen interactions in natural populations of *Linum marginale* and *Melampsora lini*. I. Patterns of resistance and racial variation in a large host population. *Evolution* 45: 205–217.

Burdon, J. J. & Marshall, D. H. (1981). Biological control and the reproductive mode of weeds. *Journal of Applied Ecology* 18: 649–658.

Burdon, J. J. & Thrall, P. H. (1999). Spatial and temporal patterns in coevolving plant and pathogen associations. *American Naturalist* 153: S15–S33.

Burdon, J. J., Ericson, L. & Müller, W. J. (1995). Temporal and spatial changes in a metapopulation of the rust pathogen *Triphragmium ulmariae* and its host, *Filipendula ulmaria*. *Journal of Ecology* 83: 979–989.

Burgman, M. A. & Lamont, B. B. (1993). A stochastic model for the viability of *Banksia cuneata* populations: environmental, demographic, and genetic effects. *Journal of Applied Ecology* 29: 719–727.

Burgman, M. A. & Lindenmayer, D. B. (1998). *Conservation biology for the Australian environment*. Chipping Norton, NSW: Surrey Beatty.

Burgman, M. A., Ferson, S. & Akçakaya, H. R. (1993). *Risk assessment in conservation biology*. London: Chapman & Hall.

Burkey, T. V. (1997). Metapopulation extinction in fragmented landscapes: using bacteria and protozoa communities as model ecosystems. *American Naturalist* 150: 568–591.

Burmaster, D. E. & Anderson, P. D. (1994). Principles of good practice for the use of Monte Carlo techniques in human health and ecological risk assessments. *Risk Analysis* 14: 477–481.

Burton, R. S. (1987). Differentiation and integration of the genome in populations of the marine copepod *Tigriopus californicus*. *Evolution* 41: 504–513.

Burton, R. S. (1990). Hybrid breakdown in developmental time in the copepod *Tigriopus californicus*. *Evolution* 44: 1814–1822.

Bush, R. M., Smouse, P. E. & Ledig, F. T. (1987). The fitness consequences of multiple-locus heterozygosity: the relationship between heterozygosity and growth rate in pitch pine (*Pinus rigida* Mill). *Evolution* 41: 787–798.

Buza, L., Young, A. G. & Thrall, P. H. (2000). Genetic erosion, inbreeding and reduced fitness in fragmented populations of the tetraploid pea *Swainsona recta*. *Biological Conservation*. 93: 177–186.

Byers, D. L. (1995). Pollen quantity and quality as explanations for low seed set in small populations exemplified by *Eupatorium* (Asteraceae). *American Journal of Botany* 82: 1000–1006.

Cabin, R. J. (1996). Genetic comparisons of seed bank and seedling populations of a desert perennial mustard, *Lesquerella fendleri*. *Evolution* 50: 1830–1841.

Caro, T. M. & Laurenson, M. K. (1994). Ecological and genetic factors in conservation: a cautionary tale. *Science* 263: 485–486.

Carr, D. E. & Dudash, M. R. (1996). Inbreeding depression in two species of *Mimulus* (Scrophulariaceae) with contrasting mating systems. *American Journal of Botany* 83: 586–593.

Carr, D. E. & Dudash, M. R. (1997). The effects of five generations of enforced selfing on potential male and female function in *Mimulus guttatus*. *Evolution* 51: 1797–1807.

Carthew, S. M. (1993). Patterns of flowering and fruit production in a natural population of *Banksia spinulosa*. *Australian Journal of Botany* **41**: 465–480.

Carthew, S. M., Ayre, D. J. & Whelan, R. J. (1988). High levels of outcrossing in populations of *Banksia spinulosa* Sm. and *Banksia paludosa* R.Br. *Australian Journal of Botany* **36**: 217–223.

Carthew, S. M., Ayre, D. J. & Whelan, R. J. (1996). Experimental confirmation of preferential outcrossing in *Banksia*. *International Journal of Plant Sciences* **157**: 615–620.

Casper, B. B. (1984). The efficiency of pollen transfer and rates of embryo initiation in *Cryptantha* (Boraginaceae). *Oecologia* **59**: 262–268.

Caswell, H. (1989). *Matrix population models: construction, analysis and interpretation.* Sunderland, MA: Sinauer Associates.

Caughley, G. (1994). Directions in conservation biology. *Journal of Animal Ecology* **63**: 215–244.

Caughley, G. & Gunn, A. (1996). *Conservation biology in theory and practice.* Cambridge, MA: Blackwell Science.

Cavener, D. R. & Clegg, M. L. (1981). Multigenic response to ethanol in *Drosophila melanogaster*. *Evolution* **35**: 1–10.

Chapman, S. B., Rose, R. J. & Clarke, R. T. (1989). The behaviour of populations of the marsh gentian (*Gentiana pneumonanthe*) – a modelling approach. *Journal of Applied Ecology* **26**: 1059–1072.

Charlesworth, D. (1991). The apparent selection on neutral marker loci in partially inbreeding populations. *Genetical Research* **57**: 159–175.

Charlesworth, D. & Charlesworth, B. (1987). Inbreeding depression and its evolutionary consequences. *Annual Review of Ecology and Systematics* **18**: 237–268.

Charlesworth, D. & Charlesworth, B. (1990). Inbreeding depression with heterozygote advantage and its effect on selection for modifiers changing the outcrossing rate. *Evolution* **44**: 870–888.

Charlesworth, D., Morgan, M. T. & Charlesworth, B. (1990). Inbreeding depression, genetic load, and the evolution of outcrossing rates in a multilocus system with no linkage. *Evolution* **44**: 1469–1489.

Chase, M. R., Kesseli, R. & Bawa, K. (1996a). Microsatellite markers for population and conservation genetics of tropical trees. *American Journal of Botany* **83**: 51–57.

Chase, M. R., Moller, C., Kesseli, R. & Bawa, K. S. (1996b). Distant gene flow in tropical trees. *Nature* **383**: 398–399.

Chepko-Sade, B. D. & Halpin, Z. T. (1987). *Mammalian dispersal patterns: the effects of social structure on population genetics.* Chicago, IL: University of Chicago Press.

Chesser, R. K. (1983). Isolation by distance: relationship to the management of genetic resources. In *Genetics and conservation: a reference for managing wild animal and plant populations*: pp. 66–77, Schonewald-Cox, C. M., Chambers, S. M., MacBryde, B. & Thomas, W. L. (eds.). Menlo Park, CA: Benjamin Cummings.

Chesson, P. L. (1982). The stabilizing effect of a random environment. *Journal of Mathematical Biology* **15**: 1–36.

Cheverud, J. M., Routman, E. J., Jaquish, C., Tardiff, S., Peterson, G., Belfiore, N. & Forman, N. (1994). Quantitative and molecular genetic variation in captive cotton-top tamarins (*Saguinus oedipus*). *Conservation Biology* **8**: 95–105.

Churchill, S. K. & Helman, P. M. (1990). Distribution of the ghost bat, *Macroderma gigas* (Chiroptera, Megadermatidae) in central and South Australia. *Australian Mammalogy* **13**: 149–156.

Clark, J. (1992). *The future for native forest logging in Australia*, Working Paper 1992/ 1. Canberra: Centre for Resource and Environmental Studies.

Clarke, G. M. (1999a). *Golden sun moth* Synemon plana *draft recovery plan*. Sydney: NSW National Parks & Wildlife Service.

Clarke, G. M. (1999b). *Genetic analysis of New South Wales populations of the endangered golden sun moth*, Synemon plana. Canberra: CSIRO Entomology.

Clarke, G. M. (1999c). *Genetic analysis of ACT populations of the endangered golden sun moth*, Synemon plana. Canberra: CSIRO Entomology.

Clarke, G. M. & O'Dwyer, C. (2000). Genetic variability and population structure of the endangered golden sun moth, *Synemon plana*. *Biological Conservation* **92**: 371–381.

Clegg, M. T., Kahler, A. L. & Allard, R. W. (1978). Estimation of life cycle components of selection in an experimental plant population. *Genetics* **89**: 765–792.

Cohen, J. T., Lampson, M. A. & Bowers, T. S. (1996). The use of two-stage Monte Carlo simulation techniques to characterize variability and uncertainty in risk analysis. *Human and Ecological Risk Assessment* **2**: 939–971.

Coimbra-Filho, A. F. & Mittermeier, R. A. (1973). Distribution and ecology of the genus *Leontopithecus* Lesson 1840 in Brazil. *Primates* **14**: 47–66.

Collins, B. G. & Rebelo, T. (1987). Pollination biology of the Proteaceae in Australia and southern Africa. *Australian Journal of Ecology* **12**: 387–421.

Collins, R. J. & Barrett, G. W. (1997). Effects of habitat fragmentation on meadow vole (*Microtus pennsylvanicus*) population dynamics in experimental landscape patches. *Landscape Ecology* **12**: 63–76.

Comins, H. N. & Hassell, M. P. (1987). The dynamics of predation and competition in patchy environments. *Theoretical Population Biology* **31**: 393–421.

Comstock, R. E. & Robinson, H. F. (1952). Estimation of average dominance of genes. In *Heterosis*: pp. 494–516, Gowan, J. W. (ed.). Ames, IA: Iowa State University Press.

Conner, R. N. & Rudolph, D. C. (1991). Forest habitat loss, fragmentation, and red-cockaded woodpecker populations. *Wilson Bulletin* **103**: 446–457.

Conner, R. N., Rudolph, D. C. & Walters, J. R. (in press). *The red-cockaded woodpecker: adaptation and conservation in a fire-climax ecosystem*. Austin, TX: University of Texas Press.

Corbet, G. B. & Hill, J. E. (1992). *The mammals of the Indomalayan region*. New York: Oxford University Press.

Cornuet, J.-M. & Luikart, G. L. (1996). Description and power analysis of two tests for detecting recent population bottlenecks from allele frequency data. *Genetics* **144**: 2001–2014.

Costa, R. & Escano, R. E. F. (1989). *Red-cockaded woodpecker: status and management in 1986*. Atlanta, GA: US Department of Agriculture, Southern Region.

Cowan, I. M. (1940). The distribution and variation in the native sheep of North America. *American Midland Naturalist* **24**: 505–580.

Coyne, J. A. & Beecham, E. (1987). Heritability of two morphological characters within and among natural populations of *Drosophila melanogaster*. *Genetics* **117**: 727–737.

Crawford, A. M., Montgomery, G. W., Pierson, C. A., Brown, T., Dodds, K. G., Sunden, S. L. F., Henry, H. M., Ede, A. J., Swarbrick, P. A., Berryman, T., Penty, J. M. & Hill, D. F. (1994). Sheep linkage mapping: nineteen linkage groups derived from the analysis of paternal half-sib families. *Genetics* **137**: 573–579.

Crawford, T. J. (1984). The estimation of neighborhood parameters for plant populations. *Heredity* **52**: 273–283.

Crnokrak, P. & Roff, D. A. (1999). Inbreeding depression in the wild. *Heredity* **83**: 260–270.

Cropper, S. (1993). *Management of endangered plants.* Melbourne: CSIRO Publishing.

Crow, J. M. & Kimura, M. (1970). *An introduction to population genetics.* New York: Harper & Row.

Crowfoot, L. (1998). Factors limiting fruit set in *Peraxilla tetrapetala* (Loranthaceae). BSc (Hons) thesis: University of Canterbury, NZ.

Cruzan, M. B. (1990). Pollen donor interactions during pollen tube growth in *Erythronium grandiflorum. American Journal of Botany* **77**: 116–122.

Culley, T. A., Weller, S. G., Sakai, A. K. & Rankin, A. E. (1999). Inbreeding depression and selfing rates in a self-compatible, hermaphroditic species, *Schiedea membranaceae* (Caryophyllaceae). *American Journal of Botany* **86**: 980–987.

Cunningham, M. & Moritz, C. (1998). Genetic effects of forest fragmentation on a rainforest restricted lizard (Scincidae, *Gnypetoscincus queenslandiae*). *Biological Conservation* **83**: 19–30.

Cunningham, R. B., Lindenmayer, D. B., Nix, H. A. & Lindenmayer, D. B. (1999). Counting birds in forests: a comparison of observers and observation methods. *Australian Journal of Ecology* **24**: 270–277.

Currens, K. P., Allendorf, F. W., Bayles, D., Bottom, D. L., Frissell, C. A., Hankin, D., Lichatowich, J. A., Trotter, P. C. & Williams, T. A. (1998). Conservation of Pacific salmon – response to Wainwright and Waples. *Conservation Biology* **12**: 1148–1149.

Czárán, T. & Bartha, S. (1992). Spatiotemporal dynamic models of plant populations and communities. *Trends in Ecology and Evolution* **7**: 38–42.

Daniels, S. J. (1997). Female dispersal and inbreeding in the red-cockaded woodpecker. MSc thesis: Virginia Polytechnic Institute and University.

Daniels, S. J. & Walters, J. R. (1999). Inbreeding depression and its effects on natal dispersal in wild birds. In *Proceedings of 22 International Ornithological Congress, Durban*: pp. 2492–2498, Adams, N. J. & Slotow, R. H. (eds.). Johannesburg: Birdlife South Africa.

Daniels, S. J. & Walters, J. R. (in press). Inbreeding depression and its effects on the natal dispersal of red-cockaded woodpeckers. *Condor.*

Darwin, C. (1859). *On the origin of species by means of natural selection or the preservation of favoured races in the struggle for life.* London: John Murray.

Darwin, C. (1876). *The effects of cross and self fertilisation in the vegetable kingdom.* London: John Murray.

Dawson, I. K., Waugh, R., Simons, A. J. & Powell, W. (1997). Simple sequence repeats provide a direct estimate of pollen-mediated gene dispersal in the tropical tree *Gliricidia sepium. Molecular Ecology* **6**: 179–183.

Day, J. R. & Possingham, H. P. (1995). A stochastic metapopulation model with variability in patch size and position. *Theoretical Population Biology* **48**: 333–360.

Dayanandan, S., Dole, J., Bawa, K. S. & Kesseli, R. (1999). Population structure delineated with microsatellite markers in fragmented populations of a tropical tree, *Carapa guianensis* (Meliaceae). *Molecular Ecology* **8**: 1585–1593.

de Lange, P. J. & Norton, D. A. (1997). *Ecology and conservation of New Zealand's loranthaceous mistletoes*. Wellington: Department of Conservation.

De Nettancourt, D. (1977). *Incompatibility in angiosperms*. Berlin: Springer-Verlag.

DeAngelis, D. L. & Gross, L. J. (1992). *Individual-based models and approaches in ecology: populations, communities and ecosystems*. New York: Chapman & Hall.

Delph, L. F. & Lloyd, D. G. (1996). Inbreeding depression in the gynodioecious shrub *Hebe subalpina* (Scrophulariaceae). *New Zealand Journal of Botany* **34**: 241–247.

DeMauro, M. M. (1993). Relationship of breeding system to rarity in the Lakeside Daisy (*Hymenoxys acaulis* var. *glabra*). *Conservation Biology* 7: 542–550.

Deng, H.-W. (1997). Decrease of developmental stability upon inbreeding in *Daphnia*. *Heredity* **78**: 182–189.

Deng, H.-W. & Lynch, M. (1996). Change of genetic architecture in response to sex. *Genetics* **143**: 203–212.

Devlin, B., Roeder, K. & Ellstrand, N. C. (1988). Fractional paternity assignment: theoretical development and comparison to other methods. *Theoretical and Applied Genetics* **74**: 369–380.

Di Rienzo, A., Peterson, A. C., Garza, J. C., Valdes, A. M., Slatkin, M. & Freimer, N. B. (1994). Mutational processes of simple-sequence repeat loci in human populations. *Proceedings of the National Academy of Sciences, USA* **91**: 3166–3170.

Diamond, J. M. & May, R. M. (1976). Island biogeography and the design of natural reserves. In *Theoretical ecology: principles and applications*: pp. 163–186, May, R. M. (ed.). Oxford: Blackwell.

Diamond, J. M., Bishop, K. D. & Van Balen, S. (1987). Bird survival in an isolated Javan woodlot: island or mirror. *Conservation Biology* **2**: 132–142.

Dietz, J. M. & Baker, A. J. (1993). Polygyny and female reproductive success in golden lion tamarins. *Animal Behaviour* **46**: 1067–1078.

Dietz, J. M., Baker, A. J. & Miglioretti, D. (1994). Seasonal variation in reproduction, juvenile growth and adult body mass in golden lion tamarins. *American Journal of Primatology* **34**: 115–132.

Dietz, J. M., Peres, C. & Pinder, L. (1997). Foraging ecology and use of space in wild golden lion tamarins (*Leontopithecus rosalia*). *American Journal of Primatology* **41**: 289–305.

Dixson, A. F., Anzenberger, G., Monteiro da Cruz, M. A. O., Patel, I. & Jeffreys, A. J. (1992). DNA fingerprinting of free-ranging groups of common marmosets (*Callithrix jacchus jacchus*) in NE Brazil. In *Paternity in primates: genetic tests and theories*: pp. 192–202, Martin, R. D., Dixson, A. F. & Wickings, E. J. (eds.). Basel: Karger.

Dizon, A. E., Lockyer, C., Perrin, W. F., DeMaster, D. P. & Sisson, J. (1992). Rethinking the stock concept: a phylogeographic approach. *Conservation Biology* **6**: 24–36.

Doak, D. F. & Mills, L. S. (1994). A useful role for theory in conservation. *Ecology* **75**: 615–626.

Dobson, A. P. & May, R. M. (1986). Disease and conservation. In *Conservation biology: the science of scarcity and diversity*: pp. 345–365, Soulé, M. E. (ed.). Sunderland, MA: Sinauer Associates.

Doebley, J., Stec, A. & Gustus, C. (1995). Teosinte branched 1 and the origin of maize: evidence for epistasis and the evolution of dominance. *Genetics* **141**: 333–

346.

Double, M. C., Dawson, D., Burke, T. & Cockburn, A. (1997). Finding the fathers in the least faithful bird: a microsatellite-based genotyping system for the Superb Fairy-Wren *Malurus cyaneus*. *Molecular Ecology* 6: 691–693.

Dow, B. D. & Ashley, M. V. (1996). Microsatellite analysis of seed dispersal and parentage of saplings in bur oak, *Quercus macrocarpa*. *Molecular Ecology* 5: 615–627.

Drechsler, M. & Wissel, C. (1998). Trade-offs between local and regional scale management of metapopulations. *Biological Conservation* 83: 31–41.

Driscoll, D. A. (1998). Genetic structure, metapopulation processes and evolution influence the conservation strategies for two endangered frog species. *Biological Conservation* 83: 43–54.

Dudash, M. R. (1990). Relative fitness of selfed and outcrossed progeny in a self-compatible, protandrous species, *Sabatia angularis* L. (Gentianaceae): a comparison in three environments. *Evolution* 44: 1129–1139.

Dudash, M. R. (1991). Plant size effects on female and male function in hermaphroditic *Sabatia angularis* (Gentianaceae). *Ecology* 72: 1004–1012.

Dudash, M. R. & Carr, D. E. (1998). Genetic underlying inbreeding depression in *Mimulus* with contrasting mating systems. *Nature* 393: 682–684.

Dudash, M. R. & Fenster, C. B. (1997). Multiyear study of pollen limitation and cost of reproduction in the iteroparous *Silene virginica*. *Ecology* 78: 484–493.

Dudash, M. R., Carr, D. E. & Fenster, C. B. (1997). Five generations of enforced selfing and outcrossing in *Mimulus guttatus*: inbreeding depression at the population and family level. *Evolution* 51: 54–65.

Dunning, J. B., Stewart, D. J., Danielson, B. J., Noon, B. R., Root, T. L., Lambertson, R. L. & Stevens, E. E. (1995). Spatially explicit population models: current forms and future uses. *Ecological Applications* 5: 3–11.

Dunstan, C. E. & Fox, B. J. (1996). The effects of fragmentation and disturbance of rainforest on ground-dwelling mammals on the Robertson Plateau, New South Wales, Australia. *Journal of Biogeography* 23: 187–201.

Durrett, R. & Levin, S. A. (1994). Stochastic spatial models: a user's guide to ecological applications. *Philosophical Transactions of the Royal Society of London B* 343: 329–350.

Dybdal, M. F. (1994). Extinction, recolonization, and the genetic structure of tidepool copepod populations. *Evolutionary Ecology* 8: 113–124.

Ebenhard, T. (1991). Colonization in metapopulations: a review of theory and observations. *Biological Journal of the Linnean Society* 42: 105–121.

Eberhardt, L. L., Knight, R. R. & Blanchard, B. M. (1986). Monitoring grizzly bear population trends. *Journal of Wildlife Management* 50: 613–618.

Edwards, E. D. (1994). *Survey of lowland grassland sites in the A.C.T. for the golden sun moth*, Synemon plana. Canberra: CSIRO Division of Entomology.

Edwards, W. & Whelan, R. J. (1995). The size, distribution and germination requirements of the soil-stored seed bank of *Grevillea barklyana* (Proteaceae). *Australian Journal of Ecology* 20: 548–555.

Effron, B. (1979). Bootstrap methods: another look at the jack-knife. *Annals of Statistics* 7: 1–26.

Ehiobu, N. G., Goddard, M. E. & Taylor, J. F. (1989). Effect of rate of inbreeding on inbreeding depression in *Drosophila melanogaster*. *Theoretical and Applied Gen-

etics **77**: 123–127.

Eldridge, K. G. & Griffin, A. R. (1983). Selfing effects in *Eucalyptus regnans*. *Silvae Genetica* **32**: 216–221.

Electricity Generating Authority of Thailand (1980). *Chiew Larn Project: environmental and ecological investigation. Final Report, vol. II: main report.* Bangkok: EGAT.

Ellegren, H., Hartmen, G., Johansson, M. & Andersson, L. (1993). Major Histocompatibility Complex monomorphism and low levels of DNA fingerprinting variability in a reintroduced and rapidly expanding population of beavers. *Proceedings of the National Academy of Sciences, USA* **90**: 8150–8153.

Ellstrand, N. C. & Elam, D. R. (1993). Population genetic consequences of small population size: implications for plant conservation. *Annual Review of Ecology and Systematics* **24**: 217–242.

El-Nahrway, M. A. & Bingham, E. T. (1989). Performance of S1 alfalfa lines from original and improved populations. *Crop Science* **29**: 920–923.

Endler, J. A. (1977). *Geographic variation, speciation, and clines.* Princeton, NJ: Princeton University Press.

Endler, J. A. (1986). *Natural selection in the wild.* Princeton, NJ: Princeton University Press.

Endo, T., Ikeo, K. & Gojobori, T. (1996). Large-scale search for genes on which positive selection may operate. *Molecular Biology and Evolution* **5**: 685–690.

Engen, S., Bakke, O. & Islam, A. (1998). Demographic and environmental stochasticity – concepts and definitions. *Biometrics* **54**: 39–45.

England, P. R., Ayre, D. J. & Whelan, R. J. (1999). Microsatellites in the Australian shrub *Grevillea macleayana*. *Molecular Ecology* **8**: 689–690.

Ennos, R. A. (1994). Estimating the relative rates of pollen and seed migration among plant populations. *Heredity* **72**: 250–259.

Environmental Systems Research Institute (1998). ARCVIEW 3.0a. Redlands, CA: Environmental Systems Research Institute.

Ericson, L., Burdon, J. J. & Müller, W. J. (1999). Spatial and temporal dynamics of epidemics of the rust fungus *Uromyces valerianae* on populations of its host, *Valeriana salina*. *Journal of Ecology* **87**: 649–658.

Eriksson, O. (1996). Regional dynamics of plants: a review of evidence for remnant, source-sink and metapopulations. *Oikos* **77**: 248–258.

Estoup, A., Tailliez, C., Cornuet, J.-M. & Solignac, M. (1995). Size homoplasy and mutational processes of interrupted microsatellites in two bee species, *Apis mellifera* and *Bombus terrestris* (Apidae). *Molecular Biology and Evolution* **12**: 1074–1084.

Estrada, A., Coates-Estrada, R., Meritt, D., Montiel, S. & Curiel, D. (1993). Patterns of frugivore species richness and abundance in forest islands and in agricultural habitats at Los Tuxtlas, Mexico. *Vegetatio* **107/108**: 245–257.

Ewens, W. J. (1979). *Mathematical population genetics.* Berlin: Springer-Verlag.

Ewens, W. J., Brockwell, P. J., Gani, J. M. & Resnick, S. I. (1987). Minimum viable population size in the presence of catastrophes. In *Viable populations for conservation*: pp. 59–68, Soulé, M. E. (ed.). Cambridge: Cambridge University Press.

Excoffier, L., Smouse, P. E. & Quattro, J. M. (1992). Analysis of molecular variance inferred from metric distances among DNA haplotypes: application to human mitocondrial DNA restriction sites. *Genetics* **131**: 479–491.

Fahrig, L. & Merriam, G. (1992). Conservation of fragmented populations. *Conservation Biology* **8**: 50–59.

Falconer, D. S. (1981). *Introduction to quantitative genetics*, 2nd edn. Harlow: Longman.

Falconer, D. S. (1989). *Introduction to quantitative genetics*, 3rd edn. Harlow: Longman.

Falconer, D. S. & Mackay, T. F. C. (1996). *Introduction to quantitative genetics*, 4th edn. Harlow: Longman.

Favre, L., Balloux, F., Goudet, J. & Perrin, N. (1997). Female-biased dispersal in the monogamous mammal *Crocidura russula*: evidence from field data and microsatellite patterns. *Proceedings of the Royal Society London B* **264**: 127–132.

Feldman, M. W., Bergman, A., Pollock, D. D. & Goldstein, D. B. (1997). Microsatellite genetic distances with range constraints: analytic description and problems of estimation. *Genetics* **145**: 207–216.

Feller, W. (1968). *An introduction to probability theory and its applications*, vol. 1. New York: John Wiley.

Felsenstein, J. (1993). PHYLIP (phylogeny inference package), version 3.5c. Seattle: University of Washington.

Fenster, C. B. (1988). Gene flow and population differentiation in *Chamaecrista fasciculata* (Leguminosae). PhD thesis: University of Chicago.

Fenster, C. B. (1991a). Gene flow in *Chamaecrista fasciculata* (Leguminosae) I. Gene dispersal. *Evolution* **45**: 398–409.

Fenster, C. B. (1991b). Gene flow in *Chamaecrista fasciculata* (Leguminosae) II. Gene establishment. *Evolution* **45**: 410–422.

Fenster, C. B. (1991c). Effect of seed parent and pollen donor on the allocation of resources to developing seeds and fruit in *Chamaecrista fasciculata*. *American Journal of Botany* **78**: 13–23.

Fenster, C. B. & Barrett, S. C. H. (1994). Inheritance of mating-system modifier genes in *Eichhornia paniculata* (Pontederiaceae). *Heredity* **72**: 433–445.

Fenster, C. B. & Dudash, M. R. (1994). Genetic considerations for plant population restoration and conservation. In *Restoration of endangered species: conceptual issues, planning and implementation*: pp. 34–62, Bowles, M. J. & Whelan, C. J. (eds.). Cambridge: Cambridge University Press.

Fenster, C. B. & Galloway, L. F. (in press a). Inbreeding and outbreeding depression in natural populations of *Chamaecrista fasciculata* (Fabaceae): consequences for conservation biology. *Conservation Biology*.

Fenster, C. B. & Galloway, L. F. (in press b). Population differentiation in an annual legume: genetic architecture. *Evolution*.

Fenster, C. B. & Galloway, L. F. (in press c). The contribution of epistasis to the evolution of natural populations: a case study of an annual plant. In *Epistasis and the evolutionary process*: Wolf, J. B., Brodie, E. D. & Wade, M. J. (eds.), pp. 232–244. Oxford: Oxford University Press.

Fenster, C. B., Galloway, L. F. & Chao, L. (1997). Epistasis and its consequences for the evolution of natural populations. *Trends in Ecology and Evolution* **12**: 282–286.

Ferson, S. (1991). RAMAS/stage 1.4 manual. Setauket, NY: Applied Biomathematics.

Ferson, S. (1996a). What Monte Carlo methods cannot do. *Human and Ecological*

Risk Assessment **2**: 990–1007.

Ferson, S. (1996*b*). Quality assurance checks on model structure in ecological risk assessment. *Human and Ecological Risk Assessment* **2**: 553–565.

Ferson, S. & Burgman, M. A. (1995). Correlations, dependency bounds, and extinction risks. *Biological Conservation* **73**: 101–105.

Ferson, S., Ginzburg, L. R. & Akçakaya, H. R. (in press). Whereof one cannot speak: when input distributions are unknown. *Risk Analysis*.

Figueroa, F., Gunther, E. & Klein, J. (1988). MHC polymorphism predating speciation. *Nature* **335**: 265–266.

Fischer, M. & Matthies, D. (1997). Mating structure and inbreeding and outbreeding depression in the rare plant *Gentianella germanica* (Gentianaceae). *American Journal of Botany* **84**: 1685–1692.

Fischer, M. & Matthies, D. (1998). Effects of population size on performance in the rare plant *Gentianella germanica*. *Journal of Ecology* **86**: 195–203.

Fisher, R. A. (1930). *The genetical theory of natural selection*. Oxford: Clarendon Press.

Fitch, W. M., Bush, R. M., Bender, C. A. & Cox, N. J. (1997). Long-term trends in the evolution of H(3) HA1 human influenza type A. *Proceedings of the National Academy of Sciences, USA* **94**: 7712–7718.

Fitzsimmons, N. N., Buskirk, S. W. & Smith, M. H. (1995). Population history, genetic variability, and horn growth in bighorn sheep. *Conservation Biology* **9**: 314–323.

Fitzsimmons, N. N., Moritz, C., Limpus, C. J., Pope, L. & Prince, R. (1997). Geographic structure of the mitochondrial and nuclear gene polymorphisms in Australian green turtle populations and male-biased gene flow. *Genetics* **147**: 1843–1854.

Forbes, S. H., Hogg, J. T., Buchanan, F. C., Crawford, A. M. & Allendorf, F. W. (1995). Microsatellite evolution in congeneric mammals: domestic and bighorn sheep. *Molecular Biology and Evolution* **12**: 1106–1113.

Ford, E. B. (1945). *Butterflies*. London: Collins.

Foré, S. A., Hickey, R. J., Vankat, J. L., Guttman, S. I. & Schaefer, R. L. (1992). Genetic structure after forest fragmentation: a landscape ecology perspective on *Acer saccharum*. *Canadian Journal of Botany* **70**: 1659–1668.

Forman, L., Kleiman, D. G., Bush, R. M., Dietz, J. M., Ballou, J. D., Phillips, L. G., Coimbra-Filho, A. F. & O'Brien, S. J. (1986). Genetic variation within and among lion tamarins. *American Journal of Physical Anthropology* **7**: 1–11.

Forman, T. T. (1996). *Land mosaics: the ecology of landscapes and regions*. New York: Cambridge University Press.

Frankel, O. H. & Soulé, M. E. (1981). *Conservation and evolution*. Cambridge: Cambridge University Press.

Frankel, O. H., Brown, A. H. D. & Burdon, J. J. (1995). *The conservation of plant biodiversity*. Cambridge: Cambridge University Press.

Frankham, R. (1995*a*). Inbreeding and extinction: a threshold effect. *Conservation Biology* **9**: 792–799.

Frankham, R. (1995*b*). Conservation genetics. *Annual Review of Genetics* **29**: 305–327.

Frankham, R. (1995*c*). Effective population size/adult population size ratios in wild-life: a review. *Genetical Research* **66**: 95–107.

Frankham, R. (1996). Relationship of genetic variation to populations size in wildlife. *Conservation Biology* **10**: 1500–1508.

Frankham, R. (1998). Inbreeding and extinction: island populations. *Conservation Biology* **12**: 665–675.

Frankham, R. & Ralls, K. (1998). Inbreeding leads to extinction. *Nature* **392**: 441–442.

Frankham, R., Lees, K., Montgomery, M., England, P. R., Lowe, E. H. & Briscoe, D. A. (1999). Do population size bottlenecks reduce evolutionary potential? *Animal Conservation* **2**: 255–260.

Franklin, I. R. (1980). Evolutionary change in small populations. In *Conservation biology: an evolutionary–ecological perspective*: pp. 135–149, Soulé, M. E. & Wilcox, B. A. (eds.). Sunderland, MA: Sinauer Associates.

Frey, H. C. & Rhodes, D. S. (1996). Characterizing, simulating, and analyzing variability and uncertainty: an illustration of methods using an air toxics emissions example. *Human and Ecological Risk Assessment* **2**: 762–797.

Fu, Y. X. & Li, W. H. (1993). Statistical tests of neutrality of mutations. *Genetics* **133**: 693–709.

Fu, Y. X., Namkoong, G. & Carlson, J. E. (1988). Comparison of breeding strategies for purging inbreeding depression via simulation. *Conservation Biology* **2**: 856–864.

Funke, O. C. (1995). Limitations of ecological risk assessment. *Human and Ecological Risk Assessment* **1**: 443–453.

Furnier, G. R., Knowles, P., Clyde, M. A. & Dancik, B. P. (1987). Effects of avian seed dispersal and genetic structure of white-bark pine populations. *Evolution* **41**: 607–612.

Gabriel, W., Lynch, M. & Bürger, R. (1993). Muller's ratchet and mutational meltdowns. *Evolution* **47**: 1744–1757.

Gagneux, P., Boesch, C. & Woodruff, D. S. (1997). Microsatellite scoring errors associated with noninvasive genotyping based on nuclear DNA amplified from shed hair. *Molecular Ecology* **6**: 861–868.

Gaines, M. S., Diffendorfer, J. E., Tamarin, R. H. & Whittam, T. S. (1997). The effects of habitat fragmentation on the genetic structure of small mammal populations. *Journal of Heredity* **88**: 294–304.

Galen, C., Plowright, R. C. & Thomson, J. D. (1985). Floral biology and regulation of seed set in the lily, *Clintonia borealis*. *American Journal of Botany* **72**: 1544–1552.

Gall, B. C. (1982). *Wildlife in the south-western slopes region of New South Wales*. Sydney: NSW National Parks and Wildlife Service.

Gall, G. A. E. (1987). Inbreeding. In *Population genetics and fishery management*: pp. 47–88, Ryman, N. & Utter, F. (eds.). Seattle, WA: Washington Sea Grant/ University of Washington Press.

Galloway, L. F. & Fenster, C. B. (1999). The role of nuclear and cytoplasmic factors in the adaptive evolution of populations of *Chamaecrista fasciculata* (Fabaceae). *Evolution* **53**: 1734–1743.

Galloway, L. F. & Fenster, C. B. (in press). Population differentiation in an annual legume: local adaptation. *Evolution*.

Garrett, M. G. & Franklin, W. L. (1988). Behavioral ecology of dispersal in the black-tailed prairie dog. *Journal of Mammalogy* **69**: 236–250.

Geiger, H. J., Smoker, W. M., Zhivotovsky, L. A. & Gharrett, A. J. (1997). Variability of family size and marine survival in pink salmon (*Oncorhynchus gorbuscha*) has implications for conservation biology and human use. *Canadian Journal of Fisheries and Aquatic Sciences* **54**: 2684–2690.

Geringer, H. (1949). Chromatid segregation of tetraploids and hexaploids. *Genetics* **22**: 665–684.

Giles, B. E. & Goudet, J. (1997*a*). A case study of genetic structure in a plant metapopulation. In *Metapopulation dynamics: ecology, genetics and evolution*: pp. 429–454, Hanski, I. A. & Gilpin, M. E. (eds.). San Diego, CA: Academic Press.

Giles, B. E. & Goudet, J. (1997*b*). Genetic differentiation in *Silene dioica* metapopulations: estimation of spatio-temporal effects in a successional plant species. *American Naturalist* **149**: 507–526.

Gill, A. M. & Williams, J. E. (1996). Fire regimes and biodiversity: the effects of fragmentation of southeastern Australian eucalypt forests by urbanisation, agriculture and pine plantations. *Forest Ecology and Management* **85**: 261–278.

Gill, D. E. (1978). The metapopulation ecology of the red-spotted newt, *Notophthalamus viridescens* (Rafinesque). *Ecological Monographs* **48**: 145–166.

Gilligan, D. M., Woodworth, L. M., Montgomery, M. E., Briscoe, D. A. & Frankham, R. (1997). Is mutation accumulation a threat to the survival of endangered populations? *Conservation Biology* **11**: 1235–1241.

Gilpin, M. E. (1991). The genetic effective size of a metapopulation. *Biological Journal of the Linnean Society* **42**: 165–175.

Gilpin, M. E. (1996). Metapopulations and wildlife conservation: approaches to modeling spatial structure. In *Metapopulations and wildlife conservation*: pp. 11–27, McCullough, D. R. (ed.). Washington, DC: Island Press.

Gilpin, M. E. & Hanski, I. A. (1991). *Metapopulation dynamics: empirical and theoretical investigations*. London: Academic Press.

Gilpin, M. E. & Soulé, M. E. (1986). Minimum viable populations: processes of species extinctions. In *Conservation biology: the science of scarcity and diversity*: pp. 19–34, Soulé, M. E. (ed.). Sunderland, MA: Sinauer Associates.

Ginsberg, J. R. & Millner-Gulland, E. J. (1994). Sex-biased harvesting and population dynamics in ungulates: implications for conservation and sustainable use. *Conservation Biology* **8**: 157–166.

Ginzburg, L. R., Slobodkin, L. B., Johnson, K. & Bindman, A. G. (1982). Quasiextinction probabilities as a measure of impact on population growth. *Risk Analysis* **2**: 171–181.

Goel, N. S. & Richter-Dyn, N. (1974). *Stochastic models in biology*. New York: Academic Press.

Gogan, P. J. P. (1990). Considerations in the reintroduction of native mammalian species to restore natural ecosystems. *Natural Areas Journal* **10**: 210–217.

Goldingay, R. L., Carthew, S. M. & Whelan, R. J. (1987). Transfer of *Banksia spinulosa* pollen by mammals: implications for pollination. *Australian Journal of Zoology* **35**: 319–325.

Goldingay, R. L., Carthew, S. M. & Whelan, R. J. (1991). The importance of non-flying mammals in pollination. *Oikos* **61**: 79–87.

Goldizen, A. W., Mendelson, J., Van Vlaardingen, M. & Terborgh, J. (1996). Saddleback tamarin (*Saguinus fuscicollis*) reproductive strategies: evidence from a thirteen-year study of a marked population. *American Journal of Primatology* **38**: 57–83.

Goldstein, D. B. & Pollock, D. D. (1997). Launching microsatellites: a review of mutation processes and methods of phylogenetic inference. *Journal of Heredity* **88**: 335–342.

Goldstein, D. B., Ruiz Linares, A. R., Cavalli-Sforza, L. L. & Feldman, M. W. (1995). An evaluation of genetic distances for use with microsatellite loci. *Genetics* **139**: 463–471.

Goldstein, D. B., Roemer, G. W., Smith, D. A., Reich, D. E., Bergman, A. & Wayne, R. K. (1999). The use of microsatellite variation to infer population structure and demographic history in a natural model system. *Genetics* **151**: 797–801.

Gompper, M. E., Stacey, P. B. & Berger, J. (1997). Conservation implications of the natural loss of lineages in wild mammals and birds. *Conservation Biology* **11**: 857–867.

Goodman, D. (1987a). Considerations of stochastic demography in the design and management of biological reserves. *Natural Resources Modelling* 1: 205–234.

Goodman, D. (1987b). The demography of chance extinction. In *Viable populations for conservation*: pp. 11–34, Soulé, M. E. (ed.). Cambridge: Cambridge University Press.

Goodman, S. J. (1997). Rst Calc: a collection of computer programs for calculating estimates of genetic differentiation from microsatellite data and determining their significance. *Molecular Ecology* **6**: 881–885.

Grant, V. (1981). *Plant speciation*, 2nd edn. New York: Columbia University Press.

Gray, A., Crawley, M. & Edwards, P. (1987). *Colonization, succession and stability.* Oxford: Blackwell.

Grenfell, B. & Harwood, J. (1997). (Meta)population dynamics of infectious diseases. *Trends in Ecology and Evolution* **12**: 395–399.

Groom, M. J. (1998). Allee effects limit population viability of an annual plant. *American Naturalist* **151**: 487–496.

Groom, M. J. & Pascual, M. A. (1997). The analysis of population persistence: an outlook on the practice of viability analysis. In *Conservation biology for the coming decade*, 2nd edn: pp. 4–27, Fiedler, P. L. & Kareiva, P. M. (eds.). New York: Chapman & Hall.

Gupta, P. K., Balyan, I. S., Sharma, P. C. & Ramesh, B. (1996). Microsatellites in plants – a new class of molecular markers. *Current Science* **70**: 45–54.

Gustafson, A. (1950). The cooperation of genotypes in barley. *Hereditas* **39**: 1–18.

Gutiérrez, R. J. & Harrison, S. (1996). Applying metapopulation theory to spotted owl management: a history and critique. In *Metapopulations and wildlife conservation*: pp. 167–185, McCullough, D. R. (ed.). Washington, DC: Island Press.

Gutiérrez-Espeleta, G. A., Kalinowski, S. T., Boyce, W. M. & Hedrick, P. W. (in press). Population structure in bighorn sheep. *Conservation Biology.*

Haas, G. N. (1997). Importance of distributional form in characterising inputs in Monte Carlo risk assessment. *Risk Analysis* **17**: 107–113.

Haig, S. M. (1998). Molecular contributions to conservation. *Ecology* **79**: 413–425.

Haig, S. M., Belthoff, J. R. & Allen, D. H. (1993a). Examination of population structure in red-cockaded woodpeckers using DNA profiles. *Evolution* **47**: 185–194.

Haig, S. M., Belthoff, J. R. & Allen, D. H. (1993b). Population viability analysis for a small population of red-cockaded woodpeckers and an evaluation of enhancement strategies. *Conservation Biology* **7**: 289–301.

Haig, S. M., Walters, J. R. & Plissner, J. H. (1994). Genetic evidence for monogamy in the cooperatively breeding red-cockaded woodpecker. *Behavioral Ecology and*

Sociobiology **34**: 295–303.

Haldane, J. B. S. (1930). Theoretical genetics of autopolyploids. *Journal of Genetics* **22**: 359–372.

Haley, C. S., Knott, S. A. & Elsen, J. M. (1994). Mapping quantitative trait loci in crosses between outbred lines using least squares. *Genetics* **136**: 1195–1207.

Hallauer, A. R. & Miranda, F. (1985). *Quantitative genetics in maize breeding*. Ames, IA: Iowa State University Press.

Halley, J. M. & Manasse, R. S. (1993). A population-dynamics model for annual plants subject to inbreeding depression. *Evolutionary Ecology* **7**: 15–24.

Hamblin, M. T. & Aquadro, C. H. (1997). Contrasting patterns of nucleotide sequence variation at the glucose dehydrogenase (*Gld*) locus in different populations of *Drosophila melanogaster*. *Genetics* **145**: 1053–1062.

Hamilton, S. & Moller, H. (1995). Can PVA models using computer packages offer useful conservation advice? Sooty Shearwaters *Puffinus griseus* in New Zealand as a case study. *Biological Conservation* **73**: 107–117.

Hamman, R. L. (1982). Induced spawning and culture of bonytail chub. *Progressive Fish-Culturist* **44**: 201–203.

Hamrick, J. L. (1992). Distribution of genetic diversity in tropical tree populations: implications for the conservation of genetic resources. In *Resolving tropical forest resource concerns through tree improvement, gene conservation and domestication of new species*: pp. 74–82, Lambeth, C. C. & Dvorak, W. (eds.). Raleigh, NC: North Carolina State University.

Hamrick, J. L. & Loveless, M. D. (1989). The genetic structure of tropical tree populations: associations with reproductive biology. In *The evolutionary ecology of plants*: pp. 129–146, Bock, J. H. & Linhart, Y. B. (eds.). Boulder, CO: Westview Press.

Hamrick, J. L., Murawski, D. A. & Nason, J. D. (1993). The influence of seed dispersal mechanisms on the genetic structure of tropical tree populations. *Vegetatio* **107/108**: 281–297.

Handel, S. N. (1983). Pollination ecology, plant population structure and gene flow. In *Pollination biology*: pp. 163–211, Real, L. A. (ed.). Orlando, FL: Academic Press.

Hanski, I. A. (1991). Single-species metapopulation dynamics: concepts, models and observations. In *Metapopulation dynamics: empirical and theoretical investigations*: pp. 17–38, Gilpin, M. E. & Hanski, I. A. (eds.). London: Academic Press.

Hanski, I. A. (1994a). Patch occupancy dynamics in fragmented landscapes. *Trends in Ecology and Evolution* **9**: 131–134.

Hanski, I. A. (1994b). Spatial scale, patchiness and population dynamics on land. *Philosophical Transactions of the Royal Society of London B* **343**: 19–25.

Hanski, I. A. (1994c). A practical model of metapopulation dynamics. *Journal of Animal Ecology* **63**: 151–162.

Hanski, I. A. (1998). Metapopulation dynamics. *Nature* **396**: 41–49.

Hanski, I. A. & Gilpin, M. E. (1991a). Metapopulation dynamics: brief history and conceptual domain. *Biological Journal of the Linnean Society* **42**: 3–16.

Hanski, I. A. & Gilpin, M. E. (1991b). *Metapopulation dynamics*. New York: Academic Press.

Hanski, I. A. & Gilpin, M. E. (1997). *Metapopulation biology: ecology, genetics and evolution*. San Diego, CA: Academic Press.

Hanski, I. A. & Simberloff, D. S. (1997). The metapopulation approach, its history, conceptual domain and application to conservation. In *Metapopulation biology: ecology, genetics and evolution*: pp. 5–26, Hanski, I. A. & Gilpin, M. E. (eds.). San Diego, CA: Academic Press.

Hanski, I. A., Kuussaari, M. & Nieminen, M. (1994). Metapopulation structure and migration in the butterfly *Melitaea cinxia*. *Ecology* **75**: 747–762.

Hanski, I. A., Moilanen, A. & Gyllenberg, M. (1996a). Minimum viable metapopulation size. *American Naturalist* **147**: 527–541.

Hanski, I. A., Moilanen, A., Pakkala, T. & Kuussaari, M. (1996b). The quantitative incidence function model and persistence of an endangered butterfly metapopulation. *Conservation Biology* **10**: 578–590.

Harcourt, A. H. (1995). Population viability estimates: theory and practice for a wild gorilla population. *Conservation Biology* **9**: 134–142.

Hard, J. J., Bradshaw, W. E. & Holzapfel, C. M. (1992). Epistasis and the genetic divergence of photoperiodism between populations of the pithcher-plant mosquito *Wyeomyia smithii*. *Genetics* **131**: 389–396.

Hard, J. J., Bradshaw, W. E. & Holzapfel, C. M. (1993). The genetic basis of photoperiodism and its evolutionary divergence among populations of the pitcher-plant mosquito, *Wyeomyia smithii*. *American Naturalist* **142**: 457–473.

Harper, J. (1977). *Population biology of plants*. London: Academic Press.

Harrison, J. L. (1956). Survival rates of Malayan rats. *Bulletin of the Raffles Museum* **27**: 5–26.

Harrison, J. L. (1958). Range of movement of some Malayan rats. *Journal of Mammalogy* **39**: 190–206.

Harrison, S. (1991). Local extinction in a metapopulation context: an empirical evaluation. *Biological Journal of the Linnean Society* **42**: 73–88.

Harrison, S. (1994). Metapopulations and conservation. In *Large-scale ecology and conservation biology*: pp. 111–128, Edwards, P. J., May, R. M. & Webb, N. (eds.). Oxford: Blackwell.

Harrison, S. & Fahrig, L. (1995). Landscape pattern and population conservation. In *Mosaic landscapes and ecological processes*: pp. 293–308, Hansson, L., Fahrig, L. & Merriam, G. (eds.). London: Chapman & Hall.

Harrison, S. & Hastings, A. (1996). Genetic and evolutionary consequences of metapopulation structure. *Trends in Ecology and Evolution* **11**: 180–183.

Harrison, S. & Quinn, J. F. (1989). Correlated environments and the persistence of metapopulations. *Oikos* **56**: 293–298.

Harriss, F. & Whelan, R. J. (1993). Selective fruit abortion in *Grevillea barklyana* (Proteaceae). *Australian Journal of Botany* **41**: 499–509.

Hartl, D. L. & Clark, A. G. (1997). *Principles of population genetics*, 3rd edn. Sunderland, MA: Sinauer Associates.

Harwood, T., Narain, S. & Edwards, E. D. (1995). *Population monitoring of the endangered moth* Synemon plana *1994–1995, York Park, Barton*. Canberra: CSIRO Division of Entomology.

Hassell, M. P. & May, R. M. (1988). Spatial heterogeneity and the dynamics of parasitoid–host systems. *Annales Zoologici Fennici* **25**: 55–61.

Hassell, M. P., Comins, H. N. & May, R. M. (1991). Spatial structure and chaos in insect population dynamics. *Nature* **353**: 255–258.

Hastings, A. & Harrison, S. (1994). Metapopulation dynamics and genetics. *Annual*

Review of Ecology and Systematics **25**: 167–188.

Hatfield, T. (1997). Genetic divergence in adaptive characters between sympatric species of stickleback. *American Naturalist* **149**: 1009–1029.

Hedrick, P. W. (1985). *Genetics of populations*. Boston, MA: Jones & Bartlett.

Hedrick, P. W. (1994). Purging inbreeding depression and the probability of extinction: full-sib mating. *Heredity* **73**: 363–372.

Hedrick, P. W. (1996). Conservation genetics and molecular techniques: a perspective. In *Molecular genetic approaches to conservation*: pp. 459–477, Smith, T. B. & Wayne, R. K. (eds.). New York: Oxford University Press.

Hedrick, P. W. (1999). Perspective: highly variable loci and their interpretation in evolution and conservation. *Evolution* **53**: 313–318.

Hedrick, P. W. & Miller, P. S. (1992). Conservation genetics: techniques and fundamentals. *Ecological Applications* **20**: 30–46.

Hedrick, P. W. & Miller, P. S. (1994). Rare alleles, MHC and captive breeding. In *Conservation genetics*: pp. 187–204, Loeschcke, V., Tomiuk, J. & Jain, S. K. (eds.). Basel: Birkhauser Verlag.

Hedrick, P. W. & Thompson, G. (1988). Maternal-foetal interactions and the maintenance of HLA polymorphism. *Genetics* **119**: 205–212.

Hedrick, P. W., Hedgecock, D. & Hamelberg, S. (1995). Effective population size in winter-run chinook salmon. *Conservation Biology* **9**: 615–624.

Hedrick, P. W., Lacy, R. C., Allendorf, F. W. & Soulé, M. E. (1996). Directions in conservation biology: comments on Caughley. *Conservation Biology* **10**: 1312–1320.

Hedrick, P. W., Miller, P. S., Geffen, E. & Wayne, R. K. (1997). Genetic evaluation of the three captive Mexican wolf lineages. *Zoo Biology* **16**: 47–69.

Hedrick, P. W., Dowling, T. E., Minckley, W. L., Tibbetts, C. A., DeMarais, B. D. & Marsh, P. C. (2000a). Establishing a captive broodstock for bonytail chub *Gila elegans*. *Journal of Heredity* **91**: 35–39.

Hedrick, P. W., Hedgecock, D., Hamelberg, S. & Croci, S. J. (2000b). The impact of supplementation in winter-run chinook salmon on effective population size. *Journal of Heredity* **91**: 112–116.

Hedrick, P. W., Rashbrook, V. K. & Hedgecock, D. (in press). Effective population size in returning winter-run chinook salmon. *Evolution*.

Herben, T. & Söderström, L. (1992). Which habitat parameters are most important for the persistence of a bryophyte species on patchy, temporary substrates? *Biological Conservation* **59**: 121–126.

Hermanutz, L., Innes, D., Denham, A. & Whelan, R. J. (1998). Very low fruit–flower ratios in *Grevillea* (Proteaceae) are independent of breeding system. *Australian Journal of Botany* **46**: 465–478.

Heschel, M. S. & Paige, K. N. (1995). Inbreeding depression, environmental stress, and population size variation in scarlet gilia (*Ipomopsis aggregata*). *Conservation Biology* **9**: 126–133.

Hess, G. (1994). Conservation corridors and contagious disease: a cautionary note. *Conservation Biology* **8**: 256–262.

Hess, G. (1996a). Disease in metapopulation models: implications for conservation. *Ecology* **77**: 1617–1632.

Hess, G. (1996b). Linking extinction to connectivity and habitat destruction in metapopulation models. *American Naturalist* **148**: 226–236.

Hewitson, H. (1997). The genetic consequences of habitat fragmentation on the Bush Rat (*Rattus fuscipes*) in a pine plantation near Tumut, NSW. BSc (Hons) thesis: Australian National University.

Hilborn, R. & Mangel, M. (1997). *The ecological detective: confronting models with data*. Princeton, NJ: Princeton University Press.

Hill, A. V. S., Jepson, A., Plebanski, M. & Gilbert, S. (1997). Genetic analysis of host–parasite coevolution in human malaria. *Philosophical Transactions of the Royal Society of London B* **352**: 1317–1325.

Hobbs, R. J. (1992). The role of corridors in conservation: solution or bandwagon? *Trends in Ecology and Evolution* **7**: 389–392.

Hoffman, F. O. & Hammonds, J. S. (1994). Propagation of uncertainty in risk assessments: the need to distinguish between uncertainty due to lack of knowledge and uncertainty due to variability. *Risk Analysis* **14**: 707–712.

Hoffmann, A. A. & Parsons, P. A. (1991). *Evolutionary genetics and environmental stress*. New York: Oxford University Press.

Hogbin, P., Ayre, D. J. & Whelan, R. J. (1998). A genetic and demographic investigation of the conservation value of the threatened shrub *Grevillea barklyana* (Proteaceae). *Heredity* **80**: 180–186.

Holderegger, R. & Schneller, J. J. (1994). Are small isolated populations of *Asplenium septentrionale* variable? *Biological Journal of the Linnean Society* **51**: 377–385.

Holsinger, K. E. & Gottlieb, L. D. (1989). The conservation of rare and endangered plants. *Trends in Ecology and Evolution* **4**: 193–194.

Holsinger, K. E. & Vitt, P. (1997). The future of conservation biology: what is a geneticist to do? In *The ecological basis for conservation: heterogeneity, ecosystems and biodiversity*: pp. 202–216, Pickett, S. T. A., Ostfeld, R. S., Shachak, M. & Likens, G. E. (eds.). New York: Chapman & Hall.

Honne, B. I. (1982). On components of the genotypic variance in autotetraploid populations. In *Proceedings of 4th Meeting of the Section Biometrics in Plant Breeding, European Association for Research on Plant Breeding*: pp. 217–226, Gallais, A. (ed.). Versailles: Institut national de la recherche agronomique (France).

Hoogland, J. L. (1992). Levels of inbreeding among prairie dogs. *American Naturalist* **139**: 591–602.

Hooper, R. G. (1983). Colony formation by red-cockaded woodpeckers: hypotheses and management implications. In *Red-cockaded woodpecker symposium II*: pp. 72–77, Wood, D. A. (ed.). Atlanta, GA: Florida Game and Freshwater Fish Commission, US Fish and Wildlife Service.

Houlden, B. A., England, P. R., Taylor, A. C., Greville, W. D. & Sherwin, W. B. (1996). Low genetic variability of the koala (*Phascolarctos cinereus*) in south eastern Australia following a severe population bottleneck. *Molecular Ecology* **5**: 269–281.

Houlden, B. A., Costello, B. H., Sharkey, D., Fowler, E. V., Melzer, A., Ellis, W., Carrick, F., Baverstock, P. R. & Elphinstone, M. S. (1999). Phylogenetic differentiation in the mitochondrial control region in the koala, *Phascolarctos cinereus* (Goldfuss 1817). *Molecular Ecology* **8**: 999–1011.

Huffaker, C. B. (1958). Experimental studies on predation: dispersion factors and predator–prey oscillations. *Hilgardia* **27**: 343–383.

Hughes, A. L. (1991). MHC polymorphism and the design of captive breeding programs. *Conservation Biology* 5: 249–251.

Hughes, A. L., Ota, T. & Nei, M. (1990). Positive Darwinian selection promotes charge profile diversity in the antigen-binding cleft of class I major-histocompatibility-complex molecules. *Molecular Biology and Evolution* 7: 515–524.

Hughes, J. B., Daily, G. C. & Ehrlich, P. R. (1997). Population diversity: its extent and extinction. *Science* 278: 689–692.

Husband, B. C. & Barrett, S. C. H. (1996). A metapopulation perspective in plant population biology. *Journal of Ecology* 84: 461–469.

Husband, B. C. & Schemske, D. W. (1996). Evolution of the magnitude and timing of inbreeding depression in plants. *Evolution* 50: 54–70.

Husband, B. C. & Schemske, D. W. (1997). The effect of inbreeding in diploid and tetraploid populations of *Epilobium angustifolium* (Onagraceae): implications for the genetic basis of inbreeding depression. *Evolution* 51: 737–746.

Imrie, B. C., Kirkman, C. T. & Ross, D. R. (1972). Computer simulation of a sporophytic self-incompatible breeding system. *Australian Journal of Biological Sciences* 25: 343–349.

Jackson, J. A. & Jackson, B. J. S. (1986). Why do red-cockaded woodpeckers need old trees? *Wildlife Society Bulletin* 14: 318–322.

Jain, S. K. (1983). Genetics of populations. In *Disturbance in ecosystems: components of response*: pp. 240–258, Mooney, H. A. & Godron, M. (eds.). New York: Springer-Verlag.

James, F. C. (1991). Signs of trouble in the largest remaining population of red-cockaded woodpeckers. *Auk* 108: 419–423.

James, F. C. (1995). The status of the red-cockaded woodpecker in 1990 and the prospect for recovery. In *Red-cockaded woodpecker: recovery, ecology, and management*: pp. 439–451, Kulhavy, D. L., Hooper, R. G. & Costa, R. (eds.). Nacogdoches, TX: College of Forestry, Stephen F. Austin State University.

James, S. H., Playford, J. & Sampson, J. F. (1991). Complex hybridity in *Isotoma petraea*. VIII. Variation for seed aborting lethal genes in the O–6 Pigeon Rock population. *Heredity* 66: 173–180.

Janzen, D. H. (1986a). *Guanacaste National Park: tropical ecological and cultural restoration*. San José, Costa Rica: EUNED.

Janzen, D. H. (1986b). Blurry catastrophes. *Oikos* 47: 1–2.

Jarne, P. & Charlesworth, D. (1993). The evolution of selfing rate in functionally hermaphroditic plants and animals. *Annual Review of Ecology and Systematics* 24: 441–466.

Jarne, P. & Delay, B. (1990). Inbreeding depression and self-fertilisation in *Lymnaea peregra* (Gastropoda: Pulmonata). *Heredity* 64: 169–175.

Jarne, P. & Lagoda, P. J. (1996). Microsatellites, from molecules to populations and back. *Trends in Ecology and Evolution* 11: 424–429.

Jennersten, O. (1988). Pollination in *Dianthus deltoides* (Caryophyllaceae): effects of habitat fragmentation on visitation and seed set. *Conservation Biology* 2: 359–366.

Jesup, D. A. & Ramey, R. R. (1995). Genetic variation of bighorn sheep as measured by blood protein electrophoresis. *Desert Bighorn Council Transactions* 39: 17–25.

Jiménez, J. A., Hughes, K. A., Alaks, G., Graham, L. & Lacy, R. C. (1994). An experimental study of inbreeding depression in a natural habitat. *Science* 266: 271–273.

Johnson, L. A. S. & Briggs, B. G. (1975). On the Proteaceae – the evolution and classification of a southern family. *Botanical Journal of the Linnean Society* **70**: 83–182.

Johnston, M. O. (1992). Effects of cross and self-fertilisation on progeny fitness in *Lobelia cardinalis* and *L. siphilitica. Evolution* **46**: 688–702.

Johnston, M. O. & Schoen, D. J. (1995). Mutation rates and dominance levels of genes affecting total fitness in two angiosperm species. *Science* **267**: 226–229.

Johnston, M. O. & Schoen, D. J. (1996). Correlated evolution of self-fertilisation and inbreeding depression: an experimental study of nine populations of *Amsinckia* (Boraginaceae). *Evolution* **50**: 1478–1491.

Jones, G. H. (1969). Further correlations between chiasmata and U-type exchanges in rye meiosis. *Chromosoma* **26**: 105–118.

Judson, O. P. (1994). The rise of the individual-based model in ecology. *Trends in Ecology and Evolution* **9**: 9–14.

Kanowski, P. J. & Boshier, D. H. (1997). Conservation of tree genetic resources *in situ*. In *Plant conservation: the* in situ *approach*: pp. 207–219, Maxted, N., Ford-Lloyd, B. V. & Hawkes, J. G. (eds.). London: Chapman & Hall.

Kapos, V. (1989). Effects of isolation on the water status of forest patches in the Brazilian Amazon. *Journal of Tropical Ecology* **5**: 173–185.

Kareiva, P. M. (1990). Population dynamics in spatially complex environments: theory and data. *Philosophical Transactions of the Royal Society of London B* **330**: 175–190.

Karlin, S. & Taylor, H. M. (1975). *A first course in stochastic processes*, 2nd edn. New York: Academic Press.

Karr, J. R. (1995). Risk assessment: we need more than an ecological veneer. *Human and Ecological Risk Assessment* **1**: 436–442.

Karron, J. D. (1987). A comparison of levels of genetic polymorphism and self-compatibility in geographically restricted and widespread plant congeners. *Evolutionary Ecology* **1**: 47–58.

Karron, J. D. (1989). Breeding systems and levels of inbreeding depression in geographically restricted and widespread species of *Astragalus* (Fabaceae). *American Journal of Botany* **76**: 331–340.

Karron, J. D. (1991). Patterns of genetic variation and breeding systems in rare plant species. In *Genetics and conservation of rare plants*: pp. 87–98, Falk, D. A. & Holsinger, K. E. (eds.). New York: Oxford University Press.

Karron, J. D., Linhart, Y. B., Chaulk, C. A. & Robertson, C. A. (1988). The genetic structure of populations of geographically restricted and widespread species of *Astragalus* (Fabaceae). *American Journal of Botany* **75**: 1114–1119.

Karron, J. D., Thumser, N. N., Tucker, R. & Hessenauer, A. J. (1995a). The influence of population density on outcrossing rates in *Mimulus ringens. Heredity* **75**: 175–180.

Karron, J. D., Tucker, R., Thumser, N. & Reinartz, J. A. (1995b). Comparison of pollinator flight movements and gene dispersal patterns in *Mimulus ringens. Heredity* **75**: 612–617.

Kearns, C. A., Inouye, D. W. & Waser, N. M. (1998). Endangered mutualisms: the conservation of plant-pollinator interactions. *Annual Review of Ecology and Systematics* **29**: 83–112.

Keast, A. (1958). The influence of ecology on variation in the mistletoe-bird

(*Dicaeum hirundinaceum*). *Emu* **58**: 195–206.

Keiding, N. (1975). Extinction and exponential growth in random environments. *Theoretical Population Biology* **8**: 49–63.

Keller, L. F. (1998). Inbreeding and its fitness effects in an insular population of song sparrows (*Melospiza melodia*). *Evolution* **52**: 240–250.

Keller, L. F., Arcese, P., Smith, J. N. M., Hochachka, W. M. & Stearns, S. C. (1994). Selection against inbred song sparrows during a natural population bottleneck. *Nature* **372**: 356–357.

Kelly, D., Ladley, J. J., Robertson, A. W., Edwards, J. & Smith, D. C. (1996). The birds and the bees. *Nature* **384**: 615.

Kelly, J. K. (1997). A test of neutrality based on interlocus associations. *Genetics* **146**: 1197–1206.

Kempenaers, B., Adriaensen, F., van Noordwijk, A. J. & Dhondt, A. A. (1996). Genetic similarity, inbreeding and hatching failure in blue tits: are unhatched eggs infertile? *Proceedings of the Royal Society of London B* **263**: 179–185.

Kemper, C. & Bell, D. T. (1985). Small mammals and habitat structure in lowland rainforest of peninsular Malaysia. *Journal of Tropical Ecology* **1**: 5–22.

Kierulff, M. C. (1993). Avaliação das populações selvagens de mico-leão dourado, *Leontopithecus rosalia*, e proposta de estratégia para sua conservação. MSc thesis: Universidade Federal de Minas Gerais.

Kierulff, M. C. & Oliveira, P. P. (1996). Re-assessing the status of conservation of the golden lion tamarin *Leontopithecus rosalia* in the wild. *Dodo, Jersey Wildlife Preservation Trust* **32**: 89–115.

Kimura, M. (1983). The neutral theory of molecular evolution. In *Evolution of genes and proteins*: pp. 208–233, Nei, M. & Koehn, R. K. (eds.). Sunderland, MA: Sinauer Associates.

King, J. C. (1955). Evidence for the integration of the gene pool from studies of DDT resistance in *Drosophila*. *Cold Spring Harbor Symposium of Quantitative Biology* **20**: 311–317.

Kirkpatrick, R. E. B., Soltis, P. S. & Soltis, D. E. (1990). Mating system and distribution of genetic variation in natural populations of *Gymnocarpium dryopteris* ssp.*disjunctum. American Journal of Botany* **77**: 1101–1110.

Kirkpatrick, J. B., McDougall, K. & Hyde, M. (1995). Australia's most threatened ecosystem – the southeastern lowland native grasslands. Chipping Norton, NSW: Surrey Beatty.

Kondo, Y., Mori, M., Muramoto, T., Yamada, J., Beckman, J. S., Simonchazottes, D., Montagutelli, X., Guenet, J. L. & Serikawa, T. (1993). DNA segments mapped by reciprocal use of microsatellite primers between mouse and rat. *Mammalian Genome* **4**: 571–576.

Kruuk, L. E. B., Fenster, C. B. & Barton, N. M. (1999). Hybridisation: a genetic perspective. *XVI International Botanical Congress Abstracts* **469**: 1059.

Kuijt, J. (1964). Critical observations on the parasitism of New World mistletoes. *Canadian Journal of Botany* **42**: 1243–1278.

Kunin, W. E. (1993). Sex and the single mustard: population density and pollinator behavior effects on seed set. *Ecology* **74**: 2145–2160.

Kunin, W. E. (1997a). Population size and density effects in pollination: pollinator foraging and plant reproductive success in experimental arrays of *Brassica kaber*. *Journal of Ecology* **85**: 225–234.

Kunin, W. E. (1997b). Density dependence in various processes. In *The biology of rarity*: pp. 150–173, Kunin, W. E. & Gaston, K. J. (eds.). London: Chapman & Hall.

Kwak, M. M. & Jennersten, O. (1991). Bumblebee visitation and seedset in *Melampyrum pratense* and *Viscaria vulgaris*: heterospecific pollen and pollen limitation. *Oecologia* 86: 99–104.

Kwak, M. M., van der Brand, C., Kremer, P. & Boerrigter, E. J. M. (1991). Visitation, flight distances and seed set in populations of the rare species *Phyteuma nigrum* (Campanulaceae). *Acta Horticulturae* 288: 303–307.

Lack, D. (1954). *The natural regulation of animal numbers.* Oxford: Clarendon Press.

Lackey, R. T. (1997). Ecological risk assessment: use, abuse, and alternatives. *Environmental Management* 21: 808–812.

Lacy, R. C. (1987). Loss of genetic diversity from managed populations: interacting effects of drift, mutation, immigration, selection, and population subdivision. *Conservation Biology* 1: 143–158.

Lacy, R. C. (1993a). Impacts of inbreeding in natural and captive populations of vertebrates: implications for conservation. *Perspectives in Biology and Medicine* 36: 480–496.

Lacy, R. C. (1993b). VORTEX: a computer simulation model for population viability analysis. *Wildlife Research* 20: 45–65.

Lacy, R. C. (1995). Clarification of genetic terms and their use in the management of captive populations. *Zoo Biology* 14: 565–578.

Lacy, R. C. (1997). Importance of genetic variation to the viability of mammalian populations. *Journal of Mammalogy* 78: 320–335.

Lacy, R. C. & Ballou, J. D. (1998). Effectiveness of selection in reducing the genetic load in populations of *Peromyscus polionotus* during generations of inbreeding. *Evolution* 52: 900–909.

Lacy, R. C. & Lindenmayer, D. B. (1995). A simulation study of the impacts of population subdivision on the mountain brushtail possum *Trichosurus caninus ogilby* (Phalangeridae, Marsupialia) in south-eastern Australia. 2. Loss of genetic variation within and between subpopulations. *Biological Conservation* 73: 131–142.

Lacy, R. C., Petric, A. M. & Warneke, M. (1993). Inbreeding and outbreeding depression in captive populations of wild animal species. In *The natural history of inbreeding and outbreeding*: pp. 352–374, Thornhill, N. W. (ed.). Chicago: University of Chicago Press.

Lacy, R. C., Hughes, K. A. & Miller, P. S. (1995). VORTEX: a stochastic simulation of the extinction process. Version 7 User's Manual. Apple Valley, MN: IUCN/SSC Conservation Breeding Specialist Group.

Ladley, J. J. & Kelly, D. (1995). Explosive New Zealand mistletoe. *Nature* 378: 766.

Ladley, J. J. & Kelly, D. (1996). Dispersal, germination and survival of New Zealand mistletoes (Loranthaceae): dependence on birds. *New Zealand Journal of Ecology* 20: 69–79.

Ladley, J. J., Kelly, D. & Robertson, A. W. (1997). Explosive flowering, nectar production, breeding systems and pollinators of New Zealand mistletoes (Loranthaceae). *New Zealand Journal of Botany* 35: 345–360.

Lamont, B. B. (1983). Germination of mistletoes. In *The biology of mistletoes*: pp. 129–144, Calder, M. & Bernhardt, P. (eds.). Sydney: Academic Press.

Lamont, B. B., Klinkhamer, P. G. L. & Witkowski, E. T. F. (1993). Population fragmentation may reduce fertility to zero in *Banksia goodii* – a demonstration of the Allee effect. *Oecologia* **94**: 446–450.

Lande, R. (1988). Genetics and demography in biological conservation. *Science* **241**: 1455–1460.

Lande, R. (1993). Risks of population extinction from demographic and environmental stochasticity and random catastrophes. *American Naturalist* **142**: 911–927.

Lande, R. (1994). Risk of population extinction from fixation of new deleterious mutations. *Evolution* **48**: 1460–1469.

Lande, R. (1995). Mutation and conservation. *Conservation Biology* **9**: 782–791.

Lande, R. (1998a). Anthropogenic, ecological, and genetic factors in extinction and conservation. *Researches in Population Ecology* **40**: 259–269.

Lande, R. (1998b). Demographic stochasticity and Allee effect on a scale with isotropic noise. *Oikos* **83**: 353–358.

Lande, R. & Barrowclough, G. F. (1987). Effective population size, genetic variation, and their use in population management. In *Viable populations for conservation*: pp. 87–123, Soulé, M. E. (ed.). Cambridge: Cambridge University Press.

Lande, R. & Orzack, S. H. (1988). Extinction dynamics of age-structured populations in a fluctuating environment. *Proceedings of the National Academy of Sciences, USA* **85**: 7418–7421.

Lande, R. & Schemske, D. W. (1985). The evolution of self-fertilization and inbreeding depression in plants. I. Genetic models. *Evolution* **39**: 24–40.

Langham, N. (1982). The ecology of the common tree shrew, *Tupaia glis* in peninsular Malaysia. *Journal of Zoology* **197**: 323–344.

Langham, N. (1983). Distribution and ecology of small mammals in three rain forest localities of peninsular Malaysia with particular reference to Kedah Peak. *Biotropica* **15**: 199–206.

Lark, K. G., Chase, K., Adler, F., Mansur, L. M. & Orf, J. H. (1995). Interactions between quantitative trait loci in soybean in which trait variation at one locus is conditional upon a specific allele at another. *Proceedings of the National Academy of Sciences, USA* **92**: 4656–4660.

Larson, A., Wake, D. B. & Yanev, K. P. (1984). Measuring gene flow among populations having high levels of genetic fragmentation. *Genetics* **106**: 293–308.

Laurance, W. F. (1991a). Ecological correlates of extinction proneness in Australian tropical rain forest mammals. *Conservation Biology* **5**: 79–89.

Laurance, W. F. (1991b). Edge effects in tropical forest fragments: application of a model for the design of nature reserves. *Biological Conservation* **57**: 205–219.

Laurance, W. F. & Bierregaard, R. O. (1997). *Tropical forest remnants: ecology, management, and conservation of fragmented communities*. Chicago: University of Chicago Press.

Lawlor, D. A., Ward, F. E., Ennis, P. D., Jackson, A. P. & Parham, P. (1988). HLA-A and B polymorphisms predate the divergence of humans and chimpanzees. *Nature* **335**: 268–270.

Leary, R. F., Allendorf, F. W. & Knudsen, K. L. (1993). Null alleles at 2 lactate dehydrogenase loci in rainbow trout are associated with decreased developmental stability. *Genetica* **89**: 3–13.

Leberg, P. L. (1993). Strategies for population reintroduction: effects of genetic variability on population growth and size. *Conservation Biology* 7: 194–199.

Ledig, F. T. & Conkle, M. T. (1983). Gene diversity and genetic structure in a narrow endemic, Torrey pine (*Pinus torreyana* Parry ex Carr.). *Evolution* 37: 79–85.

Lee, H. Y. (1967). Studies in *Swietenia* (Meliaceae): observations on the sexuality of the flowers. *Journal of the Arnold Arboretum* 48: 101–104.

Lee, R. C. & Wright, W. E. (1994). Development of human exposure-factor distributions using maximum-entropy inference. *Journal of Exposure Analysis and Environmental Epidemiology* 4: 329–341.

Leeton, P. & Fripp, Y. J. (1991). Breeding system, karyotype and variation within and between populations of *Rutidosis leptorrhynchoides* F. Muell. (Asteraceae: Inuleae). *Australian Journal of Botany* 39: 85–96.

Lefkovitch, L. P. (1965). The study of population growth in organisms grouped by stages. *Biometrics* 21: 1–18.

Legge, J. T., Roush, R., DeSalle, R., Vogler, A. P. & May, B. (1996). Genetic criteria for establishing evolutionarily significant units in Cryan's buckmoth. *Conservation Biology* 10: 85–98.

Leigh, E. G. (1981). The average lifetime of a population in a varying environment. *Journal of Theoretical Biology* 90: 213–239.

Lekagul, B. & McNeely, J. (1988). *Mammals of Thailand*, 2nd edn. Bangkok: Darnsutha Press.

Lesica, P. & Allendorf, F. W. (1992). Are small populations of plants worth preserving? *Conservation Biology* 6: 135–139.

Lesica, P. & Allendorf, F. W. (1995). When are peripheral populations valuable for conservation. *Conservation Biology* 9: 753–760.

Letcher, B. H., Priddy, J. A., Walters, J. R. & Crowder, L. B. (1998). An individual-based, spatially explicit simulation model of the population dynamics of the endangered red-cockaded woodpecker, *Picoides borealis*. *Biological Conservation* 86: 1–14.

Leung, L. K. P., Dickman, C. R. & Moore, L. A. (1993). Genetic variation in fragmented populations of an Australian rainforest rodent, *Melomys cervinipes*. *Pacific Conservation Biology* 1: 58–65.

Levenson, J. B. (1981). Woodlots as biogeographic islands in southeastern Wisconsin. In *Forest island dynamics in man-dominated landscapes*: pp. 13–39, Burgess, R. L. & Sharpe, D. M. (eds.). New York: Springer-Verlag.

Levin, D. A. (1978). Pollinator behaviour and the breeding structure of plants populations. In *The pollination of flowers by insects*: pp. 131–153, Richards, A. J. (ed.). New York: Academic Press.

Levin, D. A. (1990). The seed bank as a source of genetic novelty in plants. *American Naturalist* 135: 563–572.

Levin, D. A. & Kerster, H. W. (1969). The dependence of bee mediated pollen and gene dispersal upon plant density. *Evolution* 23: 560–571.

Levin, D. A. & Kerster, H. W. (1974). Gene flow in seed plants. *Evolutionary Biology* 7: 139–220.

Levin, D. A., Ritter, R. & Ellstrand, N. C. (1979). Protein polymorphism in the narrow endemic *Oenothera organensis*. *Evolution* 33: 534–542.

Levins, R. (1969). Some demographic and genetic consequences of environmental heterogeneity for biological control. *Bulletin of the Entomological Society of America* 15: 237–240.

Levins, R. (1970). Extinction. In *Some mathematical questions in biology: lectures on mathematics in the life sciences*, vol. 2: pp. 77–107, Grestenhaber, M. (ed.). Providence, RI: American Mathematical Society.

Lewis, O. T., Thomas, C. D., Hill, J. K., Brookes, M. I., Crane, T. P. R., Graneau, Y. A., Mallet, J. L. B. & Rose, O. C. (1997). Three ways of assessing metapopulation structure in the butterfly *Plebejus argus*. *Ecological Entomology* 22: 283–293.

Lewis, W. H. (1979). *Polyploidy: biological relevance*. New York: Plenum Press.

Lewontin, R. (1965). Selection for colonizing ability. In *The genetics of colonizing species*: pp. 77–91, Baker, H. G. & Stebbins, G. L. (eds.). New York: Academic Press.

Li, Z., Pinson, S. R. M., Park, W. D., Paterson, A. H. & Stansel, J. W. (1997). Epistasis for three grain components in rice (*Oryza sativa* L.). *Genetics* 145: 453–456.

Lindenmayer, D. B. (1995). Disturbance, forest wildlife conservation and a conservative basis for forest management in the mountain ash forests of Victoria. *Forest Ecology and Management* 74: 223–231.

Lindenmayer, D. B. & Possingham, H. P. (1996). Ranking conservation and timber management options for Leadbeater's Possum in southeastern Australia using population viability analysis. *Conservation Biology* 10: 235–251.

Lindenmayer, D. B., Cunningham, R. B., Donnelly, C. F., Triggs, B. J. & Belvedere, M. (1994). The diversity, abundance and microhabitat requirements of terrestrial mammals in contiguous forests and retained linear strips in the montane ash forests of the central highlands of Victoria. *Forest Ecology and Management* 67: 113–133.

Lindenmayer, D. B., Pope, M. L., Cunningham, R. B., Donnelly, C. F. & Nix, H. A. (1996). Roosting in the Sulphur-Crested Cockatoo (*Cacatua galerita*). *Emu* 96: 209–212.

Lindenmayer, D. B., Cunningham, R. B., Nix, H. A., Lindenmayer, D. B., McKenzie, S., McGregor, C., Pope, M. L. & Incoll, R. D. (1997a). *Counting birds in forests: a comparison of observers and observation methods*, CRES Working Paper 1997/6. Canberra: CRES.

Lindenmayer, D. B., Cunningham, R. B., Pope, M. L., Donnelly, C. F., Nix, H. A. & Incoll, R. D. (1997b). *The Tumut fragmentation experiment in south-eastern Australia: the effects of landscape context and fragmentation on arboreal marsupials*, CRES Working Paper 1997/4. Canberra: CRES.

Lindenmayer, D. B., Cunningham, R. B., Pope, M. L. & Donnelly, C. F. (1998a). *A field-based quasi-experiment to examine the response of mammals to landscape context and habitat fragmentation*, CRES Working Paper 1998/1. Canberra: CRES.

Lindenmayer, D. B., McCarthy, M. A. & Pope, M. L. (1998b). A test of Hanski's simple model for metapopulation model in a fragmented ecosystem. *Oikos* 84: 99–109.

Lindenmayer, D. B., Incoll, R. D., Cunningham, R. B., Pope, M. L., Donnelly, C. F., MacGregor, C. I., Tribolet, C. & Triggs, B. E. (1999a). Comparison of hairtube types for the detection of mammals. *Wildlife Research* 26: 745–753.

Lindenmayer, D. B., Cunningham, R. B., Pope, M. L. & Donnelly, C. F. (1999b). The Tumut fragmentation experiment in south-eastern Australia: the effects of landscape context and fragmentation of arboreal marsupials. *Ecological Applications* 9: 594–611.

Lindenmayer, D. B., Cunningham, R. B., Pope, M. L. & Donnelly, C. F. (1999c). A field-based experiment to examine the response of mammals to landscape context and habitat fragmentation. *Biological Conservation* **88**: 387–403.

Lindenmayer, D. B., Pope, M. L. & Cunningham, R. B. (1999*d*). Roads and nest predation: an experimental study in a modified forest ecosystem. *Emu* **99**: 148–152.

Litt, M. & Luty, J. A. (1989). A hypervariable microsatellite revealed by in vitro amplification of a dinucleotide repeat within the cardiac muscle actin gene. *American Journal of Human Genetics* **44**: 397–401.

Lively, C. M., Craddock, C. & Vrijenhoek, R. C. (1990). The red queen hypothesis supported by parasitism in sexual and clonal fish. *Nature* **344**: 864–866.

Loebel, D. A., Nurthen, R. K., Frankham, R., Briscoe, D. A. & Craven, D. (1992). Modeling problems in conservation genetics using captive *Drosophila* populations: consequences of equalizing founder representation. *Zoo Biology* **11**: 319–332.

Loeschcke, V., Tomiuk, J. & Jain, S. K. (1994). *Conservation genetics*. Basel: Birkhäuser Verlag.

Loiselle, B. A., Sork, V. L., Nason, J. D. & Graham, C. (1995). Spatial genetic structure of a tropical understory shrub, *Psychotria officinalis* (Rubiaceae). *American Journal of Botany* **82**: 1420–1425.

Lorenz, G. C. & Barrett, G. W. (1990). Influence of simulated landscape corridors on house mouse (*Mus musculus*) dispersal. *American Midland Naturalist* **123**: 348–356.

Louda, S. M., Kendall, D., Connor, J. & Simberloff, D. S. (1997). Ecological effects of an insect introduced for the biological control of weeds. *Science* **277**: 1088–1090.

Lovejoy, T. E. & Bierregaard, R. O. (1990). Central Amazonian forests and the Minimum Critical Size of Ecosystems Project. In *Four neotropical rainforests*: pp. 60–71, Gentry, A. H. (ed.). New Haven, CT: Yale University Press.

Lovejoy, T. E., Bierregaard, R. O., Rylands, A. B., Malcolm, J. R., Quintella, C. E., Harper, L. H., Brown, K. S., Powell, A. H., Powell, G. V. N., Schubart, H. O. R. & Hays, M. B. (1986). Edge and other effects of isolation on Amazon forest fragments. In *Conservation biology: the science of scarcity and diversity*: pp. 257–285, Soulé, M. E. (ed.). Sunderland, MA: Sinauer Associates.

Loveless, M. D. (1992). Isozyme variation in tropical trees: patterns of genetic organization. *New Forests* **6**: 67–94.

Loveless, M. D. & Hamrick, J. L. (1984). Ecological determinants of genetic structure in plant populations. *Annual Review of Ecology and Systematics* **15**: 65–95.

Ludwig, D. (1975). Persistence of dynamical systems under random perturbations. *Society for Industrial and Applied Mathematics Review* **17**: 605–640.

Ludwig, D. (1976). A singular perturbation problem in the theory of population extinction. *Society for Industrial and Applied Mathematics Proceedings* **10**: 87–104.

Ludwig, D. (1996). The distribution of population survival times. *American Naturalist* **147**: 506–526.

Ludwig, D. (1999). Is it meaningful to estimate a probability of extinction? *Ecology* **80**: 298–310.

Luikart, G. L. & England, P. R. (1999). Statistical analysis of microsatellite DNA data. *Trends in Ecology and Evolution* **14**: 253–255.

Luikart, G. L., Sherwin, W. B., Steele, B. M. & Allendorf, F. W. (1998). Usefulness of molecular markers for detecting population bottlenecks via monitoring genetic change. *Molecular Ecology* 7: 963–974.

Lynam, A. J. (1995). Effects of habitat fragmentation on the distributional patterns of small mammals in a tropical forest in Thailand. PhD thesis: University of California, San Diego.

Lynam, A. J. (1997). Rapid decline of small mammal diversity in monsoon evergreen forest fragments in Thailand. In *Tropical forest remnants: ecology, management and conservation of fragmented communities*: pp. 222–240, Laurance, W. F. & Bierregaard, R. O. (eds.). Chicago, IL: University of Chicago Press.

Lynch, M. (1988). The rate of polygenic mutation. *Genetical Research* 51: 127–148.

Lynch, M. (1991). The genetic interpretation of inbreeding and outbreeding depression. *Evolution* 45: 622–629.

Lynch, M. (1996). A quantitative-genetic perspective on conservation issues. In *Conservation genetics: case histories from nature*: pp. 471–501, Avise, J. C. & Hamrick, J. L. (eds.). New York: Chapman & Hall.

Lynch, M. & Crease, T. J. (1990). The analysis of population survey data on DNA sequence variation. *Molecular Biology and Evolution* 7: 377–394.

Lynch, M. & Gabriel, W. (1990). Mutation load and the survival of small populations. *Evolution* 44: 1725–1737.

Lynch, M. & Walsh, B. (1998). *Genetics and analysis of quantitative traits*. Sunderland, MA: Sinauer Associates.

Lynch, M., Conery, J. & Bürger, R. (1995a). Mutation accumulation and the extinction of small populations. *American Naturalist* 146: 489–518.

Lynch, M., Conery, J. & Bürger, R. (1995b). Mutational meltdowns in sexual populations. *Evolution* 49: 1067–1080.

Lynch, M., Pfrender, M., Spitze, K., Lehman, N., Hicks, J., Allen, D., Latta, L., Ottene, M., Bogue, F. & Colbourne, J. (1999). The quantitative and molecular genetic architecture of a subdivided species. *Evolution* 53: 100–110.

Lyons, E. E., Waser, N. M., Price, M. V., Antonovics, J. & Motten, A. F. (1989). Sources of variation in plant reproductive success and implications for concepts of sexual selection. *American Naturalist* 134: 409–433.

MacArthur, R. H. & Wilson, E. O. (1967). *The theory of island biogeography*. Princeton, NJ: Princeton University Press.

Mace, G. M. & Lande, R. (1991). Assessing extinction threats: towards a reevaluation of IUCN threatened species categories. *Conservation Biology* 5: 148–157.

Mackay, T. F. C. (1981). Genetic variation in varying environments. *Genetical Research* 37: 79–93.

Madsen, T., Stille, B. & Shine, R. (1996). Inbreeding depression in an isolated colony of adders, *Vipera brevis*. *Biological Conservation* 75: 113–118.

Magnanini, A. (1977). Progress in the development of Poço das Antas Biological Reserve for *Leontopithecus rosalia* in Brazil. In *The biology and conservation of the Callitrichidae*: pp. 131–136, Kleiman, D. G. (ed.). Washington, DC: Smithsonian Institution Press.

Maguire, L. A. (1986). Using decision analysis to manage endangered species populations. *Journal of Environmental Management* 22: 345–360.

Makinson, R. O. (1999). *Grevillea*. Melbourne: CSIRO Publishing.

Mangel, M. & Tier, C. (1994). Four facts every conservation biologist should know about persistence. *Ecology* 75: 607–614.

Margules, C. R., Mikovits, G. A. & Smith, G. T. (1994). Contrasting the effects of habitat fragmentation on the scorpion *Cercophonius squama* and amphipod *Arcitalitrus sylvaticus*. *Ecology* **75**: 2033–2042.

Martin, P. & Bateson, P. (1993). *Measuring behaviour*. Cambridge: Cambridge University Press.

Masterson, J. (1994). Stomatal size in fossil plants: evidence for polyploidy in majority of angiosperms. *Science* **264**: 421–424.

Mayo, O. (1987). *The theory of plant breeding*, 2nd edn. Oxford: Clarendon Press.

McCall, C., Waller, D. M. & Mitchell-Olds, T. (1994). Effects of serial inbreeding on fitness components in *Impatiens capensis*. *Evolution* **48**: 818–827.

McCarthy, M. A. (1996). Modelling extinction dynamics of the Helmeted Honeyeater: effects of demography, stochasticity, inbreeding and spatial structure. *Ecological Modelling* **85**: 151–163.

McCarthy, M. A. & Lindenmayer, D. B. (1999). Incorporating metapopulation dynamics of Greater Gliders into reserve design in disturbed landscapes. *Ecology* **80**: 651–667.

McCarthy, M. A., Burgman, M. A. & Ferson, S. (1995). Sensitivity analysis for models of population viability. *Biological Conservation* **73**: 93–100.

McCauley, D. E. (1991). Genetic consequences of local population extinction and recolonisation. *Trends in Ecology and Evolution* **6**: 5–8.

McCauley, D. E. (1994). Contrasting the distribution of chloroplast DNA and allozyme polymorphism among local populations of *Silene alba*: implications for studies of gene flow in plants. *Proceedings of the National Academy of Sciences USA* **91**: 8127–8131.

McCauley, D. E. (1997a). A population of populations. *Trends in Ecology and Evolution* **12**: 241–242.

McCauley, D. E. (1997b). The relative contributions of seed and pollen movement to local genetic structure of *Silene alba*. *Journal of Heredity* **88**: 257–263.

McCauley, D. E., Raveill, J. & Antonovics, J. (1995). Local founding events as determinants of genetic structure in a plant metapopulation. *Heredity* **75**: 630–636.

McCauley, D. E., Stevens, J. E. & Peroni, P. A. (1996). The spatial distribution of chloroplast DNA and allozyme polymorphisms within a population of *Silene alba* (Caryophyllaceae). *American Journal of Botany* **83**: 727–731.

McClenaghan, L. R. & Beauchamp, A. C. (1986). Low genic differentiation among isolated populations of the californian fan palm (*Washingtonia filifera*). *Evolution* **40**: 315–322.

McCullagh, P. & Nelder, J. A. (1989). *Generalised linear models*, 2nd edn. New York: Chapman & Hall.

McCullough, D. R. (1996). Introduction. In *Metapopulations and wildlife conservation*: pp. 1–10, McCullough, D. R. (ed.). Washington, DC: Island Press.

McDonald, J. H. (1996). Detecting non-neutral heterogeneity across a region of DNA sequence in the ratio of polymorphism to divergence. *Molecular Biology and Evolution* **13**: 253–260.

McDonald, J. H. & Kreitman, M. (1991). Adaptive protein evolution at the *Adh* locus in *Drosophila*. *Nature* **351**: 652–654.

McFarlane, D. J., George, R. J. & Farrington, P. (1993). Changes in the hydrologic cycle. In *Reintegrating fragmented landscapes: towards sustainable production and nature conservation*: pp. 146–186, Hobbs, R.J. & Saunders, D.A. (eds.). New York: Springer-Verlag.

McFarquhar, A. M. & Robertson, F. W. (1963). The lack of evidence for co-adaptation in crosses between geographical races of *Drosophila subobscura* Coll. *Genetical Research* **4**: 104–131.

McGillivray, D. J. (1993). *Grevillea*. Melbourne: Melbourne University Press.

McKechnie, S. W., Halford, M. M., McColl, G. & Hoffmann, A. A. (1998). Both allelic variation and expression of nuclear and cytoplasmic transcripts are closely associated with thermal phenotype in *Drosophila*. *Proceedings of the National Academy of Sciences, USA* **95**: 2423–2428.

McKenzie, J. A. & Batterham, P. (1998). Predicting insecticide resistance – mutagenesis, selection and response. *Philosophical Transactions of the Royal Society of London B* **353**: 1729–1734.

McKenzie, J. A. & Parsons, P. A. (1974). Microdifferentiation in a natural population of *Drosophila melanogaster* to alcohol in the environment. *Genetics* **77**: 385–394.

McKenzie, J. A., McKechnie, S. W. & Batterham, P. (1994). Perturbation of gene frequencies in a natural population of *Drosophila melanogaster* – evidence for selection at the *ADH* locus. *Genetica* **92**: 187–196.

McNeill, J. (1977). The biology of Canadian weeds. 25. *Silene alba* (Miller) E. H. L. Krause. *Canadian Journal of Plant Science* **57**: 1103–1114.

Meagher, T. R., Antonovics, J. & Primack, R. (1978). Experimental ecological genetics in *Plantago* III. Genetic variation and demography in relation to survival of *Plantago cordata*, a rare species. *Biological Conservation* **14**: 243–257.

Medway, L. (1983). *The wild mammals of Malaya (Peninsular Malaysia) and Singapore*, 2nd edn with corrns. Kuala Lumpur: Oxford University Press.

Menges, E. S. (1990). Population viability analysis for a rare plant. *Conservation Biology* **4**: 52–62.

Menges, E. S. (1991a). Seed germination percentage increases with population size in a fragmented prairie species. *Conservation Biology* **5**: 158–164.

Menges, E. S. (1991b). The application of minimum viable population theory to plants. In *Genetics and conservation of rare plants*: pp. 45–61, Falk, D. A. & Holsinger, K. E. (eds.). Oxford: Oxford University Press.

Menges, E. S. (1992). Stochastic modeling of extinction in plant populations. In *Conservation biology: the theory and practice of nature conservation, preservation, and management*: pp. 253–276, Fiedler, P. L. & Jain, S. K. (eds.). New York: Chapman & Hall.

Menges, E. S. & Dolan, R. W. (1998). Demographic viability of populations of *Silene regia* in midwestern prairies: relationships with fire management, genetic variation, geographic location, population size and isolation. *Journal of Ecology* **86**: 63–78.

Merrell, P. (1995). Legal issues of ecological risk assessment. *Human and Ecological Risk Assessment* **1**: 454–458.

Merriam, G. & Wegner, J. F. (1992). Local extinctions, habitat fragmentation, and ecotones. In *Landscape boundaries: consequences for biotic diversity and ecological flows*: pp. 151–169, Hansen, A. J. & di Castor, F. (eds.). New York: Springer-Verlag.

Miller, P. S. (1994). Is Inbreeding depression more severe in a stressful environment? *Zoo Biology* **13**: 195–208.

Millner-Gulland, E. J. (1997). A stochastic dynamic programming model for the management of the saiga antelope. *Ecological Applications* **7**: 130–142.

Mills, L. S. (1996). Cheetah extinction: genetics or extrinsic factors? *Conservation Biology* **10**: 315.

Mills, L. S. & Allendorf, F. W. (1996). The one-migrant-per-generation rule in conservation and management. *Conservation Biology* **10**: 1509–1518.

Mills, L. S. & Smouse, P. E. (1994). Demographic consequences of inbreeding in remnant populations. *American Naturalist* **144**: 412–431.

Minckley, W. L., Buth, D. G. & Mayden, R. L. (1989). Origin of brood stock and allozyme variation in hatchery-reared bonytail, an endangered North American cyprinid fish. *Transactions of the American Fisheries Society* **118**: 131–137.

Mitchell-Olds, T. (1995). The molecular basis of quantitative genetic variation in natural populations. *Trends in Ecology and Evolution* **8**: 324–328.

Mitton, J. B. (1998). *Selection in natural populations.* New York: Oxford University Press.

Moilanen, A. & Hanski, I. A. (1998). Metapopulation dynamics: effects of habitat quality and landscape structure. *Ecology* **79**: 2503–2515.

Moilanen, A., Smith, A. T. & Hanski, I. A. (1998). Long-term dynamics in a metapopulation of the American pika. *American Naturalist* **152**: 530–542.

Molina-Freaner, F. & Jain, S. K. (1993). Inbreeding effects in a gynodioecious population of the colonizing species *Trifolium hirtum* All. *Evolution* **47**: 1472–1479.

Moll, R. H., Lindsey, M. F. & Robinson, H. F. (1965). Estimates of genetic variances and level of dominance in maize. *Genetics* **49**: 411–423.

Montllor, R. M. & Bernays, E. A. (1993). Invertebrate predators and caterpillar foraging. In *Caterpillars: ecological and evolutionary constraints on foraging*: pp. 170–202, Stamp, N. E. & Casey, T. M. (eds.). London: Chapman & Hall.

Moore, J. J. (1993). Inbreeding and outbreeding in primates: what's wrong with the dispersing sex? In *The natural history of inbreeding and outbreeding*: pp. 392–426, Thornhill, N. W. (ed.). Chicago, IL: University of Chicago Press.

Moran, G. F. & Hopper, S. D. (1983). Genetic diversity and the insular population structure of the rare granite rock species, *Eucalyptus caesia* Benth. *Australian Journal of Botany* **31**: 161–172.

Morgan, J. W. (1995a). Ecological studies of the endangered *Rutidosis leptorrhynchoides*. I. Seed production, soil seed bank dynamics, population density and their effect on recruitment. *Australian Journal of Botany* **43**: 1–11.

Morgan, J. W. (1995b). Ecological studies of the endangered *Rutidosis leptorrhynchoides*. II. Patterns of seedling emergence and survival in a native grassland. *Australian Journal of Botany* **43**: 13–24.

Morgan, J. W. (1999). Effects of population size on seed production and germinability in an endangered, fragmented grassland plant. *Conservation Biology* **13**: 266–273.

Morgante, M., Pfeiffer, A., Costacurta, A. & Olivieri, A. M. (1996). Molecular tools for population and ecological genetics in coniferous trees. *Phyton* **36**: 129–138.

Morin, P. A. & Woodruff, D. S. (1992). Paternity exclusion using multiple hypervariable microsatellite loci amplified from nuclear DNA of hair cells. In *Paternity in primates: genetic tests and theories*: pp. 63–81, Martin, R. D., Dixson, A. F. & Wickings, E. J. (eds.). Basel: Karger.

Morin, P. A. & Woodruff, D. S. (1996). Noninvasive genotyping for vertebrate conservation. In *Molecular genetic approaches to conservation*: pp. 298–313, Smith, T.

B. & Wayne, R. K. (eds.). New York: Oxford University Press.

Morin, P. A., Moore, J. J., Chakraborty, R., Jin, L., Goodall, J. & Woodruff, D. S. (1993). Kin selection, social structure, gene flow and the evolution of chimpanzees. *Science* **265**: 1193–1201.

Moritz, C. (1994*a*). Applications of mitochondrial DNA analysis to conservation: a critical review. *Molecular Ecology* **3**: 401–411.

Moritz, C. (1994*b*). Defining 'evolutionarily significant units' for conservation. *Trends in Ecology and Evolution* **9**: 373–375.

Moritz, C. (1999*a*). A molecular perspective on the conservation of diversity. In *The biology of biodiversity*: pp. 21–34, Kato, S. (ed.). Tokyo: Springer-Verlag.

Moritz, C. (1999*b*). Conservation units and translocations: strategies for conserving evolutionary processes. *Hereditas* **130**: 217–228.

Moritz, C., Worthington-Wilmer, J., Pope, L., Sherwin, W. B. & Taylor, A. C. (1996). Applications of genetics to the conservation and management of Australian fauna: four case studies from Queensland. In *Molecular genetic approaches in conservation*: pp. 442–456, Smith, T. B. & Wayne, R. K. (eds.). New York: Oxford University Press.

Morris, W. F. (1993). Predicting the consequences of plant spacing and biased movement for pollen dispersal by honey bees. *Ecology* **74**: 493–500.

Mundy, N. I., Winchell, C. S., Burr, T. & Woodruff, D. S. (1997). Microsatellite variation and microevolution in the critically endangered San Clemente Island loggerhead shrike (*Lanius ludovicianus mearnsi*). *Proceedings of the Royal Society of London B* **264**: 869–875.

Munte, A., Aguade, M. & Segarre, C. (1997). Divergence of the *yellow* gene between *Drosophila melanogaster* and *D. subobscura*: recombination rate, codon bias, and synonymous substitutions. *Genetics* **147**: 165–175.

Murawski, D. A. & Hamrick, J. L. (1991). The effect of the density of flowering individuals on the mating systems of nine tropical tree species. *Heredity* **67**: 167–174.

Murcia, C. (1995). Edge effects in fragmented forests: implications for conservation. *Trends in Ecology and Evolution* **10**: 58–62.

Muse, S. V. (1996). Estimating synonymous and non-synonymous substitution rates. *Molecular Biology and Evolution* **13**: 105–114.

Mutikainen, P. & Delph, L. F. (1998). Inbreeding depression in gynodioecious *Lobelia siphilitica*: among-family differences override between-morph differences. *Evolution* **52**: 1572–1582.

Nakasathien, S. (1988). The first wildlife rescue operation in Thailand. *Thai Journal of Forestry* **7**: 250–265.

Nakasathien, S. (1989). Chiew Larn Dam wildlife rescue operation. *Oryx* **23**: 146–154.

Nason, J. & Ellstrand, N. C. (1995). Lifetime estimates of biparental inbreeding depression in the self-incompatible annual plant *Raphanus sativus*. *Evolution* **49**: 307–316.

Nason, J. D. & Hamrick, J. L. (1997). Reproductive and genetic consequences of forest fragmentation: two case studies of neotropical canopy trees. *Journal of Heredity* **88**: 264–276.

Nason, J. D., Herre, E. A. & Hamrick, J. L. (1998). The breeding structure of a tropical keystone plant resource. *Nature* **391**: 685–687.

Nauta, M. J. & Weissing, F. J. (1996). Constraints on allele size at microsatellite loci: implications for genetic differentiation. *Genetics* **143**: 1021–1032.

Nei, M. (1978). Estimation of average heterozygosity and genetic distances from a small number of individuals. *Genetics* **89**: 583–590.

Nei, M. (1987). *Molecular evolutionary genetics.* New York: Columbia University Press.

Nei, M., Maruyama, T. & Chakraborty, R. (1975). The bottleneck effect and genetic variability in populations. *Evolution* **29**: 1–10.

Neigel, J. E. (1996). Estimation of effective population size and migration parameters from genetic data. In *Molecular genetic approaches in conservation*: pp. 329–346, Smith, T. B. & Wayne, R. K. (eds.). New York: Oxford University Press.

Neil, D. & Fogarty, P. (1991). Land use and sediment yield on the southern tablelands of New South Wales. *Australian Journal of Soil and Water Conservation* **4**: 33–39.

Newman, D. & Pilson, D. (1997). Increased probability of extinction due to decreased genetic effective population size: experimental populations of *Clarkia pulchella. Evolution* **51**: 354–362.

Newton, A. C., Cornelius, J. P., Baker, P., Gillies, A. C. M., Hernandez, M., Ramnarine, S., Mesen, J. F., Watt, A. D. & MacLellan, A. (1996). Mahogany as a genetic resource. *Botanical Journal of the Linnean Society* **122**: 61–73.

Neyman, P. F. (1977). Aspects of the ecology and social organisation of free-ranging cotton-top tamarins (*Saguinus oedipus*) and the conservation status of the species. In *The biology and conservation of the Callitrichidae*: pp. 39–69, Kleiman, D. G. (ed.). Washington, DC: Smithsonian Institution Press.

Noon, B. R. & McKelvey, K. S. (1996). A common framework for conservation planning: linking individual and metapopulation models. In *Metapopulations and wildlife conservation*: pp. 139–165, McCullough, D. R. (ed.). Washington, DC: Island Press.

Norman, J. K., Sakai, A. K., Weller, S. G. & Dawson, T. E. (1995). Inbreeding depression in morphological and physiological traits of *Schiedea lydgatei* (Caryophyllaceae) in two environments. *Evolution* **49**: 297–306.

Norton, D. A. (1991). *Trilepidea adamsii*: an obituary for a species. *Conservation Biology* **5**: 52–57.

Norton, D. A. & Ladley, J. J. (1998). Establishment and early growth of *Alepis flavida* in relation to *Nothofagus solandri* branch size. *New Zealand Journal of Botany* **36**: 213–217.

Norton, D. A. & Reid, N. (1997). Lessons in ecosystem management from management of threatened and pest Loranthaceous mistletoes in New Zealand and Australia. *Conservation Biology* **11**: 759–769.

Norton, D. A., Hobbs, R. J. & Atkins, L. (1995). Fragmentation, disturbance, and plant distribution: mistletoes in woodland remnants in the Western Australian wheatbelt. *Conservation Biology* **9**: 426–438.

Nunney, L. (1993). The influence of mating system and overlapping generations on effective population size. *Evolution* **47**: 1329–1341.

Nunney, L. & Campbell, K. A. (1993). Assessing minimum viable population size: demography meets population genetics. *Trends in Ecology and Evolution* **8**: 234–239.

Nunney, L. & Elam, D. R. (1994). Estimating the effective population size of conserved populations. *Conservation Biology* **8**: 175–184.

O'Brien, S. J., Roelke, M. E., Marker, A., Newman, C. A., Winkler, D., Meltzer, L., Colly, J. F., Bush, M. & Wildt, D. E. (1985). Genetic basis for species vulnerability in the cheetah. *Science* **227**: 1428–1434.

O'Brien, S. J., Wildt, D. E., Bush, M., Caro, T. M., Fitzgibbon, C., Aggundey, I. & Leakey, R. E. (1987). East African cheetahs: evidence for two population bottlenecks? *Proceedings of the National Academy of Sciences, USA* **84**: 508–511.

Ogle, C. & Wilson, P. R. (1985). Where have all the mistletoes gone? *Forest and Bird* **237**: 10–13.

Olde, P. M. & Marriott, N. R. (1994). The *Grevillea* book, vol. 1. Kenthurst, NSW: Kangaroo Press.

Olde, P. M. & Marriott, N. R. (1995). The *Grevillea* book, vols. 2 and 3. Kenthurst, NSW: Kangaroo Press.

Olivieri, I., Couvet, D. & Gouyon, P. H. (1990). The genetics of transient populations: research at the metapopulation level. *Trends in Ecology and Evolution* **5**: 207–210.

Olivieri, I., Michalakis, Y. & Gouyon, P. H. (1995). Metapopulation genetics and the evolution of dispersal. *American Naturalist* **146**: 202–228.

Oostermeijer, J. G. B. (1996a). Population size, genetic variation, and related parameters in framented plant populations: a case study. In *Species survival in fragmented landscapes*: pp. 61–68, Settele, J., Margules, C. R., Poschlod, P. & Henle, K. (eds.). Dordrecht: Kluwer.

Oostermeijer, J. G. B. (1996b). Population viability of the rare *Gentiana pneumonanthe*: the relative importance of demography, genetics and reproductive biology. PhD thesis: University of Amsterdam.

Oostermeijer, J. G. B., den Nijs, J. C. M., Raijmann, L. E. L. & Menken, S. B. J. (1992). Population biology and management of the Marsh Gentian (*Gentiana pneumonanthe* L.), a rare species in The Netherlands. *Botanical Journal of the Linnean Society* **108**: 117–130.

Oostermeijer, J. G. B., van Eijck, M. W. & den Nijs, J. C. M. (1994a). Offspring fitness in relation to population size and genetic variation in the rare perennial plant species *Gentiana pneumonanthe* (Gentianaceae). *Oecologia* **97**: 289–296.

Oostermeijer, J. G. B., 't Veer, R. & den Nijs, J. C. M. (1994b). Population structure of the rare, long-lived perennial *Gentiana pneumonanthe* in relation to vegetation and management in The Netherlands. *Journal of Applied Ecology* **31**: 428–438.

Oostermeijer, J. G. B., van Eijck, M. W., van Leeuwen, N. C. & den Nijs, J. C. M. (1995). Analysis of the relationship between allozyme heterozygosity and fitness in the rare *Gentiana pneumonanthe* L. *Journal of Evolutionary Biology* **8**: 739–759.

Oostermeijer, J. G. B., Brugman, M. L., de Boer, E. R. & den Nijs, J. C. M. (1996). Temporal and spatial variation in the demography of *Gentiana pneumonanthe*, a rare perennial herb. *Journal of Ecology* **84**: 153–166.

Oostermeijer, J. G. B., Luijten, S. H., Krenova, Z. V. & den Nijs, J. C. M. (1998). Relationships between population and habitat characteristics and reproduction of the rare *Gentiana pneumonanthe* L. *Conservation Biology* **12**: 1042–1053.

Oreskes, N., Shrader-Frechette, K. & Belitz, K. (1994). Verification, validation, and confirmation of numerical models in the earth sciences. *Science* **263**: 641–646.

Orr, H. A. (1995). The population genetics of speciation: the evolution of hybrid compatibilities. *Genetics* **139**: 1805–1813.

Ota, T. (1993). DISPAN: Genetic distance and phylogenetic analysis. Pennsylvania State University: Institute of molecular evolutionary genetics.

Ouborg, N. J. (1993*a*). Isolation, population size and extinction: the classical and metapopulation approaches applied to vascular plants along the Dutch Rhine system. *Oikos* **66**: 298–308.

Ouborg, N. J. (1993*b*). On the relative contribution of genetic erosion to the chance of population extinction. PhD thesis: University of Utrecht.

Ouborg, N. J. & van Treuren, R. (1994). The significance of genetic erosion in the process of extinction. IV. Inbreeding load and heterosis in relation to population size in the mint *Salvia pratensis*. *Evolution* **48**: 996–1008.

Ouborg, N. J., van Treuren, R. & Van Damme, J. M. M. (1991). The significance of genetic erosion in the process of extinction. II. Morphological variation and fitness components in populations varying in size of *Salvia pratensis* L. and *Scabiosa columbaria* L. *Oecologia* **86**: 359–367.

Overton, J. M. (1994). Dispersal and infection in mistletoe metapopulations. *Journal of Ecology* **82**: 711–723.

Owen, H. J. & Norton, D. A. (1995). The diet of introduced brushtail possums *Trichosurus vulpecula* in a low-diversity New Zealand *Nothofagus* forest and possible implications for conservation management. *Biological Conservation* **71**: 339–345.

Paetkau, D., Shields, G. F. & Strobeck, C. (1998). Gene flow between insular coastal and interior populations of brown bears in Alaska. *Molecular Ecology* **7**: 1283–1292.

Paillat, G. & Butet, A. (1996). Spatial dynamics of the bank vole (*Clethrionomys glareolus*) in a fragmented landscape. *Acta Oecologica* **17**: 553–559.

Palapoli, M. F. & Wu, C. I. (1994). Genetics of hybrid male sterility between *Drosophila* sibling species: a complex web of epistasis is revealed in interspecific studies. *Genetics* **138**: 329–341.

Parker, J. S. & Wilby, A. S. (1989). Extreme chromosomal heterogeneity in a small-island population of *Rumex acetosa*. *Heredity* **62**: 133–140.

Parker, M. A. (1991). Outbreeding depression in a selfing annual. *Evolution* **46**: 837–841.

Parsons, P. A. (1983). *The evolutionary biology of colonizing species*. Cambridge: Cambridge University Press.

Pascual, M. A., Kareiva, P. M. & Hilborn, R. (1997). The influence of model structure on conclusions about the viability and harvesting of Serengeti Wildebeest. *Conservation Biology* **11**: 966–976.

Patrick, B. H. & Dugdale, J. S. (1997). Mistletoe moths. In *Ecology and conservation of New Zealand's Loranthaceous mistletoes*: pp. 125–132, de Lange, P. J. & Norton, D. A. (eds.). Wellington: Department of Conservation.

Payne, J., Francis, C. M. & Phillipps, K. (1985). *A field guide to the mammals of Borneo*. Koto Kinabalu, Malaysia: The Sabah Society.

Peakall, R. & Smouse, P. E. (1998). GenAlEX: Genetic Analysis in Excel. Support software for the national workshop on 'Genetic Analysis for Population and Biogeographic Studies', July 1998. Canberra: Australian National University.

Peakall, R. & Sydes, M. A. (1996). Determining priorities for achieving practical outcomes from the genetic studies of rare plants. In *Back from the brink: refining the threatened species recovery process*: pp. 119–129, Stephens, S. & Maxwell, S. (eds.). Chipping Norton, NSW: Surrey Beatty.

Peakall, R., Smouse, P. E. & Huff, D. R. (1995). Evolutionary implications of allozyme and RAPD variation in diploid populations of Buffalo grass (*Buchloë dactyloides* Nutt. Engelm.). *Molecular Ecology* **4**: 135–147.

Peakall, R., Gilmore, S., Keys, W., Morgante, M. & Rafalski, A. (1998). Cross-species amplification of Soybean (*Glycine max*) simple-sequence-repeats (SSRs) within the genus and other legume genera: implications for the transferability of SSRs in plants. *Molecular Biology and Evolution* **15**: 1257–1287.

Petanidou, T., den Nijs, J. C. M., Oostermeijer, J. G. B. & Ellis-Adam, A. C. (1995). Pollination ecology and patch-dependent reproductive success of the rare perennial *Gentiana pneumonanthe* L. *New Phytologist* **129**: 155–163.

Pimm, S. L., Jones, H. L. & Diamond, J. M. (1988). On the risk of extinction. *American Naturalist* **132**: 757–785.

Pimm, S. L., Gittleman, J. L., McCracken, G. F. & Gilpin, M. E. (1989). Plausible alternatives to bottlenecks to explain reduced genetic diversity. *Trends in Ecology and Evolution* **4**: 176–177.

Polans, N. O. & Allard, R. W. (1989). An experimental evaluation of the recovery potential of rye grass populations from genetic stress resulting from restriction of population size. *Evolution* **43**: 1320–1324.

Pope, T. R. (1996). Socioecology, population fragmentation, and patterns of genetic loss in endangered primates. In *Conservation genetics: case histories from nature*: pp. 119–159. Avise, J. C. & Hamrick, J. L. (eds.). New York: Chapman & Hall.

Possingham, H. P. (1996). Decision theory and biodiversity management: how to manage a metapopulation. In *Proceedings of the Nicholson Centenary Conference 1995*: pp. 391–398, Floyd, R. B., Sheppard, A. W. & De Barro, P. J. (eds.). Melbourne: CSIRO Publishing.

Possingham, H. P. (1997). State-dependent decision analysis for conservation biology. In *The ecological basis for conservation: heterogeneity, ecosystems and biodiversity*: pp. 298–304, Pickett, S. T. A., Ostfeld, R. S., Shachak, M. & Likens, G. E. (eds.). New York: Chapman & Hall.

Possingham, H. P. & Davies, I. (1995). ALEX: A model for the viability analysis of spatially structured populations. *Biological Conservation* **73**: 143–150.

Possingham, H. P. & Tuck, G. (1998). Fire management strategies that minimise the probability of population extinction for early and mid-successional species. In *Proceedings of the conference on statistics in ecology and environmental monitoring 2. Decision-making and risk assessment in biology*: pp. 157–167, Fletcher, D., Kavalieris, L. & Manly, B. B. (eds.). Otago: University of Otago.

Possingham, H. P., Lindenmayer, D. B. & Norton, T. W. (1993). A framework for the improved management of threatened species based on population viability analysis. *Pacific Conservation Biology* **1**: 39–45.

Possingham, H. P., Lindenmayer, D. B., Norton, T. W. & Davies, I. (1994). Metapopulation viability analysis of the greater glider *Petauroides volans* in a wood production area. *Biological Conservation* **70**: 227–236.

Postel, S. L., Daily, G. C. & Ehrlich, P. R. (1996). Human appropriation of renewable fresh water. *Science* **271**: 785–788.

Powell, A. H. & Powell, G. V. N. (1987). Population dynamics of male euglossine bees in Amazonian forest fragments. *Biotropica* **19**: 176–179.

Power, M. & Adams, S. M. (1997). Perspectives of the scientific community on the status of ecological risk assessment. *Environmental Management* **21**: 803–830.

Pray, L. A. & Goodnight, C. J. (1995). Genetic variation in inbreeding depression in the red flour beetle *Tribolium castaneum*. *Evolution* **49**: 179–188.

Pray, L. A., Schwartz, J. M., Goodnight, C. J. & Stevens, L. (1994). Environmental dependency of inbreeding depression: implications for conservation biology. *Conservation Biology* **8**: 562–568.

Price, M. V. & Gilpin, M. E. (1996). Modelers, mammalogists, and metapopulations: designing Stephens' kangaroo rat reserves. In *Metapopulations and wildlife conservation*: pp. 217–240, McCullough, D. R. (ed.). Washington, DC: Island Press.

Price, M. V. & Kelly, P. A. (1994). An age-structured demographic model for the endangered Stephens' kangaroo rat. *Conservation Biology* **8**: 810–821.

Price, M. V. & Waser, N. M. (1979). Pollen dispersal and optimal outcrossing in *Delphinium nelsonii*. *Nature* **277**: 294–297.

Prober, S. M. & Brown, A. H. D. (1994). Conservation of Grassy White woodlands: population genetics and fragmentation of *Eucalyptus albens*. *Conservation Biology* **8**: 1003–1013.

Prout, T. & Barker, J. S. F. (1989). Ecological aspects of the heritability of body size in *Drosophila buzzatii*. *Genetics* **123**: 803–813.

Pusey, A. & Wolf, M. (1996). Inbreeding avoidance in animals. *Trends in Ecology and Evolution* **11**: 201–206.

Quammen, D. (1996). *The song of the dodo: island biogeography in an age of extinction*. New York: Scribner.

Rabinowitz, D. (1981). Seven forms of rarity. In *The biological aspects of rare plant conservation*: pp. 205–217, Synge, H. (ed.). Chichester: John Wiley.

Raijmann, L. E. L., van Leeuwen, N. C., Kersten, R., Oostermeijer, J. G. B., den Nijs, J. C. M. & Menken, S. B. J. (1994). Genetic variation and outcrossing rate in relation to population size in *Gentiana pneumonanthe* L. *Conservation Biology* **8**: 1014–1025.

Ralls, K. & Ballou, J. D. (1982). Effects of inbreeding on infant mortality in captive primates. *International Journal of Primatology* **3**: 491–505.

Ralls, K. & Ballou, J. D. (1983). Extinction: lessons from zoos. In *Genetics and conservation: a reference for managing wild animal and plant populations*: pp. 164–184, Schonewald-Cox, C. M., Chambers, S. M., MacBryde, B. & Thomas, W. L. (eds.). Menlo Park, CA: Benjamin Cummings.

Ralls, K. & Starfield, A. M. (1995). Choosing a management stratgey: two structured decision-making methods for evaluating the predictions of stochastic simulation models. *Conservation Biology* **9**: 175–181.

Ralls, K., Ballou, J. D. & Templeton, A. R. (1988). Estimates of lethal equivalents and the cost of inbreeding in mammals. *Conservation Biology* **2**: 185–193.

Ramey, R. R. (1995). Mitochondrial DNA variation, population structure, and evolution of mountain sheep in the southwestern United States and Mexico. *Molecular Ecology* **4**: 429–439.

Rand, D. M. (1996). Neutrality tests of molecular markers and the connection between DNA polymorphism, demography, and conservation biology. *Conservation Biology* **10**: 665–671.

Ranney, J. W., Burner, M. C. & Levenson, J. B. (1981). The importance of edge in the structure and dynamics of forest islands. In *Forest island dynamics in man-dominated landscapes*: pp. 67–95, Burgess, R. L. & Sharpe, D. M. (eds.). New York: Springer-Verlag.

Rathke, B. (1983). Competition and facilitation among plants for pollination. In *Pollination biology*: pp. 305–329, Real, L. A. (ed.). Orlando, FL: Academic Press.

Rausher, M. D. (1979). Larval habitat suitability and oviposition preference in three related butterflies. *Ecology* **60**: 503–511.

Raw, A. (1989). The dispersal of euglossine bees between isolated patches of eastern Brazilian wet forest (Hymneoptera, Apidae). *Revista Brasilia Entomologia* **33**: 103–107.

Raymond, M. & Rousset, F. (1995). GENEPOP (version 1.2): population genetics software for exact tests and ecumenicism. *Journal of Heredity* **86**: 248–249.

Reed, J. M., Doerr, P. D. & Walters, J. R. (1988). Minimum viable population size of the red-cockaded woodpecker. *Journal of Wildlife Management* **52**: 385–391.

Reed, J. M., Murphy, D. D. & Brussard, P. F. (1998). Efficacy of population viability analysis. *Wildlife Society Bulletin* **26**: 244–251.

Reed, J. M., Walters, J. R., Emigh, T. E. & Seaman, D. E. (1993). Effective population size in red-cockaded woodpeckers: population and model differences. *Conservation Biology* **7**: 302–308.

Rees, H. (1961). Genotypic control of chromosome form and behaviour. *Botanical Review* **27**: 288–318.

Reeve, H. K., Westneat, D. F., Noon, W. A., Sherman, P. W. & Aquadro, C. F. (1990). DNA "fingerprinting" reveals high levels of inbreeding in colonies of the eusocial naked mole-rat. *Proceedings of the National Academy of Sciences, USA* **87**: 2496–2500.

Reinartz, J. A. & Les, D. H. (1994). Bottleneck-induced dissolution of self-incompatibility and breeding consequences in *Aster furcatus* (Asteraceae). *American Journal of Botany* **81**: 446–455.

Rhymer, J. M. & Simberloff, D. (1996). Extinction by hybridization and introgression. *Annual Review of Ecology and Systematics* **27**: 83–109.

Rice, K. & Jain, S. K. (1985). Plant population genetics and evolution in disturbed environments. In *The ecology of natural disturbance and patch dynamics*: pp. 287–303, Pickett, S. T. A. & White, P. (eds.). Orlando, FL: Academic Press.

Richards, C. M. (1997). The ecological genetics of population establishment in *Silene alba*. PhD thesis: Duke University.

Richards, C. M. (2000). Gene flow and genetic rescue in a plant metapopulation. *American Naturalist* **155**: 383–394.

Richards, C. M., Church, S. & McCauley, D. E. (1999a). The influence of population size and isolation on gene flow by pollen. *Evolution* **53**: 63–73.

Richards, P. W. (1996). *The tropical rain forest*, 2nd edn. Cambridge: Cambridge University Press.

Richards, S. A., Possingham, H. P. & Tizard, J. (1999b). Optimal fire management for maintaining community diversity. *Ecological Applications* **9**: 880–892.

Richardson, M. B. G., Ayre, D. J. & Whelan, R. J. (in press). Pollinator behaviour, mate choice and the realised mating systems of *Grevillea mucronulata* and *Grevillea sphacelata*. *Australian Journal of Botany*.

Richman, A. D. & Kohn, J. R. (1996). Learning from rejection: the evolutionary biology of single-locus incompatibility. *Trends in Ecology and Evolution* **11**: 497–

502.

Richter-Dyn, N. & Goel, N. S. (1972). On the extinction of a colonizing species. *Theoretical Population Biology* **3**: 406–433.

Rieseberg, L. H. (1997). Hybrid origins of plant species. *Annual Review of Ecology and Systematics* **28**: 359–389.

Rieseberg, L. H., Sinervo, B., Linder, C. R., Ungerer, M. C. & Arias, D. M. (1995). Role of gene interactions in hybrid speciation: evidence from ancient and experimental hybrids. *Science* **272**: 741–745.

Rishworth, C. & Tanton, M. T. (1995). Diet of the Common Wombat, *Vombatus ursinus*, in plantations of *Pinus radiata*. *Wildlife Research* **22**: 333–339.

Rishworth, C., McIlroy, J. & Tanton, M. T. (1995). Factors affecting population densities of the Common Wombat, *Vombatus ursinus*, in plantations of *Pinus radiata*. *Forest Ecology and Management* **76**: 11–19.

Ritland, K. (1989). Genetic differentiation , diversity and inbreeding in the mountain monkeyflower (*Mimulus caespitosus*) of the Washington Cacades. *Canadian Journal of Botany* **67**: 2017–2024.

Ritland, K. (1996). A marker-based method for inferences about quantitative inheritance in natural populations. *Evolution* **50**: 1062–1073.

Robertson, A. (1962). Selection for heterozygotes in small populations. *Genetics* **47**: 1291–1300.

Robertson, A. W., Kelly, D., Ladley, J. J. & Sparrow, A. D. (1999). Effects of pollinator loss on endemic New Zealand mistletoes (Loranthaceae). *Conservation Biology* **13**: 499–508.

Robinson, G. R., Holt, R. D., Gaines, M. S., Hamburg, S. P., Johnson, M. L., Fitch, H. S. & Martinko, E. A. (1992). Diverse and contrasting effects of habitat fragmentation. *Science* **257**: 524–526.

Robinson, N. A., Sherwin, W. B. & Murray, N. D. (1993). Use of VNTR loci to reveal population structure in the eastern barred bandicoot, *Perameles gunn*. *Molecular Ecology* **2**: 195–207.

Rogers, A. R. & Harpending, H. C. (1983). Population structure and quantitative characters. *Genetics* **105**: 985–1002.

Ronfort, J. (1999). The mutation load under tetrasomic inheritance and its consequences for the evolution of the selfing rate in autotetraploid species. *Genetical Research* **74**: 31–42.

Rossetto, M., Weaver, P. K. & Dixon, K. W. (1995). Use of RAPD analysis in devising conservation strategies for the rare and endangered *Grevillea scapigera* (Proteaceae). *Molecular Ecology* **4**: 321–329.

Rousset, F. & Raymond, M. (1997). Statistical analyses of population genetic data: new tools, old concepts. *Trends in Ecology and Evolution* **12**: 313–317.

Routman, E. J. & Cheverud, J. M. (1997). Gene effects on a quantitative trait: two-locus epistatic effects measured at microsatellite markers and at estimated QTL. *Evolution* **51**: 1654–1662.

Ruckelshaus, M. H., Hartway, C. & Kareiva, P. M. (1997). Assessing the data requirements of spatially explicit dispersal models. *Conservation Biology* **11**: 1298–1306.

Rudd, R. L. (1979). Niche dimension in the bamboo mouse, *Chiropodomys gliroides* (Rodentia; Muridae). *Malayan Nature Journal* **32**: 347–349.

Ruesink, J. L., Parker, I. M., Groom, M. J. & Kareiva, P. M. (1995). Reducing the risks of nonindigenous species introductions: guilty until proven innocent. *Bioscience* **45**: 465–477.

Ruscoe, W. A., Mather, P. B. & Wilson, J. (1998). Genetic structure in populations of *Rattus sordidus* in sugarcane-growing districts of Queensland, Australia. *Journal of Mammalogy* **79**: 612–623.

Sabelis, M. W. & Diekmann, O. (1988). Overall population stability despite local extinction: the stabilizing influence of prey dispersal from predator-invaded patches. *Theoretical Population Biology* **34**: 169–176.

Saccheri, I., Kuussaari, M., Kankare, M., Vikman, P., Fortelius, W. & Hanski, I. A. (1998). Inbreeding and extinction in a butterfly metapopulation. *Nature* **392**: 491–494.

Sæther, B. E., Engen, S., Islam, A., McCleery, R. & Christopher, P. (1998). Environmental stochasticity and extinction risk in a population of a small songbird, the great tit. *American Naturalist* **151**: 441–450.

Sakai, A. K., Karoly, K. & Weller, S. G. (1989). Inbreeding depression in *Schiedea globosa* and *S. salicaria* (Caryophyllaceae), subdioecious and gynodioecious Hawaiian species. *American Journal of Botany* **76**: 437–444.

Sakai, A. K., Weller, S. G., Chen, M. L., Chou, S. H. & Tasonot, C. (1997). Evolution of gynodioecy and maintenance of females: the role of inbreeding depression, outcrossing rates, and resource allocation in *Schiedea adamantis* (Caryophyllaceae). *Evolution* **51**: 724–736.

Sambrook, J., Fritsch, E. F. & Maniatis, T. (1989). *Molecular cloning: a laboratory manual*. New York: Cold Spring Harbor Laboratory Press.

Sanjayan, M. A., Crooks, K., Zegers, G. & Foran, D. (1996). Genetic variation and the immune response in natural populations of pocket gophers. *Conservation Biology* **10**: 1519–1527.

Sarre, S. (1995). Mitochondrial DNA variation among populations of *Oedura reticulata* (Gekkonidae) in remnant vegetation: implications for metapopulation structure and population decline. *Molecular Ecology* **4**: 395–405.

SAS (1997). SAS/STAT® software: changes and enhancements through release 6.12. Cary, NC: SAS Institute.

Saunders, D. A., Arnold, G. W., Burbridge, A. A. & Hopkins, A. J. (1987). *Nature conservation: the role of remnants of native vegetation*. Chipping Norton, NSW: Surrey Beatty.

Saunders, D. A., Hobbs, R. J. & Margules, C. R. (1991). Biological consequences of ecosystem fragmentation: a review. *Conservation Biology* **5**: 18–32.

Savage, A., Giraldo, L. H., Soto, L. H. & Snowdon, C. T. (1996). Demography, group composition, and dispersal in wild cotton-top tamarin (*Saguinus oedipus*) groups. *American Journal of Primatology* **38**: 85–100.

Savolainen, O. & Hedrick, P. W. (1995). Heterozygosity and fitness: no association in Scots pine. *Genetics* **140**: 755–766.

Scarlett, N. H. & Parsons, R. F. (1990). Conservation biology of the southern Australian daisy *Rutidosis leptorrhynchoides*. In *Management and conservation of small populations*: pp. 195–205, Clark, T. W. & Seebeck, J. H. (eds.). Chicago, IL: Chicago Zoological Society.

Schaal, B. A. & Levin, D. A. (1976). The demographic genetics of *Liatris cylindracea*. *American Naturalist* **110**: 191–206.

Schemske, D. W. (1983). Breeding system and habitat effects on fitness components in three neotropical *Costus* (Zingiberaceae). *Evolution* **37**: 523–539.

Schemske, D. W. & Lande, R. (1985). The evolution of self-fertilization and inbreeding depression in plants. II. Empirical observations. *Evolution* **39**: 41–52.

Schemske, D. W., Husband, B. C., Ruckelshaus, M. H., Goodwillie, C., Parker, I. M. & Bishop, J. G. (1994). Evaluating approaches to the conservation of rare and endangered plants. *Ecology* **75**: 584–606.

Schierenbeck, K. A., Skupski, M., Lieberman, D. & Lieberman, M. (1997). Population structure and genetic diversity in four tropical tree species in Costa Rica. *Molecular Ecology* **6**: 137–144.

Schmidt-Adam, G., Young, A. G. & Murray, B. G. (in press). Low outcrossing rates and shift in pollinators in New Zealand Pohutukawa (*Metrosideros excelsa*) (Myrtaceae). *American Journal of Botany*.

Schmitt, J. & Gamble, S. E. (1990). The effect of distance from the parental site on offspring performance and inbreeding depression in *Impatiens capensis*: a test of the local adaptation hypothesis. *Evolution* **44**: 269–278.

Schnabel, A., Nason, J. D. & Hamrick, J. L. (1998). Understanding the population genetic structure of *Gleditsia triacanthos* L.: seed dispersal and variation in female reproductive success. *Molecular Ecology* **7**: 819–832.

Schneider, C. J., Smith, T. B., Larison, B. & Moritz, C. (1999). A test of alternative models of diversification in tropical rainforests: ecological gradients versus rainforest refugia. *Proceedings of the National Academy of Sciences, USA* **96**: 13869–13873.

Schneider, S., Kueffer, J., Roessli, D. & Excoffier, L. (1997). ARLEQUIN v. 1.1. A software for population genetic data analysis. University of Geneva: Genetics and Biometry Laboratory.

Schoen, D. J. (1983). Relative fitnesses of selfed and outcrossed progeny in *Gilia achilleifolia* (Polemoniaceae). *Evolution* **37**: 292–301.

Schoen, D. J. & Brown, A. H. D. (1991). Intraspecific variation in population gene diversity and effective population size correlates with the mating system in plants. *Proceedings of the National Academy of Sciences, USA* **88**: 4494–4497.

Schonewald-Cox, C. M., Chambers, S. M., MacBryde, B. & Thomas, W. L. (1983). *Genetics and conservation: a reference for managing wild animal and plant populations.* Menlo Park, CA: Benjamin Cummings.

Schwartz, M. K., Tallmon, D. A. & Luikart, G. L. (1998). Review of DNA-based census and effective population size estimates. *Animal Conservation* **1**: 293–299.

Scott, J. K. (1980). Estimation of the outcrossing rate for *Banksia attenuata* R.Br. and *Bansksia menziesii* R.Br. (Proteaceae). *Australian Journal of Botany* **28**: 53–59.

Scott, J. M., Mountainspring, S., Kepler, C. B., Jacobi, J. D., Burr, T. A. & Giffin, J. G. (1984). Annual variation in the distribution, abundance, and habitat response of the palila (*Loxioides bailleui*). *Auk* **101**: 647–664.

Scotts, D. J. & Craig, S. A. (1988). Improved hair-sampling tube for the detection of small mammals. *Australian Wildlife Research* **15**: 469–472.

Seal, U. S. (1992). What is CBSG? What do we do? *Conservation Breeding Specialist Group News* **3**: 1.

Serikawa, T., Kuramoto, T., Hilbert, P., Mori, M., Yamada, J., Dubay, C. J., Lindpainter, K., Ganten, D., Guenet, J. L., Lathrop, G. M. & Beckmann, J. S. (1992). Rat gene mapping using PCR-analyzed microsatellites. *Genetics* **131**: 701–721.

Sgro, C. & Hoffmann, A. A. (1998). Effects of stress on the expression of additive genetic variation for fecundity in *Drosophila melanogaster*. *Genetical Research* **72**: 13.

Shaffer, M. L. (1981). Minimum population sizes for species conservation. *Bioscience* **31**: 131–134.

Shaffer, M. L. (1987). Minimum viable populations: coping with uncertainty. In *Viable populations for conservation*: pp. 69–86, Soulé, M. E. (ed.). Cambridge: Cambridge University Press.

Shea, K. & NCEAS Working Group on Population Management (1998). Management of populations in conservation, harvesting and control. *Trends in Ecology and Evolution* **13**: 371–375.

Sherwin, W. B. & Murray, N. D. (1990). Population and conservation genetics of marsupials. *Australian Journal of Zoology* **37**: 161–180.

Sherwin, W. B., Murray, N. D., Graves, J. A. M. & Brown, P. R. (1991). Measurement of genetic variation in endangered populations: Bandicoots (Marsupialia: Peramelidae) as an example. *Conservation Biology* **5**: 103–108.

Sherwin, W. B., Timms, P., Wilcken, J. & Houlden, B. A. (in press). Genetics of koalas: an analysis and conservation implications. *Conservation Biology*.

Shykoff, J. A. & Bucheli, E. (1995). Pollinator visitation patterns, floral rewards and the probability of transmission of *Microbotryum violaceum*: a venereal disease in plants. *Journal of Ecology* **83**: 189–198.

Sih, A. & Baltus, M. (1987). Patch size, pollinator behavior, and pollinator limitation in catnip. *Ecology* **68**: 1679–1690.

Siikamaki, P. & Lammi, A. (1998). Fluctuating asymmetry in central and marginal populations of *Lychnis viscaria* in relation to genetic and environmental factors. *Evolution* **52**: 1285–1292.

Simberloff, D. S. (1988). The contribution of population and community biology to conservation science. *Annual Review of Ecology and Systematics* **19**: 473–511.

Simmons, M. J. & Crow, J. F. (1977). Mutations affecting fitness in *Drosophila* populations. *Annual Review of Genetics* **11**: 49–78.

Simonsen, K. L., Churchill, G. A. & Aqaudro, C. F. (1995). Properties of statistical tests of neutrality for DNA polymorphism. *Genetics* **141**: 413–429.

Sjögren, P. (1991a). Extinction and isolation gradients in metapopulations: the case of the pool frog (*Rana lessonae*). In *Metapopulation dynamics: empirical and theoretical investigations*: pp. 135–147, Gilpin, M. E. & Hanski, I. A. (eds.). London: Academic Press.

Sjögren, P. (1991b). Genetic variation in relation to demography of peripheral pool frog populations (*Rana lessonae*). *Evolutionary Ecology* **5**: 248–271.

Sjögren-Gulve, P. & Ray, C. (1996). Using logistic regression to model metapopulation dynamics: large-scale forestry extirpates the pool frog. In *Metapopulations and wildlife conservation*: pp. 111–137, McCullough, D. R. (ed.). Washington, DC: Island Press.

Skole, D. & Tucker, C. (1993). Tropical deforestation and habitat fragmentation in the Amazon: satellite data from 1978–1988. *Science* **260**: 1905–1910.

Slade, R. W. (1992). Limited MHC polymorphism in the southern elephant seal: implications for MHC evolution and marine mammal population biology. *Proceedings of the Royal Society of London B* **249**: 163–171.

Slatkin, M. (1977). Gene flow and genetic drift in a species subject to frequent local extinctions. *Theoretical Population Biology* **12**: 253–262.

Slatkin, M. (1995). A measure of population subdivision based on microsatellite allele frequencies. *Genetics* **139**: 457–462.

Smith, J. M. (1982). *Ecological comparisons between pine plantations and native forests, Clouds Creek, New South Wales,* Research Series in Applied Geography. Armidale, NSW: University of New England.

Smith, T. B. & Wayne, R. K. (1996). *Molecular genetic approaches in conservation.* New York: Oxford University Press.

Smith, T. B., Wayne, R. K., Girmam, D. J. & Bruford, M. W. (1997). A role for ecotones in generation of rainforest diversity. *Science* **276**: 1855–1857.

Smouse, P. E. & Peakall, R. (1999). Spatial autocorrelation analysis of multi-allele and multi-locus genetic micro-structure. *Heredity* **82**: 561–573.

Sokal, R. R. & Rohlf, F. J. (1995). *Biometry,* 3rd edn. San Francisco, CA: W. H. Freeman.

Soltis, D. E. & Soltis, P. S. (1989). Genetic consequences of autopolyploidy in *Tolmiea* (Saxifragaceae). *Evolution* **43**: 586–594.

Soltis, D. E. & Soltis, P. S. (1992). The distribution of selfing rates in homosporous ferns. *American Journal of Botany* **79**: 97–100.

Soltis, P. S. & Soltis, D. E. (1990). Evolution of inbreeding and outcrossing in ferns and fern-allies. *Plant Species Biology* **5**: 1–11.

Sorensen, F. C. (1999). Relationship between self-fertility, allocation of growth and inbreeding depression in three coniferous species. *Evolution* **53**: 417–425.

Soulé, M. E. (1976). Allozyme variation, its determinants in space and time. In *Molecular evolution*: pp. 60–77, Ayala, F. J. (ed.). Sunderland, MA: Sinauer Associates.

Soulé, M. E. (1980). Thresholds for survival: maintaining fitness and evolutionary potential. In *Conservation biology: an evolutionary–ecological perspective*: pp. 151–169, Soulé, M. E. & Wilcox, B. A. (eds.). Sunderland, MA: Sinauer Associates.

Soulé, M. E. (1985). What is conservation biology? *Bioscience* **11**: 727–734.

Soulé, M. E. (1986a). Conservation biology and the "real world". In *Conservation biology: the science of scarcity and diversity*: pp. 1–12, Soulé, M. E. (ed.). Sunderland, MA: Sinauer Associates.

Soulé, M. E. (1986b). The fitness and viability of populations. In *Conservation biology: the science of scarcity and diversity*: pp. 13–18, Soulé, M. E. (ed.). Sunderland, MA: Sinauer Associates.

Soulé, M. E. (1987). *Viable populations for conservation.* Cambridge: Cambridge University Press.

Soulé, M. E. & Gilpin, M. E. (1991). The theory of wildlife corridor capability. In *Nature conservation*, vol. 2, *The role of corridors*: pp. 3–8, Saunders, D. A. & Hobbs, R. J. (eds.). Chipping Norton, NSW: Surrey Beatty.

Soulé, M. E. & Mills, L. S. (1998). Population genetics: no need to isolate genetics. *Science* **242**: 1658–1659.

Soulé, M. E. & Simberloff, D. S. (1986). What do genetics and ecology tell us about the design of nature reserves? *Biological Conservation* **35**: 19–40.

Soulé, M. E. & Wilcox, B. A. (1980a). Conservation biology: its scope and its challenge. In *Conservation biology: an evolutionary–ecological perspective*: pp. 1–8, Soulé, M. E. & Wilcox, B. A. (eds.). Sunderland, MA: Sinauer Associates.

Soulé, M. E. & Wilcox, B. A. (1980b). *Conservation biology: an evolutionary–ecological perspective.* Sunderland, MA: Sinauer Associates.

Spielman, D. & Frankham, R. (1992). Modeling problems in conservation genetics using captive *Drosophila* populations: improvement of reproductive fitness due to immigration of one individual into small partially inbred populations. *Zoo Biology* 11: 343–351.

Srikwan, S. (1998). Genetic erosion in small mammal populations following rain forest fragmentation in Thailand. PhD thesis: University of California, San Diego.

Srikwan, S., Field, D. & Woodruff, D. S. (1996). Genotyping free-ranging rodents with heterologous PCR primer pairs for hypervariable nuclear microsatellite loci. *Journal of the Science Society of Thailand* 22: 267–274.

Stacy, E. A., Hamrick, J. L., Nason, J. D., Hubbell, S. P., Foster, R. B. & Condit, R. (1996). Pollen dispersal in low density populations of three neotropical tree species. *American Naturalist* 148: 275–298.

Stamps, J. A., Buechner, M. & Krishnan, V. V. (1987). The effects of edge permeability and habitat geometry on emigration from patches of habitat. *American Naturalist* 129: 533–552.

Stangel, P. W. & Dixon, P. M. (1995). Associations between fluctuating asymmetry and heterozygosity in the red-cockaded woodpecker. In *Red-cockaded woodpecker: recovery, ecology, and management*: pp. 239–247, Kulhavy, D. L., Hooper, R. G. & Costa, R. (eds.). Nacogdoches, TX: College of Forestry, Stephen F. Austin State University.

Stangel, P. W., Lennartz, M. R. & Smith, M. H. (1992). Genetic variation and population structure of red-cockaded woodpeckers. *Conservation Biology* 6: 283–292.

Steffen, P., Eggen, A., Dietz, A. B., Womack, J. E., Stranzinger, G. & Fries, R. (1993). Isolation and mapping of polymorphic microsatellites in cattle. *Animal Genetics* 24: 121–124.

Stevens, J. P. & Bougourd, S. M. (1988). Inbreeding depression and the outcrossing rate in natural populations of *Allium schoenoprasum* L. (wild chives). *Heredity* 60: 257–261.

Stith, B. M., Fitzpatrick, J. W., Woolfenden, G. E. & Pranty, B. (1996). Classification and conservation of metapopulations: a case study of the Florida scrub jay. In *Metapopulations and wildlife conservation*: pp. 187–215, McCullough, D. R. (ed.). Washington, DC: Island Press.

Styles, B. T. (1972). The flower biology of the Meliaceae and its bearing on tree breeding. *Silvae Genetica* 21: 175–182.

Styles, B. T. (1981). Swietenioideae. In *Flora Neotropica*, Monograph no. 28, *Meliaceae*: pp. 359–418, Pennington, T. D., Styles, B. T. & Taylor, D. A. H. (eds.). New York: New York Botanical Garden.

Suckling, G. C. (1978). A hair sampling tube for the detection of small mammals in trees. *Australian Wildlife Research* 5: 249–252.

Suckling, G. C. (1982). Value of reserved habitat for mammal conservation in plantations. *Australian Forestry* 45: 19–27.

Suckling, G. C. & Heislers, A. (1978). Populations of small mammals in *Radiata* pine plantations and eucalypt forests of north-eastern Victoria. *Australian Wildlife Research* 5: 305–315.

Sun, M. (1996). Effects of population size, mating system, and evolutionary origin on genetic diversity in *Spiranthes sinensis* and *S. hongongensis*. *Conservation Biology* 10: 785–795.

Sun, M. & Corke, H. (1992). Population genetics of colonizing success of weedy rye in northern California. *Theoretical and Applied Genetics* **83**: 321–329.

Taggart, J. B., McNally, S. F. & Sharp, P. M. (1990). Genetic variability and differentiation among founder populations of the pitcher plant (*Sarracenia purpurea* L.) in Ireland. *Heredity* **64**: 177–183.

Tajima, F. (1989). Statistical method for testing the neutral mutation hypothesis by DNA polymorphism. *Genetics* **123**: 585–595.

Tanaka, Y. (1997). Extinction of populations due to inbreeding depression with demographic disturbances. *Researches in Population Ecology* **39**: 57–66.

Tautz, D. (1989). Hypervariability of simple sequences as a general source of polymorphic DNA. *Nucleic Acids Research* **17**: 6462–6471.

Taylor, A. C., Sherwin, W. B. & Wayne, R. K. (1994). The use of simple sequence loci to measure genetic variation in bottlenecked species: the decline of the northern hairy-nosed wombat (*Lasiorhinus krefftii*). *Molecular Ecology* **3**: 277–290.

Taylor, A. D. (1988). Large-scale spatial structure and population dynamics in arthropod predator–prey systems. *Annales Zoologici Fennici* **25**: 63–74.

Taylor, A. D. (1990). Metapopulations, dispersal, and predator–prey dynamics: an overview. *Ecology* **71**: 429–436.

Taylor, B. L. (1995). The reliability of using population viability analysis for risk classification of species. *Conservation Biology* **9**: 551–558.

Taylor, B. L. (1997). Defining "population" to meet management objectives for marine mammals. In *Molecular genetics of marine mammals*, Special publication no. 3: pp. 49–65, Dizon, A. E., Chivers, S. J. & Perrin, W. F. (eds.). Lawrence, KS: Society for Marine Mammalogy.

Taylor, B. L., Chivers, S. J. & Dizon, A. E. (1997). Using statistical power to interpret genetic data to define management units for marine mammals. In *Molecular genetics of marine mammals*, Special publication no. 3: pp. 347–364, Dizon, A. E., Chivers, S. J. & Perrin, W. F. (eds.). Lawrence, KS: Society for Marine Mammalogy.

Taylor, G. & Whelan, R. J. (1988). Can honeybees pollinate *Grevillea*? *Australian Zoologist* **24**: 193–196.

Tease, C. & Jones, G. H. (1976). Chromosome-specific control of chiasma formation in *Crepis capillaris*. *Chromosoma* **57**: 33–49.

Templeton, A. R. & Levin, D. A. (1979). Evolutionary consequences of seed banks. *American Naturalist* **114**: 232–249.

Templeton, A. R. & Read, B. (1984). Factors eliminating inbreeding depression in a captive herd of Speke's gazelle (*Gazella spekei*). *Zoo Biology* **3**: 177–199.

Templeton, A. R. & Read, B. (1994). Inbreeding: one word, several meanings, much confusion. In *Conservation genetics*: pp. 91–106, Loeschcke, V., Tomiuk, J. & Jain, S. K. (eds.). Basel: Birkhäuser Verlag.

Templeton, A. R., Sing, C. F. & Brokaw, B. (1976). The unit of selection in *Drosophila mercatorum*. I. The interaction of selection and meiosis in parthenogenetic strains. *Genetics* **82**: 349–376.

The Commonwealth of Australia & Department of Natural Resources and Environment (1997). Comprehensive Regional Assessment – Biodiversity. Central Highlands of Victoria. Canberra: The Commonwealth of Australia and Department of Natural Resources and Environment.

Thomas, C. D. & Hanski, I. A. (1997). Butterfly metapopulations. In *Metapopulation biology: ecology, genetics, and evolution*: pp. 359–386, Hanski, I. A. & Gilpin, M. E. (eds.). San Diego, CA: Academic Press.

Thompson, J. N. (1994). *The coevolutionary process*. Chicago, IL: University of Chicago Press.

Thompson, J. N. (1997). Evaluating the dynamics of coevolution among geographically structured populations. *Ecology* 78: 1619–1623.

Thrall, P. H. & Antonovics, J. (1995). Theoretical and empirical studies of metapopulations: population and genetic dynamics of the *Silene–Ustilago* system. *Canadian Journal of Botany* 73: S1249–S1258.

Thrall, P. H. & Burdon, J. J. (1997). Host–pathogen dynamics in a metapopulation context: the ecological and evolutionary consequences of being spatial. *Journal of Ecology* 85: 743–753.

Thrall, P. H. & Burdon, J. J. (1999). The spatial scale of pathogen dispersal: consequences for disease dynamics and persistence. *Evolutionary Ecology Research* 1: 681–701.

Thrall, P. H. & Jarosz, A. M. (1994a). Host–pathogen dynamics in experimental populations of *Silene alba* and *Ustilago violacea*. I. Ecological and genetic determinants of disease spread. *Journal of Ecology* 82: 549–559.

Thrall, P. H. & Jarosz, A. M. (1994b). Host–pathogen dynamics in experimental populations of *Silene alba* and *Ustilago violacea*. II. Experimental tests of theoretical models. *Journal of Ecology* 82: 561–570.

Thrall, P. H., Richards, C. M., McCauley, D. E. & Antonovics, J. (1998). Metapopulation collapse: the consequences of limited gene-flow in spatially structured populations. In *Modelling spatiotemporal dynamics in ecology*: pp. 83–104, Bascompte, J. & Solé, R. V. (eds.). Berlin: Springer-Verlag.

Tilman, D. & Kareiva, P. M. (1997). *Spatial ecology: the role of space in population dynamics and interspecific interactions*. Princeton, NJ: Princeton University Press.

Tilman, D., May, R. M., Lehman, C. L. & Nowak, M. A. (1994). Habitat destruction and the extinction debt. *Nature* 371: 65–66.

Tonsor, S. J., Kalisz, S., Fisher, J. & Holtsford, T. P. (1993). A life-history based study of population genetic structure: seed bank to adults in *Plantago lanceolata*. *Evolution* 47: 833–843.

Triggs, S. J., Powlesland, R. G. & Daugherty, C. H. (1989). Genetic variation and conservation of kakapo (*Stigops habroptilus*: Psittaciformes). *Conservation Biology* 3: 92–96.

Tuljapurkar, S. D. (1982). Population dynamics in variable environments. III. Evolutionary dynamics of r-selection. *Theoretical Population Biology* 21: 141–165.

Tuljapurkar, S. D. (1989). An uncertain life: demography in random environments. *Theoretical Population Biology* 35: 227–294.

Tyndale-Biscoe, C. H. & Smith, R. F. (1969a). Studies of the marsupial glider, *Schoinobates volans* (Kerr). II. Population structure and regulatory mechanisms. *Journal of Animal Ecology* 38: 637–650.

Tyndale-Biscoe, C. H. & Smith, R. F. (1969b). Studies of the marsupial glider, *Schoinobates volans* (Kerr). III. Response to habitat destruction. *Journal of Animal Ecology* 38: 651–659.

Valdes, A. M., Slatkin, M. & Freimer, N. B. (1993). Allele frequencies at microsatellite loci – the stepwise mutation model revisited. *Genetics* 133: 737–749.

Van Dongen, S. (1995). How should we bootstrap allozyme data? *Heredity* **74**: 445–447.

Van Dongen, S., Backeljau, T., Matthysen, E. & Dhondt, A. A. (1998). Genetic population structure of the winter moth (*Operophthera brumata* L.) (Lepidoptera, Geometridae) in a fragmented landscape. *Heredity* **80**: 92–100.

van Treuren, R. (1993). The significance of genetic erosion for the extinction of locally endangered plant populations. PhD thesis: State University of Groningen.

van Treuren, R., Bijlsma, R., Ouborg, N. J. & van Delden, W. (1991). The significance of genetic erosion in the process of extinction. I. Genetic differentiation in *Salvia pratensis* and *Scabiosa columbaria* in relation to population size. *Heredity* **66**: 181–189.

van Treuren, R., Bijlsma, R., Ouborg, N. J. & Kwak, M. M. (1994). Outcrossing rates in *Scabiosa columbaria*: effects of density and differential success of self- and cross-pollination. *Journal of Evolutionary Biology* **7**: 287–302.

Varvio, S., Chakraborty, R. & Nei, M. (1986). Genetic variation in subdivided populations and conservation genetics. *Heredity* **57**: 189–198.

Vaughton, G. (1995). No evidence for selective fruit abortion in the Australian shrub *Grevillea barklyana* (Proteaceae). *International Journal of Plant Sciences* **156**: 417–424.

Vaughton, G. (1996). Pollination disruption by European honeybees in the Australian bird-pollinated shrub *Grevillea barklyana* (Proteaceae). *Plant Systematics and Evolution* **200**: 89–100.

Vaughton, G. (1998). Soil seed bank dynamics in the rare, obligate seeding shrub, *Grevillea barklyana* (Proteaceae). *Australian Journal of Ecology* **23**: 375–384.

Vaughton, G. & Carthew, S. M. (1993). Evidence for selective fruit abortion in *Banksia spinulosa* (Proteaceae). *Biological Journal of the Linnean Society* **50**: 35–46.

Vekemans, X., Schierup, M. H. & Christiansen, F. B. (1998). Mate availability and fecundity selection in multi-allelic self-incompatibility systems in plants. *Evolution* **52**: 19–29.

Verboom, J. & Metz, J. A. J. (1991). Linking local and regional dynamics in stochastic metapopulation models. *Biological Journal of the Linnean Society* **42**: 39–55.

Vitousek, P. M. (1994). Beyond global warming. *Ecology* **75**: 1861.

Vitousek, P. M., D'Antonio, C. M., Loope, L. L. & Westbrooks, R. (1996). Biological invasions as global environmental change. *Bioscience* **84**: 468–478.

Vitousek, P. M., Mooney, H. A., Lubchenco, J. & Melillo, J. M. (1997). Human domination of earth's ecosystems. *Science* **277**: 494–499.

Vogler, A. P. & DeSalle, R. (1994). Diagnosing units of conservation management. *Conservation Biology* **8**: 354–363.

Vrijenhoek, R. C. (1996). Conservation genetics of North America desert fishes. In *Conservation genetics: case histories from nature*: pp. 367–397, Avise, J. C. & Hamrick, J. L. (eds.). New York: Chapman & Hall.

Vucetich, J. A., Waite, T. A. & Nunney, L. (1997). Fluctuating population size and the ratio of effective to census population size. *Evolution* **51**: 2017–2021.

Wade, M. J. & McCauley, D. E. (1988). Extinction and colonization: their effects on the genetic differentiation of local populations. *Evolution* **42**: 995–1005.

Wainwright, T. C. & Waples, R. S. (1998). Prioritizing Pacific salmon stocks for conservation – response to Allendorf *et al. Conservation Biology* **12**: 1144–1147.

Walker, H. P. & Bawa, K. (1996). Effects of forest fragmentation on genetic diversity and mating systems in a tropical tree, *Pithecellobium elegans. Conservation Biology* **10**: 757–768.

Walker, S. & Rabinowitz, A. (1992). The small mammal community of a dry tropical forest in central Thailand. *Journal of Tropical Ecology* **8**: 57–71.

Wallace, B. (1953). On coadaptation in *Drosophila. American Naturalist* **87**: 343–358.

Wallace, B. & Vetukhiv, M. (1955). Adaptive organization of the gene pools of *Drosophila* populations. *Cold Spring Harbor Symposium on Quantitative Biology* **20**: 303–309.

Walters, J. R. (1990). Red-cockaded woodpeckers: a 'primitive' cooperative breeder. In *Cooperative breeding in birds: long-term studies of ecology and behavior*: pp. 67–101, Stacey, P. B. & Koenig, W. D. (eds.). Cambridge: Cambridge University Press.

Walters, J. R. (1991). Application of ecological principles to the management of endangered species: the case of the red-cockaded woodpecker. *Annual Review of Ecology and Systematics* **22**: 505–523.

Walters, J. R., Doerr, P. D. & Carter, J. H. (1988*a*). The cooperative breeding system of the red-cockaded woodpecker. *Ethology* **78**: 275–305.

Walters, J. R., Hansen, S. K., Carter, J. H. & Manor, P. D. (1988*b*). Long-distance dispersal of an adult red-cockaded woodpecker. *Wilson Bulletin* **100**: 494–496.

Walters, J. R., Copeyon, C. K. & Carter, J. H. (1992). Test of the ecological basis of cooperative breeding in red-cockaded woodpeckers. *Auk* **109**: 90–97.

Wardle, J. A. (1984). *The New Zealand beeches: ecology, utilization and management.* Christchurch: New Zealand Forest Service.

Warneke, R. M. (1971). Field study of the Bush Rat (*Rattus fuscipes*). *Wildlife Contributions Victoria* **14**: 1–115.

Warren-Hicks, W. J. & Moore, D. R. J. (1998). *Uncertainty analysis in ecological risk assessment.* Pensacola, FL: Society of Environmental Toxicology and Chemistry (SETAC).

Waser, N. M. (1993). Population structure, optimal outbreeding, and assortative mating in angiosperms. In *The natural history of inbreeding and outbreeding, theoretical and empirical perspectives*: pp. 173–199, Thornhill, N. W. (ed.). Chicago, IL: University of Chicago Press.

Waser, N. M. & Price, M. V. (1989). Optimal outcrossing in *Ipomopsis aggregata*: seed set and offspring fitness. *Evolution* **43**: 1097–1109.

Waser, N. M. & Price, M. V. (1994). Crossing distance effects in *Delphinium nelsonii*: outbreeding and inbreeding depression in progeny fitness. *Evolution* **48**: 842–852.

Waser, P. M. & Strobeck, C. (1998). Genetic signatures of interpopulation dispersal. *Trends in Ecology and Evolution* **13**: 43–44.

Watt, W. B. (1985). Bioenergetics and evolutionary genetics: opportunities for new synthesis. *American Naturalist* **125**: 118–143.

Watt, W. B., Cassin, R. C. & Swan, M. S. (1983). Adaptation at specific loci. III. Field behaviour and survivorship differences among *Colias* PGI genotypes are predictable from in vitro biochemistry. *Genetics* **103**: 725–739.

Wauters, L. A., Hutchinson, Y., Parkin, D. T. & Dhondt, A. A. (1994). The effects of habitat fragmentation on demography and on the loss of genetic variation in the red squirrel. *Proceedings of the Royal Society of London B* **255**: 107–111.

Webb, C. J. & Bawa, K. S. (1983). Pollen dispersal by hummingbirds and butterflies: a comparative study of two lowland tropical plants. *Evolution* **37**: 1258–1270.

Weber, J. L. (1990). Human DNA polymorphisms and methods of analysis. *Current Opinion in Biotechnology* **1**: 166–171.

Weber, J. L. & May, P. E. (1989). Abundant class of human DNA polymorphisms which can be typed using the polymerase chain reaction. *American Journal of Human Genetics* **44**: 338–396.

Weddell, B. J. (1991). Distribution and movements of Columbian ground squirrels (*Spermophilus columbianus* (Ord)): are habitat patches like islands? *Journal of Biogeography* **18**: 385–394.

Wegner, J. F. & Merriam, G. (1979). Movements by birds and small mammals between a wood and adjoining farmland habitats. *Journal of Applied Ecology* **16**: 349–357.

Wehausen, J. D. (1991). Some potentially adaptive characters of mountain sheep populations in the Owens Valley region. In *History of water*, vol. 3: *Eastern Sierra Nevada, Owens Valley, White-Inyo Mountains*: pp. 124–135, Hall, C. A., Doyle-Jones, V. & Widawski, B. (eds.). Bishop, CA: White Mountain Research Station.

Wehausen, J. D. & Ramey, R. R. (1993). A morphometric reevaluation of the peninsular bighorn subspecies. *Desert Bighorn Council Transactions* **37**: 1–10.

Weir, B. S. & Cockerham, C. C. (1984). Estimating F-statistics for the analysis of population structure. *Evolution* **38**: 1358–1370.

Weller, S. G. (1994). The relationship of rarity to plant reproductive biology. In *Restoration of endangered species: conceptual issues, planning and implementation*: pp. 90–117, Bowles, M. J. & Whelan, C. J. (eds.). Cambridge: Cambridge University Press.

Westemeier, R. L., Brawn, J. D., Simpson, S. A., Esker, T. L., Jansen, R. W., Walk, J. W., Kershner, E. L., Bouzat, J. L. & Paige, K. N. (1998). Tracking the long-term decline and recovery of an isolated population. *Science* **282**: 1695–1698.

Wethington, A. R. & Dillon, R. T. (1997). Selfing, outcrossing, and mixed mating in the freshwater snail *Physa heterostropha*: lifetime fitness and inbreeding depression. *Invertebrate Biology* **116**: 192–199.

Whelan, R. J. (1995). *The ecology of fire*. Cambridge: Cambridge University Press.

Whelan, R. J. & Goldingay, R. L. (1986). Do pollinators influence seed set in *Banksia spinulosa* Sm. and *B. paludosa* R.Br.? *Australian Journal of Ecology* **11**: 181–186.

Whelan, R. J. & Goldingay, R. L. (1989). Factors affecting fruit set in *Telopea speciosissima* (Proteaceae): the importance of pollen limitation. *Journal of Ecology* **77**: 1123–1134.

Whelan, R. J., de Jong, N. & von der Burg, S. (1998). Variation in bradyspory and seedling recruitment without fire among populations of *Banksia serrata* (Proteaceae). *Australian Journal of Ecology* **23**: 121–128.

White, G. & Powell, W. (1997). Isolation and characterisation of microsatellite loci in *Swietenia humilis* (Meliaceae): an endangered tropical hardwood species. *Molecular Ecology* **6**: 851–860.

White, G. M., Boshier, D. H. & Powell, W. (1999). Genetic variation within a fragmented population of *Swietenia humilis* Zucc. *Molecular Ecology* **8**: 1899–1909.

Whitlock, M. C. & McCauley, D. E. (1990). Some population genetic consequences of colony formation and extinction: genetic correlations within founding groups. *Evolution* **44**: 1717–1724.

Whitlock, M. C., Phillips, P. C., Moore, F. G. G. & Tonsor, S. J. (1995). Multiple fitness peaks and epistasis. *Annual Review of Ecology and Systematics* **26**: 601–629.

Whitmore, T. C. (1984). *Tropical rain forests of the Far East*, 2nd edn. Oxford: Oxford University Press.

Whitmore, T. C. (1990). *An introduction to tropical rainforests*. Oxford: Oxford University Press.

Wiens, J. A. (1996). Wildlife in patchy environments: metapopulations, mosaics, and management. In *Metapopulations and wildlife conservation*: pp. 53–84, McCullough, D. R. (ed.). Washington, DC: Island Press.

Wiens, J. A. (1997). Metapopulation dynamics and landscape ecology. In *Metapopulation biology: ecology, genetics and evolution*: pp. 43–62, Hanski, I. A. & Gilpin, M. E. (eds.). New York: Academic Press.

Wilcove, D. S., McLellan, C. H. & Dobson, A. P. (1986). Habitat fragmentation in the temperate zone. In *Conservation biology: the science of scarcity and diversity*: pp. 237–256, Soulé, M. E. (ed.). Sunderland, MA: Sinauer Associates.

Wilcove, D. S., Rothstein, D., Dubow, J., Phillips, A. & Losos, E. (1998). Quantifying threats to imperiled species in the United States. *Bioscience* **48**: 607–616.

Wilkens, H. (1971). Genetic interpretation of regressive evolutionary processes. *Evolution* **25**: 530–540.

Williams, C. G. & Savolainen, O. (1996). Inbreeding depression in conifers: implications for breeding strategy. *Forest Science* **42**: 102–117.

Williams, J. E., Bowman, D. B., Brooks, J. E., Echelle, A. A., Edwards, R. J., Hendrickson, D. A. & Landye, J. J. (1985). Endangered aquatic ecosystems in North American deserts, with a list of vanishing fishes of the region. *Journal of the Arizona–Nevada Academy of Sciences* **20**: 1–62.

Williams, J. G. K., Kubelik, A. R., Livak, K. J., Rafalski, J. A. & Tingey, S. V. (1990). DNA polymorphisms amplified by arbitrary primers are useful as genetic markers. *Nucleic Acids Research* **18**: 6531–6535.

Williams, P. A. & Karl, B. J. (1996). Fleshy fruits of indigenous and adventive plants in the diet of birds in forest remnants, Nelson, New Zealand. *New Zealand Journal of Ecology* **20**: 127–145.

Williams, W. & Brown, A. G. (1956). Genetic response to selection in cultivated plants: gene frequencies in *Prunus avium*. *Heredity* **10**: 237–245.

Willis, J. H. (1993). Effects of different levels of inbreeding on fitness components in *Mimulus guttatus*. *Evolution* **47**: 864–876.

Willis, K. & Wiese, R. J. (1997). Elimination of inbreeding depression from captive populations: Speke's gazelle revisited. *Zoo Biology* **16**: 9–16.

Wilson, D. S. (1992). Complex interactions in metacommunities, with implications for biodiversity and higher levels of selection. *Ecology* **73**: 1984–2000.

Wilson, H. B. & Hassell, M. P. (1997). Host–parasitoid spatial models: the interplay of demographic stochasticity and dynamics. *Proceedings of the Royal Society of London B* **264**: 1189–1195.

Wilson, P. R. (1984). The effects of possums on mistletoe on Mt Misery, Nelson Lakes National Park. In *Protection and parks: essays in the preservation of natural values in protected areas*: pp. 53–60, Dingwall, P. R. (ed.). Wellington: Department of Lands and Survey.

Wolfe, L. M. (1993). Inbreeding depression in *Hydrophyllum appendiculatum*: role of maternal effects and crowding, and parent mating history. *Evolution* **47**: 374–386.

Wolff, K. & Haeck, J. (1990). Genetic analysis of ecologically relevant morphological variation in *Plantago lanceolata* L. VI. The relation between allozyme heterozygosity and some fitness components. *Journal of Evolutionary Biology* **3**: 243–255.

Woodruff, D. S. (1989). The problems of conserving genes and species. In *Conservation for the twenty-first century*: pp. 76–78, Pearl, M. (ed.). New York: Oxford University Press.

Woodruff, D. S. (1990). Genetics and demography in the conservation of biodiversity. *Journal of the Science Society of Thailand* **16**: 117–132.

Woodruff, D. S. (1992). Genetics and the conservation of animals in fragmented habitats. In *In harmony with nature, Proceedings of the International Conference on Tropical Biodiversity*: pp. 258–272, Kheong, Y. S. & Win, L. S. (eds.). Kuala Lumpur: Malay Nature Society.

Woodruff, D. S. (1993). Non-invasive genotyping of primates. *Primates* **34**: 233–246.

Woodworth, L. M., Montgomery, M. E., Nurthen, R. K., Briscoe, D. A. & Frankham, R. (1994). Modelling problems in conservation genetics using *Drosophila*: consequences of fluctuating populations sizes. *Molecular Ecology* **3**: 393–399.

Worthington-Wilmer, J., Hall, L., Barratt, E. & Moritz, C. (1999). Genetic structure and male-mediated gene flow in the ghost bat (*Macroderma gigas*). *Evolution* **53**: 1582–1591.

Wright, S. (1931). Evolution in Mendelian populations. *Genetics* **16**: 97–159.

Wright, S. (1940). Breeding structure of populations in relation to speciation. *American Naturalist* **74**: 232–248.

Wright, S. (1943). Isolation by distance. *Genetics* **28**: 114–138.

Wright, S. (1946). Isolation by distance under diverse systems of mating. *Genetics* **31**: 39–59.

Wright, S. (1951). The genetical structure of populations. *Annals of Eugenics* **15**: 323–354.

Wright, S. (1964). The distribution of self-incompatability in populations. *Evolution* **18**: 609–619.

Wright, S. (1969). *Evolution and the genetics of populations*, vol. 2, *The theory of gene frequencies*. Chicago, IL: University of Chicago Press.

Wright, S. (1977). *Evolution and the genetics of populations*, vol. 3, *Experimental results and evolutionary deductions*. Chicago, IL: University of Chicago Press.

Wright, S. (1978). *Evolution and the genetics of populations*, vol. 4, *Variability within and among natural populations*. Chicago, IL: University of Chicago Press.

Yamazaki, K., Boyse, E. A., Mike, V., Thaler, H. T., Mathieson, B. J., Abbot, J., Boyse, J., Zayas, Z. A. & Thomas, L. (1976). Control of mating preferences in mice by genes in the major histocompatibility complex. *Journal of Experimental Medicine* **144**: 1324–1335.

Yeh, F. C., Yang, R. C., Boyle, T., Ye, Z. H. & Mao, J. X. (1997). POPGENE, the user-friendly shareware for population genetic analysis. Alberta, Canada: Molecular Biology and Biotechnology Centre, University of Alberta.

Young, A. G. & Brown, A. H. D. (1998). Comparative analysis of the mating system of the rare woodland shrub *Daviesia suaveolens* and its common congener *D.*

mimosoides. Heredity **80**: 374–381.

Young, A. G. & Brown, A. H. D. (1999). Paternal bottlenecks in fragmented populations of the grassland daisy, *Rutidosis leptorrhynchoides. Genetical Research* **73**: 111–117.

Young, A. G. & Merriam, H. G. (1994). Effects of forest fragmentation on the spatial genetic structure of *Acer saccharum* Marsh. (sugar maple) populations. *Heredity* **72**: 201–208.

Young, A. G. & Murray, B. G. (in press). Genetic bottlenecks and dysgenic gene flow into re-established populations of the grassland daisy *Rutidosis leptorrhynchoides. Australian Journal of Botany.*

Young, A. G., Merriam, H. G. & Warwick, S. I. (1993). The effects of forest fragmentation on genetic variation in *Acer saccharum* Marsh. (sugar maple) populations. *Heredity* **71**: 227–289.

Young, A. G., Boyle, T. & Brown, A. H. D. (1996). The population genetic consequences of habitat fragmentation for plants. *Trends in Ecology and Evolution* **11**: 413–418.

Young, A. G., Brown, A. H. D. & Zich, F. C. (1999). Genetic structure of fragmented populations of the endangered grassland daisy, *Rutidosis leptorrhynchoides. Conservation Biology* **13**: 256–265.

Ziehe, M. & Roberds, J. H. (1989). Inbreeding depression due to overdominance in partially self-fertilising populations. *Genetics* **121**: 861–868.

Index

Note: page numbers in *italics* refer to figures and tables.

424 | *Index*

bottlenecks (*cont.*)
  gene selection 23
  genetic load reduction 41
  population fitness 36
box, white 86

Callitrichidae 204, 205, 206
Canis lupus baileyi (Mexican gray wolf) 122
Chamaecrista fasciculata
  fitness 50, *51*, 52
  inbreeding and outbreeding depression
    49–50, *51, 52*
  population crossing 49–50
Chiew Larn reservoir (Thailand) 152, *153*
chimerism, tamarins 204
Chiropodomys gliroides (pencil-tail tree
    mouse) 149, 152, *154, 155*
  abundance 168, *169*
  allelic diversity *159, 160,* 164
  demographic collapse 169
  effective population size 165
  generation time 158
  genetic erosion 169
  heterozygosity *159, 162,* 165
  microsatellite loci *159, 160, 162, 164*–5
chloroplast DNA markers 278–9
chub, bonytail
  additional wild fish in broodstock 121–2
  allozyme genetic data 120
  broodstock establishment 119–22
  egg production 120
  F_1 progeny 119, 120
  genetic variation loss 121
  genotypes 120–1
  mitochondrial DNA information 120,
    121
  proportion of egg fertilisation 120
Clarkia pulchella 276
coadapted gene complexes 46, 48
  disruption 46, 47, 48
coevolutionary processes, conservation
    92–3
colonisation
  Allee effects 288
  colonist exports 288
  compressed founder populations 168
  effective population size 273
  episodes 272
  founding cohort
    relatedness 283
    selective forces 273

  viability 274
  inbreeding 283
    depression 288
  island biogeography 89
  plants 272
  probability and patch occupancy 83
  processes 76, 78
    metapopulation 77, 78
    preservation 79
  rate 89
  relatedness of founders 283
  Silene alba 277, 279, 283
  Synemon plana 223–4
  stages 287–8
  strategies 272
community effects, metapopulation
    studies 85–6
community structure, fragment size 85
connectivity 91, 94
conservation biology 1–2
  applied science 114
  molecular techniques 113–15
  population genetics 113–15
  studies 79, 80
conservation forecasting, genetics 14–16
conservation value, metapopulation
    paradigm 88
contribution to next generation 116
corridors, demographic sinks 91
cross-fertilisation in small populations 237

daisy *see Rutidosis leptorrhynchoides* (daisy)
Daphnia, genetic variation 14
decision theory, PVA 111
declining-population paradigm 3, 58, 71,
    98–9
  false dichotomy from small-population
    paradigm 114
deforestation 5
demographic data, population viability
    analysis 99
demographic factors 3
demographic processes 2
demographic sinks 91
demographic stochasticity 60–1, 64, 66–7
demographic variance 61
demography 2, 3–4
  genetic interactions 273–4
deterministic decline 65–6
deterministic processes, population
    viability analysis 100, 101